Recent Advances in Artificial Life

ADVANCES IN NATURAL COMPUTATION

Published

Vol. 1: Applications of Multi-Objective Evolutionary Algorithms
Eds: Carlos A. Coello Coello (CINVESTAV-IPN, Mexico) and
Gary B. Lamont (Air Force Institute of Technology, USA)

Vol. 2: Recent Advances in Simulated Evolution and Learning
Eds: Kay Chen Tan (National University of Singapore, Singapore),
Meng Hiot Lim (Nanyang Technological University, Singapore),
Xin Yao (University of Birmingham, UK) and
Lipo Wang (Nanyang Technological University, Singapore)

Recent Advances in Artificial Life

Advances in Natural Computation – Vol. 3

Sydney, Australia 5 – 8 December 2005

editors

H. A. Abbass
University of New South Wales, Australia

T. Bossomaier
Charles Sturt University, Australia

J. Wiles
The University of Queensland, Austalia

World Scientific

NEW JERSEY • LONDON • SINGAPORE • BEIJING • SHANGHAI • HONG KONG • TAIPEI • CHENNAI

Published by

World Scientific Publishing Co. Pte. Ltd.

5 Toh Tuck Link, Singapore 596224

USA office: 27 Warren Street, Suite 401-402, Hackensack, NJ 07601

UK office: 57 Shelton Street, Covent Garden, London WC2H 9HE

British Library Cataloguing-in-Publication Data
A catalogue record for this book is available from the British Library.

RECENT ADVANCES IN ARTIFICIAL LIFE
Advances in Natural Computation — Vol. 3

ISBN 981-256-615-5

Printed in Singapore by World Scientific Printers (S) Pte Ltd

Preface

This book arose from the set of papers that were submitted and accepted at the Australian Conference on Artificial Life (ACAL'05), 5-9 December, 2005. ACAL'05 is the second in a series of Australian artificial life conferences at which the advances in Artificial Life are reported. All papers submitted to the conference were peer reviewed by at least two experts in the relevant area and only those that were accepted are included in this proceedings. ACAL'05 received 40 papers, out of which 25 papers only were selected to appear in this book. The conference attracted submissions from Australia, China, Japan, New Zealand, the Netherlands, UK, and the USA. The conference hosted a number of events including the Second Workshop on Patterns. ACAL'05 witnessed a number of invited speakers including Prof. David Goldberg, from the University of Illinois, Urbana-Champaign, and Prof. Mark Bedau, from Reed College.

We wish to acknowledge the role of the advisory and program committees of the conference. The advisory committee included: Mark Bedau (Reed, USA); Eric Bonabeau (Icosystem, USA); David Fogel (Natural Selection, USA); Peter Stadler (Leipzig, Germany); Masanori Sugisaka (Oita, Japan).

The program committee included: Alan Blair (UNSW, Australia); Xiaodong Li (RMIT, Australia); Stephen Chalup (Newcastle, Australia); Xavier Llor (UIUC, USA); Tan Kay Chen (NUS, Singapore); Frederic Maire (QUT, Australia); Vic Ciesielski (RMIT, Australia); Bob Mckay (UNSW@ADFA, Australia); David Cornforth (CSU, Australia); Naoki Mori (Osaka Prefecture University, Japan); Alan Dorine (Monash, Australia); Akira Namatame (Defence Academy, Japan); Daryl Essam (UNSW@ADFA, Australia); Chrystopher Nehaniv (Herts, UK); David Green (Monash, Australia); David Newth (CSIRO, Australia); Tim Hendtlass (Swinburne, Australia); Stefano Nolfi (CNR, Italy); Christian Jacob (Calgary, Canada); Marcus Randall (Bond, Australia); Ray Jarvis (Monash, Australia); Alex Ryan (DSTO, Australia); Graham Kendall (Nottingham, UK); Russell Standish (UNSW, Australia); Kevin Korb (Monash, Australia); Jason Teo (Universiti Malaysia Sabah, Malaysia).

The editors also wish to acknowledge the efforts of the enthusiastic team of World Scientific Publishing, who worked hard to get high quality manuscript and to make it available for the authors at the conference. We also wish to thank the series editor, Prof. Xin Yao, for accepting our proposal and making this publication possible.

ACAL runs biannually and we look forward to ACAL 2007. We hope that this event benefited all researchers who attended the conference.

Hussein A. Abbass, Terry Bossomaier, and Janet Wiles (Editors)

Contents

Chapter 1

Recreating Large-Scale Evolutionary Phenomena

P.-M. Agapow
VieDigitale Limited,
20 Matthias Ct,
119 Church Rd.
Richmond, London TW10 6LL, United Kingdom
E-mail: agapow@bbsrc.ac.uk

Any study of the evolutionary past is hampered by the size and type of events involved. Change on such a grand scale cannot be observed directly or manipulated experimentally, and so to date it has been deduced from clues left in the present day. This limitation can be overcome by replaying the course of evolution and observing what results and whether it matches what we know of present day biology. Here I present MeSA, a sophisticated framework for the simulation of large-scale evolution, and demonstrate how this approach can be used to investigate key innovations, a putative cause of patterns of biodiversity.

1.1 Introduction

It has often been said about evolution that "the present is the key to the past" [368], that studying observable microevolutionary [1] process and facets of extant biology can tell us about the the history of life on Earth. Given our necessarily limited view of paleontological life (due to the sparseness of the fossil record, and the subsequent difficulty in deducing the paths of evolution) until recently this forensic approach dominated evolutionary discussion. The unknowable past was a subject for study, but not a source of information in itself.

It is obvious that studying past via the present has limitations. It assumes that the evolutionary game has not changed substantially over the course of life's history, that while the fine details of forces and patterns behind biodiversification may vary, the broad picture is the same. This may not be true, with possible explosions of organismal variety occurring in the past [154] and anthropogenic extinctions shaping the present [315]. Second, the scale and size of macroevolution (potentially trillions of organisms over millions of years) defies direct examination and manipulation. In contrast, the scale of most contemporary ecological observations is less than a single square meter over less than a year. The gulf between the two prevents all but the most general conclusions.

Conversely, there is a great deal that a "knowable past" can tell us about the present. Evolutionary history is a smoking gun. Its course and shape are a rich trove of information about how organisms arise, prosper and die. In addition, it can serve as a large pool of samples for statistical testing of hypotheses, each diverging population acting as a replicate. Thus by examining the family tree sketched by evolution, it is possible (for example) to study what characteristics make organisms more susceptible to extinction, how body-size changes when the environment changes, what organismal characteristics co-evolve. In this way phylogenies have been used for insights into epidemiology [319], conservation [228], development [96] and many other fields [182].

Unfortunately, there are several obstructions in the path of such investigation. Different workers use a wide variety of metrics or tests to assess how organismal traits change and correlate. It is unclear which tests are best, how tests perform under different evolutionary scenarios [314], or even what such measurements actually mean in terms of the behaviour of real organisms. Given the large number of possible interactions between organisms and their environment, macroevolution may be irreducibly complex [152]. Certainly, no general analytical formulation can possibly capture its many properties. To complicate matters further, phylogenies are at best only estimates of the actual pattern of evolution and may be inaccurate or even incomplete. In summary, the current analysis of large-scale evolution is unable to cope with the complexities of real life.

[1]Some specialised terms are required for this paper and they are grouped here for convenience. *Microevolution* refers to those processes that shape and shape a population from generation to generation. *Macroevolution* in contrast describes the longer term behaviour that leads to the formation of species and higher groups of organisms. A *phylogeny* is a evolutionary family tree, a branching pattern that describes how species give rise to other species or die out. A *trait* is here used to abstractly refer to some heritable organismal trait, including bodyform, behaviour, senses *etc.*. A *key innovation* is the evolution of a novel organismal trait within a species that grants it an advantage over other species in terms of giving rise to other species.

Below, I present a general simulation solution to the problem of macroevolutionary analysis, and demonstrate its application on a perplexing issue in macroevolution, the causes of diversity.

1.2 A framework for recreating evolution

If analysis is problematic, the generating system is complex and inference uncertain, how can we study macroevolution? An alternative approach is to replay the process evolution. Put in its simplest form: by observing the results of a model that recreates macroevolution and comparing them to what we know of contemporary biology, conclusion can be made about the verisimilutude of the model. Further, the statistical support for these conclusions can be made by compiling the results of many runs. Use of a synthetic model also allows total knowledge of the system unlike real world data.

To this end, I present MeSA (Macroevolutionary Analysis and Simulation), a portable and extensible software program and framework for macroevolutionary investigation. Evolutionary hypotheses may be tested by recreating them within MeSA's simulation framework and analysing the results for comparison to the products of terrestrial evolution, contemporary organisms, and what knowledge we have of their ancestors.

1.2.1 *Simulation*

MeSA incorporates a discrete-event simulation of the salient events of macroevolution – the creation and extinction of species, and organismal trait evolution within species – which it uses to recreate and grow synthetic phylogenies. The factors controlling these events are expressed as a set of rules. Each rule defines the resultant event, its instantaneous rate, and how this rate might vary as a result of the current state of the entire evolutionary system or the particular species for which the rate is being calculated.For example, speciation can be invoked as a number of rule types including Markovian / equal rates (the rate is constant for all species at all times), latency (the rate falls immediately after speciation for a period and by an amount calculated by a constant function), dependent (the rate varies according to one or more trait values of the possessing species), density dependence (speciation rate alters according the total number of species in the system) and so on. A similar variety of rules exists for extinction, and also extends to a variety of mass extinction rules that may randomly or selectively target species.

Organismal traits may be discrete or continuous and may evolve in a punctuational or saltational manner (*i.e.* where change occurs only at speciation events or in a smooth continuous manner respectively). The direction and velocity of trait evolution can be brownian, directed brownian or dependent on other trait values. Where traits are evolving punctuationally, trait evolution of the two new daughter taxa can be set to be symmetrical (i.e., they inherit traits under the same rules) or asymmetrical (i.e., they inherit different rules). Asymmetrical trait inheritance can be used to model situations like master-slave evolution, where a single rapidly speciating master species gives rise repeatedly to slowly speciating slave species, as is seen in RNA viruses [103].

In order to simulate different periods of evolution in which conditions may change, rules can be grouped together in epochs. An epoch is a period within which those rules hold until an end-condition is reached. These conditions can be related to time or number of species. Thus periods of climatic change or the intervals between mass extinction can be simulated.

A discrete event simulation implies that no change occurs in the system between events. Several evolutionary models may violate this assumption. Examples include anagenesis or phyletic transformation, where taxa traits vary continuously across time between speciations and extinctions. In many such cases, this change does not effect other entities or events in the simulation and MeSA sums the change and updates their state at logical points in the simulation (*i.e.* speciation, extinction, the end of the simulation). In other cases these continuous changes interact with and affect the probability of other events. Examples include speciation with latency (where the probability of speciation changes with increasing taxa age), species senescence and stability (where vulnerability to extinction changes with increasing taxa age), or other cases where the probability of extinction or speciation is dependent on traits that vary continuously through time. Wholly realistic simulations of such systems are difficult to construct and computationally expensive. Where necessary, MeSA can discretize these continuous changes across time into a series of small differences that approximate smooth and continuous change.

1.2.2 *Data analysis and manipulation*

A wide range of macroevolutionary analyses are built into MeSA. While outside the scope of the current discussion, these include measures of phylogeny shape and balance, distance and diversity. To facilitate further analyses, MeSA can also report various information gathered across the nodes of a tree including the total number of taxa, the number of extant taxa, the age, time since speciation, number of children, number of siblings and number of ancestors ultimately subtended. These analyses (where appropriate) can be calculated over all nodes, or selectively over terminal nodes, internal nodes or extant taxa. Also, investigation of the effects of inaccuracy on tests often requires distortions in phylogenetic data to represent these uncertainties. To this end, MeSA allows systematic distortion of data, including equalising or randomising phylogenies branch lengths, randomising character states, and reducing phylogenies from paleontological to neontological forms.

A thorough comparison of a range of evolutionary scenarios can prove to be a long and tedious task. The detail and repetition required may discourage researchers from testing as many cases as necessary or lead to the making of mistakes. To this end, MeSA allows all the actions of the program – data manipulation, analysis, simulation and saving of results – to be linked together in a queue and executed as one. Furthermore, such actions can be embedded in loops that repeat those actions a given number of times or over the available set of trees. Thus a user can program the queue to repeatedly generate trees under a variety of conditions and then loop over those trees analysing them with the same set of metrics. While it is not as powerful or as flexible as these as full scripting approach, I feel its simplicity and robustness make up for this fact.

1.2.3 *Practical issues*

Finally, *MeSA* solves some prosaic issues *en passant*. Different file formats are often used for storing phylogenetic and organismal information and the (often manual) task of converting between them can be so error-prone and laborious as to prevent the practical use of datasets. Analytical tests are scattered across a variety of software packages (on a number of different platforms), hampering comparison. MeSA uses NEXUS [255] as its primary data format, because it is the most prevalent. Furthermore MeSA can read and write CAIC [318; 8], PHYLIP [124] and simple tab-delimited format files. Combined with existing translators this allows the program to easily interface with most forms of data.

MeSA may be found on the web at `<http://www.agapow.net/software/mesa>`. As it is written in standard C++ and eschews a graphical interface, it is highly portable across platforms and fast.

1.3 Life out of balance

Biology is skewed. It is clear from many studies that biodiversity and phylogenies show a non-random distribution of species. Some groups of organisms are richly fertile, having divided repeatedly to give rise to many, many ancestral species (*e.g.* beetles). Other groups of similar or greater age are depauperate, with few ancestors (*e.g.* pangolins, tuataras). Different taxa seem to follow different macroevolutionary clocks of wildly varying speeds. The meaning of this has long been a puzzle of great controversy (*e.g.* [107; 170; 345; 287; 314; 283; 353]). Even when confounding problems - such as artefacts from reconstruction, our limited sampling of the planets evolutionary history, and the random nature of evolution - are accounted for, this pattern remains. Persistent and unidentified factors are systematically distorting the shape of evolution.

1.3.1 *Key innovations*

The search for putative causes of this imbalance has a substantial recent history (*e.g.*, [29; 97]). However, given the wide array of processes that could operate at a macroevolutionary level and the aforementioned complexity of macroevolution, interpretation is necessarily complex. In response, a bewildering array of methodologies has arisen and created a spectrum of sometimes contradictory hypotheses about the factors promoting speciation and diversity. Confusion over the factors behind imbalance has thus been augmented by confusion over which methodologies are best under which circumstances.

One plausible explanation is the theory of key innovations. This argues that the chance evolution of certain novel morphological and functional traits is associated with new lineages and rapid diversification [401] and may be what creates major taxonomic groups [240]. Put another way, the "discovery" of a novel strategy, bodyform, behaviour *etc.* by a species gives it – and its descendants – a decisive advantage over other species. This places the idea at the heart of macroevolution, but the paucity of evidence (or approaches for examining it) makes it still "just potentially interesting" [97]. For example, one putative innovation in the plants is a shift in the mode of seed dispersal. Logically, the various patterns of dispersal (*e.g.*, animals carriers, wind, etc.) should have a great effect on the isolation of individual plants and subsequent possibilities for speciation [386]. However several studies failed to show any relationship (*e.g.*, [333; 387]), while one indicated a possible complex interaction [97].

1.3.2 *Methods*

Conventional forensic analysis having failed, here I attack the problem from the reverse perspective. Rather than search hopefully for correlations between any particular putative key traits and imbalance, instead the shape and imbalance of published plant, arthropod and vertebrate phylogenies shall be compared to those generated by MeSA under a range of evolutionary models in which speciation is determined by key innovations or by other factors (species age or random chance). The aim is to determine which scenarios yield phylogenies of realistic shape.

Phylogeny shape and imbalance is assessed with recently-developed imbalance measures [317; 9] that are based on the degree of asymmetry in individual nodes. The technical details of these metrics can be found in referenced publications, but briefly imbalance is calculated as the discrepancy between diversification in sister species. For example, a parent species gives rise to two daughter species that (between them) give rise to 10 contemporary species. Maximal imbalance occurs when the split of descendants between the daughters is 1 and 9 (*i.e.*, one daughter is the ancestor to all but one of the eventual descendants). Maximal balance occurs when the daughters give rise to the same number of descendants. An imbalance metric is calculated by dividing the observed imbalance by the maximal possible imbalance such that 1 is maximal imbalance, 0 is minimal and 0.5 the expected value under a markovian model (*i.e.*, when all species divide at the same rate).

Figure 1.1 shows the imbalance signature of a dataset of 208 phylogenies collected from the literature [52]; while Figure 1.2 shows the signatures for arthropods, plants and vertebrates separately. Note that imbalance is significantly greater than the 0.5 value expected under random speciation ($p < 0.0001$); in fact it rises significantly when the older, more ancestral species are tested (data not shown). Further, the imbalance signatures of different types of organisms cannot be distinguished statistically, either across all nodes (comparison of slopes $p > 0.9$; comparison of intercepts $p > 0.4$).

Can key innovations explain these patterns? A number of possible scenarios were constructed in MeSA. In each of these 100 phylogenies were grown until they contained 1000 species:

Key innovation In this model, a discrete trait X has two states, 0 and 1, that confer very different speciation rates. The ancestral state, 0, gives a speciation rate of 0.01, whereas the derived state, 1 (a key innovation), gives a tenfold higher speciation rate of 0.1. X was set to switch between states 0 and 1 with a per-lineage rate of 0.0005 per unit time. In these models, the overall speciation rate was initially low but rose as X changed to a 1 in some lineages.

Gradual trait-dependent Here, a continuous trait X evolves by gradual brownian motion, with the speciation rate being directly proportional to X. The starting value of was 100 and the instantaneous

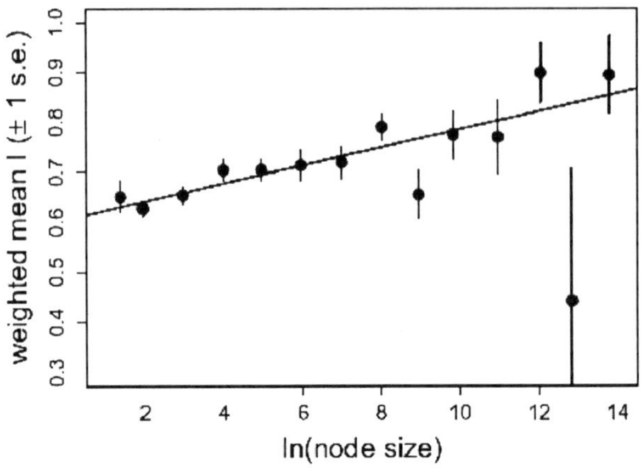

Fig. 1.1 Imbalance signature for combined phylogenies. Error bars are 1 standard error. Filled circles = arthropods; hollow circles = plants; hollow triangles = vertebrates. There are no significant differences among the three lines.

rate of speciation was determined as $a \cdot X + c$ where a and c are 0.0001. Thus, given the initial value of X, the speciation rate was low. There was no upper bound set on X, but if X evolved to a negative value it was set to zero

Punctuational trait-dependent Speciation rates were determined by the values of a punctuationally evolving trait, *i.e.* one that changed only at speciation time. In other respects, this was set as per gradual trait-dependent model. Previous research [362] has shown that punctuational change can lead to greater imbalance than gradual change.This is because lineages can be "stuck" with values of X associated with low speciation rates - their values of X cannot change until they speciate, which they are unlikely to do.

While to some extent the exact parameter values (and functions) chosen for these three models are arbitrary, they can still be interpreted as the general trends that would be seen in these models. This has been confirmed by experimentation (data not shown). Figure 1.3 shows their imbalance signatures. While details differ among the models and depend upon parameter values, in every case, the trend is for the larger (*i.e.* deeper and older) nodes to show an increasing imbalance. In contrast the smaller nodes are indistinguishable from the null model expectation of 0.5, very different from the pattern seen in the empirical data. In retrospect this is unsurprising. As time increases, the possible divergence in speciation rate between descendants increases. Species that speciate rapidly give rise to more species

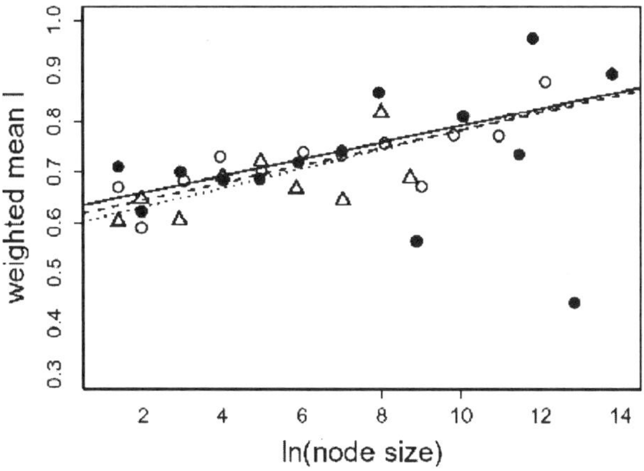

Fig. 1.2 Imbalance signature for select phylogenies. Error bars are 1 standard error. Filled circles = arthropods; hollow circles = plants; hollow triangles = vertebrates. There are no significant differences among the three lines.

that speciate rapidly, giving the characteristic "ski-jump" profile seen in the plots.

1.4 Discussion

Although the main point of this paper is a description of the approach, the biological findings deserve some consideration. While they do not dispose of key innovations as a possible cause of evolutionary imbalance, it is now arguable that key innovations are a seductively simple idea with many problems. If there were such a things as innovations, then species should be engaged in an "innovation arms-race" with the velocity of macroevolution increasing steadily through time as more innovative traits are discovered and spread. There is no evidence for this, with recent studies showing rates of speciation have little or no heritability [91; 332]. If key innovations are occurring, they are soon lost or lose their adaptive edge.

Perhaps the apparent advantages some organism enjoy over others can be explained not in terms of the organism but their environment. If a species is in the right situation to exploit a new environment, a new niche, a change in climate *etc.*, it and some descendants may enjoy a macroevolutionary advantage before the status quo is restored. This may help to explain why despite much research the list of demonstrated correlates of diversity is short. Also, some other different models of speciation may fit the observed data better than innovations. Patency models [316] describe

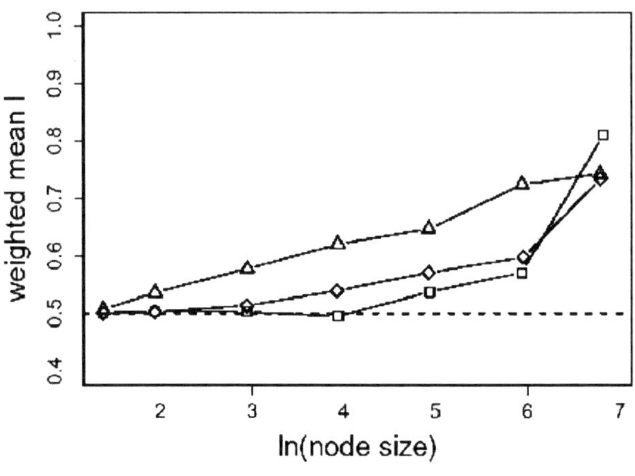

Fig. 1.3 Imbalance signatures from simulations. Diamonds = key innovations; triangles = gradual evolution model; inverted triangles = punctuational model. Points are averages from 100 simulations of each model. Bars showing 1 standard error would be hidden by the plotting symbols except at the largest node size.

situations where newly created species have a tendency to speciate again, creating rapidly forming chains of species. Other researchers have recently proposed models based on an abstract "niche-space" in which each species occupies a position in a multidimensional grid where the axes are characteristics of the species or environment in occupies [131]. Speciation occurs by species colonizing randomly generated new niches in the space. This produces a similar effect to patency as above because, as the space becomes occupied, the species radiates out from its origin into unpopulated niche-space. It is unclear as yet exactly what realworld properties these models represent. However they do apparently have certain features in common with "real" macroevolution: (1) closely related species can differ markedly in speciation rate, and (2) these rates are not strongly heritable.

As mentioned, phylogenies are but estimates and sometimes bad estimates. Could a systematic bias in constructing phylogenies have created the patterned seen? While there have been some reports linking poor quality data to imbalance, no such link was found in a more recent, larger survey [362]. Finally, one recent study has provocatively suggested that some imbalance is the result of human biases in categorization. As always, much work remains to be done.

Within biological rsearch there is a long history of model systems, using small manipulatable surrogates for complex realities. This of course has been very sucessful, with evolution being studied by manipulating bacterial populations in a test-tube, studying ecological interactions in small controlled environments, and animal behaviour in artificial populations of rodents and birds. Yet, there are many questions that cannot be handled by these approaches. Processes on the scale of months and metres cannot tell us about millions of years and continents. Bacterial evolution cannot tell us about megafauna, microevolution cannot tell us about macroevolution.

This, I think, demonstrates a clear need for artificial life and complex simulation approaches like MeSA in the biological sciences. Such techniques should not be seen as something fundamentally different, but as an extension of the use of model systems. While these methods are not unknown in biology, they are as yet underused. With them we can study problems at scales that were not previously accessible, recreating rather than (over)simplifying scenarios. Further, they have certain advantages. By directly recreating experimental questions rather than adapting pre-existing biology, no factor is silently incorporated into the experiment. Thus the effects of hidden states and variables are avoided. Finally, experiments can be easily replicated for statistical analysis, which cannot be done with many biological experiments. Note these advantages do not eliminate the need for rigor. If phenomena are observed in a synthetic model, researchers must always question whether these are the result of peculiarities of the model including implementation details. This, arguably, is the greatest barrier facing the wider use of artificial life systems.

As regards MeSA, it has already proven highly useful in exploring a wide variety of macroevolutionary questions, including the assessing methods to measuring phylogenetic imbalance [317], extrapolating the impact of current extinction trends [315] and patterns of viral evolution. Obviously the simulation framework could be extended to encompass other modes of extinction and speciation. The aforementioned niche-space model is an obvious candidate. Another direction for expansion would the development of modes including abiotic, non-evolving factors such as climate, latitude and resources. Such data is often available from stratigraphic records, and may allowing the testing of hypotheses that relate diversity to extrinsic (environmental) as opposed to intrinsic (organismal) factors.

A valid criticism of artificial life and simulation models, where attempts are made to replicate realworld biological phenomena, is that goodness-of-fit can be hard to measure. When one researcher sees punctuated equilibrium in a model, another may find only a fleeting resemblance. The experiments above escape this problem thanks to well defined metrics for imbalance. For less strictly defined phenomena assessing a match can be highly subjective. More sophisticated statistical models (such as Bayesian inference) would be useful. In this way, the match between synthetic model and realworld can be properly calculated.

Acknowledgement

Some of this work was published in [316] but is included here for clarity. I am grateful to the Institute for Animal Health for their support during the preparation of the manuscript. The early part of this work was supported by the Natural Environment Research Council (U.K.) through grant GR3/11526.

Chapter 2

Neural Evolution for Collision Detection & Resolution in a 2D Free Flight Environment

S. Alam[1], M. McPartland[1], M. Barlow[1], P. Lindsay[2], and H. A. Abbass[1]

[1]*ARC Center for Complex Systems, The Artificial Life & Adaptive Robotics Laboratory, Australian Defence Force Academy, University of New South Wales, Canberra, Australia*
{z3147403,z3153140,spike,abbass}@itee. adfa. edu. au

[2]*ARC Center for Complex Systems, School of Information Technology and Electrical Engineering. The University of Queensland St Lucia Queensland 4072 Australia*
pal@itee. uq. edu. au

During the last decade, Air Traffic movements worldwide have experienced a tremendous growth with new concepts such as Free Flight. Under Free Flight, current procedures of Airways and Waypoints for maintaining separation wouldn't be there. In the absence of Airway structure and ground based tactical support, automated conflict detection and resolution tools will be required to ensure safe and smooth flow of Air traffic. The main challenge is to develop robust and efficient conflict detection and resolution algorithms to achieve real time performance for complex scenarios of conflicts in a Free Flight Airspace. This paper investigates preliminary design and implementation issues in two dimension application of evolutionary techniques for collision detection and resolution. The preliminary results demonstrate that an artificial neural network (ANN) using evolutionary techniques can be trained not only follow optimum trajectories, but also to detect and avoid collisions in two dimensions.

2.1 Background

Collision detection and resolution (CD&R) is a fundamental problem to many mission critical & real time applications. It is also of prime importance in Air Traffic Control, Vehicle Navigation, Robotics, and Maritime operations. A literature review revealed that most of the CD&R techniques discussed assumes the availability of intent information of all the other aircrafts in the vicinity for the purpose of trajectory prediction. However in the real life scenario it's highly unlikely that long term intent information can be made available to predict conflicts with accuracy. Free flight concept lets the pilot decide the course, altitude, air speed and other parameters dynamically in real time. This makes it extremely hard for conventional CD&R algorithms to work in such environment. Use of linear programming techniques were discussed in the SOM (Stream Operation Manager) model [291], it suggests techniques for automated integration of aircraft separation, merging and stream management using linear programming techniques. The SOM input requires quite a few data to be input before hand including static Aircraft performance data, which given the dynamic nature of a flight plan may change dramatically en-route. One of the key parameters given as input in case of a conflict resolution is pilot preference, which may override the Aircraft performance envelop while negotiating a conflict with other aircraft compromising the safety. Another mathematical modeling technique, multi-point boundary value problem with ordinary differential equations, to be solved numerically with the multiple shooting method was discussed by Rainer Lachner (Collision Avoidance as a Differential Game) [230]. However the unavoidable analytical or numerical calculation of hundreds, thousands or even ten thousands of optimal trajectories to obtain the optimal strategies representation is generally a difficult task given the time criticality and lack of high computing power onboard.

A geometrical approach to the problem of CD&R was investigated by Ki-Yin Chang and Gene Parberry [75] by using 4 Geometry Maze routing algorithm (A modified version of 2D Maze Lee Algorithm) to obtain the particular route of each naval ship that have potential to collide, which is detected by simulating the particular routes with ship domains. The algorithm provides linear time complexity and guarantees to find an optimal path if exists. However the algorithm is suitable for navigational situations at sea characterised by slow cruise speed and large time window for heading change. It assumes that trajectory and speed of the naval ships remains unchanged which is highly unlikely in a free flight environment. A rule based approach for solving CD&R problem is described by I. Hwang and C. Tomlin (Protocol Based CD&R for ATC control) [203]. It uses multiple conflict detection model and detects collision in 2D horizontal plane. It uses rule based approach for conflict resolution. The algorithm discussed by the authors is robust to uncertainties in aircraft position, heading & velocity. However the experiments performed by the authors using fixed 20 minutes lookahead time window for conflict prediction gave rise to lots of false alerts and act as a limiting factor to extend the model further. Free flight model where uncertainty in trajectory in inherent makes the detection of conflict between aircrafts very complex task, A. Chakravarthy and D. Ghose [74] proposed collision cone approach as an aid to collision detection and avoidance between irregularly shaped moving objects with unknown trajectories. This mathematical model was restricted to CD&R between two objects only and is discussed mathematically with out any simulation and experimental results. This uncertainty in intent information and its complexity in detecting conflict in a free routing environment was also investigated by Prandini, M. [309]. The 2-D, two Aircraft CD&R algorithm uses probabilistic framework, thus allowing uncertainty in the aircraft positions to be explicitly taken in to account when detecting a potential conflict. The authors use the fight plans of the two aircraft, generate pairs of aircraft trajectories over a 20 minutes time horizon according to the discretized version of the stochastic differential equation, and do the computations for conflict detection. High false alert rate(18%) shows that algorithm needs further improvements and sensitivity analysis on the part of crossing angles, minimum deterministic distance, and time of minimum distance, in order to set a value for the threshold which is appropriate for the typically encountered path configuration. Many of these 2D CD&R algorithms are justifiable for a completely known environment. A partially known dynamic environment like a free flight airspace where long term trajectories cannot be predicted, requires an entirely different approach. An evolutionary algorithm based model may handle the flexibility required in free routing model and may handle additional constraints. These facts and also the relative simplicity of dealing with the two dimensional case have caused our approach proposed in this paper to focus on 2D conflict detection, and horizontal resolution maneuvers.

2.2 Modelling the Problem

It is assumed for modeling the problem that two Aircraft are flying at a constant speed and altitude, in a 2-D free flight environment. The Aircrafts (referred to as agents from here onwards) explore the environment trying to reach their destination in a given time interval. The agents have to minimize the off track error (The difference between the planned trajectory and actual trajectory) and to detect & resolve collision with other agents. Our objective is to build and train an artificial neural network (ANN) applying Evolutionary techniques which are able to modify the heading of the Agents to avoid the conflict while keeping the agent nearest to its optimal trajectory and directing them to reach their destinations. The experimental environment (airspace) is made up of 2-dimensional cells and is discrete. The preliminary experiment used a 10x10 environment with 2 agents. For experimental purpose the starting position of an agent A and the starting position of an agent B are chosen such that their mission trajectories cross each other ensuring a collision scenario, if the agents maintain planned paths and do not use avoidance.

Each Agent has proximity sensors for the neighboring eight cells and they also emit probability signals into the surrounding cells for likelihood of occupation on their next move. These signals are directly proportional to the Euclidian Distance from the cell position to the destination. Any obstacles like terrain proximity, Bad weather; Special Use Airspace is read by 8 obstacle sensor. The ANN has to deal with the issue of Agents reaching the edges of the discrete environment, for that a wrap-around environment was implemented. An Agent can wrap-around the environment from left to right, top to bottom as well as the four diagonals. For example if an Agent is at position (0,10) and decides to move East, the Agent's new position is (29,10). As wrap-around behavior is not desired, Agents that perform it are penalized. In conceptual terms wrapping around the environment is a greater distance, and therefore time, than moving one cell within the environment's boundaries. Agents' movements are updated asynchronously to ensure no particular Agent is biased.

At each time step the environment is updated according to the following steps.

(1) **Sense:** Compute the Euclidian distance to destination from each neighboring cell.
Compute the probability of collision for each neighboring cell based on the other Agent proximity signals.
Compute the probability of collision for each neighboring cell for the presence of Obstacles.

(2) **Make a decision:** Based on the objective function Set the inputs to the ANN (proximity signals, obstacles, distance)

(3) **Move:** Update the Agent's position based on the ANN output.

(4) **Update:** Update the neighboring cells of the new position for probability of occupancy in the next move.

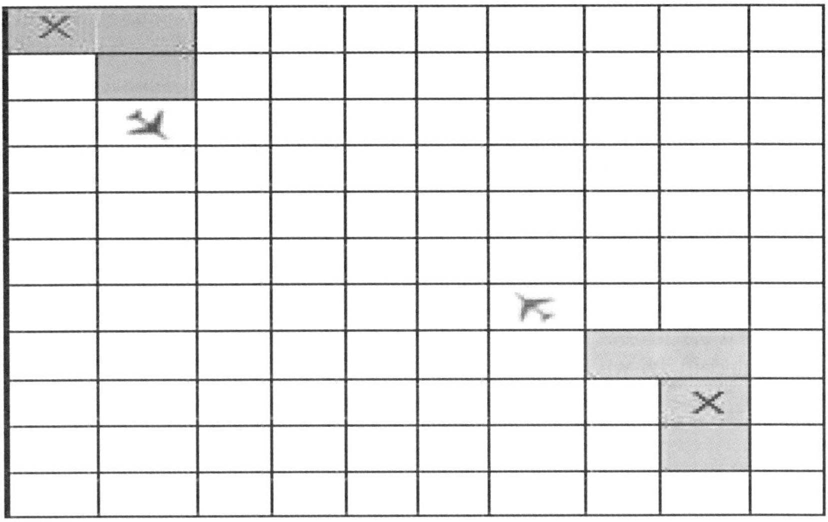

Fig. 2.1 Representation of the cellular environment with two Agents, their destination targets and their paths. X marks the destination target of the Agents; the shaded area represents the paths of the Agents.

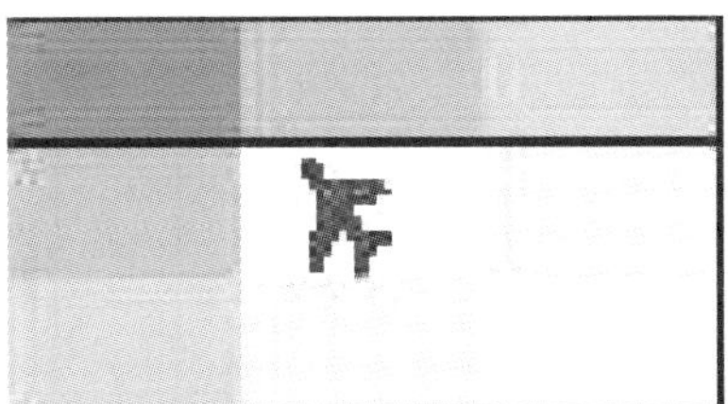

Fig. 2.2 Representation of the cell occupancy signals model based on the destination target as shown in Figure 1, the depth of shading is indicative of the distance to the target.

Each Agent emit occupancy signals for each 8 adjacent neighboring cell at every time step indicating the normalized Euclidean distance to its destination according to the following equation.

$$Ph_i = \frac{\sqrt{(X1_i - X2_i) + (Y1_i - Y2_i)}}{\sum_{n=1}^{8} \sqrt{(X1_n - X2_n) + (Y1_n - Y2_n)}} \tag{2.1}$$

After each step the occupancy signals are re-assigned. The Agents are equipped with 18 sensors: (a) 1 distance to destination sensor; (b) 9 proximity sensors; and, (c) 8 obstacle sensors. The distance sensor indicates the Euclidean distance from the current position to the Agent's destination position. The proximity sensors detect the combined Agent's occupancy signal values in the adjacent cells to the current position. The obstacle sensor's act similarly to the signal values, by detecting obstacles in adjacent cells (for the preliminary experiments, the Agent's are the only obstacles and are represented by 1. 0 if present and 0. 0 if not). At the end of each run, the Euclidian distance from the agent's end position to destination is computed for fitness. The collision counter maintains the number of times an agent collided with another. In this preliminary experimentation, if the agent reaches its destination then its position is not updated further.

2.2.1 *Collision Detection*

Collisions among the agents are detected according to the following rules as shown in Fig 3:

(1) Agent A and B occupy the same cell
(2) Agent A and B have switched cells and,
(3) Agent A and B's paths have crossed over in the same time step.

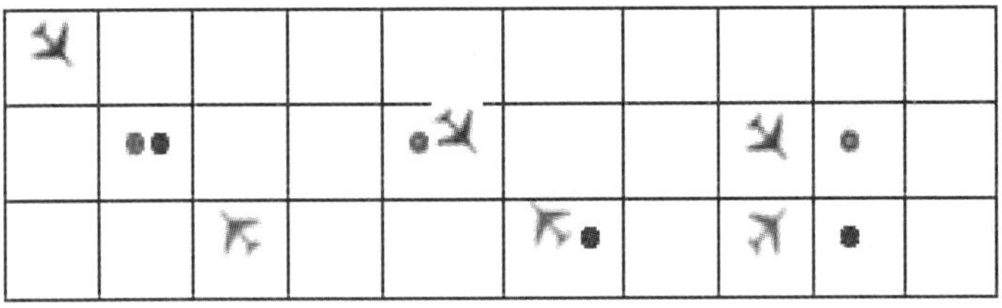

Fig. 2.3 Collision Scenarios between two agents in a 2D environment

2.3 The Neural Network Structure

A three layer ANN architecture is used (Fig 4). The input layer has 18 inputs based on the Agent sensors as mentioned in the section above, the middle (hidden) layer contains a fixed number of nodes and are varied as 2, 5, 10 and 12. The third layer, the output layer has three binary outputs which denote the direction of movement i.e. $2^3 = 8$ possible moves.

The ANN topology that was used was a feed forward network with input to output connections and recurrence on the hidden neurons (see Fig 4). Recurrent connections were used to attempt to stop the Agents from getting 'stuck' in a two-step movement (moving back and forth in two cells only). This problem will occur, particularly in feed-forward topologies, when the inputs to the network are identical in the cells. For example if the ANN decides to move to the cell directly East, then based on the new inputs for the Easterly cell decides to move back to the original cell, the original inputs are identical to the first time the Agent was here (assuming another Agent is not nearby). Using recurrent connections adds a dimension of time to the ANN, so this problem is less likely to occur. Note if the context neurons are not being used by the ANN this problem may still occur. Classical back propagation cannot be used in our case because conflict free trajectories are not known in advance. We have used the Self-Adaptive Pareto Artificial Neural Network (SPANN-R) algorithm [1; 380] for evolving the weights of the ANN.

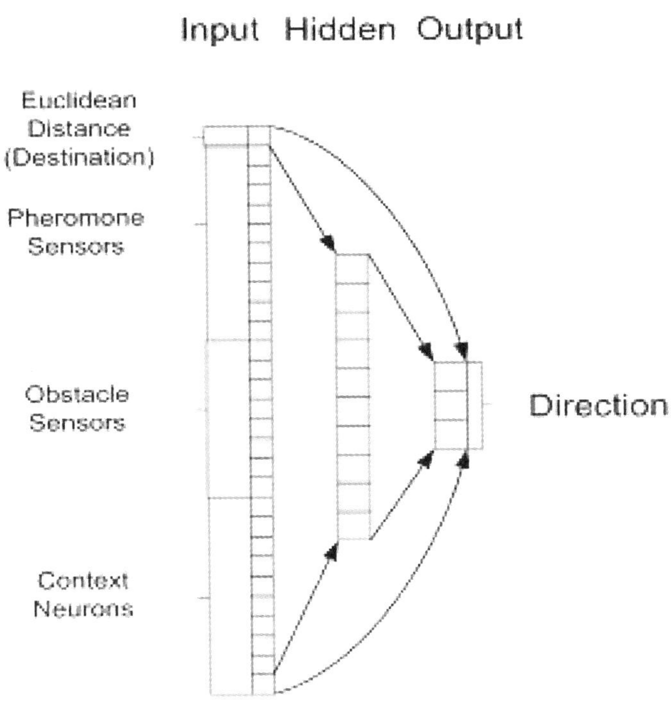

Fig. 2.4 Type 3 - ANN topology with recurrence in the hidden nodes

2.4 Preliminary Experimental Setup

Two fitness functions were used. The first fitness function attempts to minimize the off track distance travel by the Agents and reduce the time to find the target destination represented by the following equation.

$$f1 = \sum_{n=1}^{N} D(Cur_n, Dest_n) + T_n + \alpha.P_n \tag{2.2}$$

where N is the number of Agents, D is the Euclidean distance between the current position of the Agent and its destination, T is the time the Agent found its destination (this will be the total time steps if the Agent does not reach its destination). T can be seen as a penalty term which is used to set a pressure on each Agent to get from origin to destination in shortest possible time. If the Agent is not able to find the destination at the end of a run, a larger value is assigned for T. α is the wrap around penalty and P is the number of times the agent wrapped around the environment, the more times the agent wraps around an environment, the higher is the P value.

The second fitness function is the total number of collisions that occurred in the run.

$$f2 = C \tag{2.3}$$

where C is the total number of collisions detected in the run.

The evolutionary runs were performed on a population size of 100 chromosomes for 1000 generations. The initialization of the ANN weights in SPANN-R is done using a Gaussian N(0,1). The crossover rate and mutation rate are assigned random values according to a uniform distribution between [0, 1].

These functions were designed to guide the evolution to avoid the other Agents as well as to find their targets in the shortest possible path. Direct encoding is used here as it is easy to implement and simple to understand encoding scheme. The mutation rate and crossover rate are assigned when the random initial population of potential solutions is created. However,the representation scheme of SPANN-R algorithm allows for the self adaptation of crossover and mutation rate during optimization. It is recognized that there are issues with the design in that it does not promote generalization of the networks (i.e. changing the environment will cause unexpected results). The preliminary experiments were designed to test the initial theory of target finding and collision avoidance behavior in a static scenario.

Table 2.1 Environment Parameter set up for preliminary runs

Parameter	Value
World Dimensions	10x10
Number of Agents	2
Agent A Origin	1,1
Agent A Destination	8,8
Agent B Origin	8,2
Agent B Destination	2,8
Generations	1000
Time Steps	40
Population size	100
Hidden Nodes	2,5,10,12

2.4.1 *Preliminary Results*

The preliminary experiments and results show that an ANN can be trained for target finding and collision avoidance. The results are recorded from the best performing ANN in fitness 1 with the lowest scoring fitness 2; or, in other words we only consider solutions where no collisions occurred. From the results below it can be seen that an ANN with 10 hidden nodes has a better chance of finding a good solution as seen by the low average of the solution set. It is also seen that ANNs with hidden nodes as low as 2 are still able to find a good solution but it seems that the ruggedness of the landscape is high. The best overall solution is found in the experiment with 12 hidden nodes (value of 55) which is the lowest possible value that can be obtained without a collision occurring. From these results it was decided to use 10 hidden nodes for the main experiments.

The implementation of the wrap around behavior adds additional complexity to the problem. From observation of initial generations the agents were found to wrap around the environment and find their destination positions in a couple of steps. Due to the wrap around penalty, the evolution eventually finds better solutions which do not perform the wrap around behavior. This adds a level of complexity to the solution space.

2.5 Main Experiment Setup

For the main experiment, the Agents have their mission trajectories embedded in them and one of the components of the modified fitness function is to try to minimize the off track movement in the environment while still trying to reach the destination. The trajectories are generated from an equation of an ellipse whose major and minor axis points are given.

Table 2.2 Preliminary results with various hidden units and random seeds. In all results, there was no collision

RandSeed	2 Nodes	5 Nodes	10 Nodes	12 Nodes
101	56. 00	128. 00	56. 00	122. 42
102	128. 00	115. 94	68. 00	68. 00
103	102. 35	128. 00	55. 00	68. 00
104	105. 76	119. 89	68. 00	120. 31
105	69. 23	69. 00	128. 00	55. 00
106	128. 00	56. 00	56. 00	55. 00
107	128. 00	88. 63	115. 94	76. 21
108	118. 24	104. 76	69. 23	128. 00
109	104. 76	56. 00	68. 00	115. 43
110	120. 45	83. 65	58. 00	68. 00

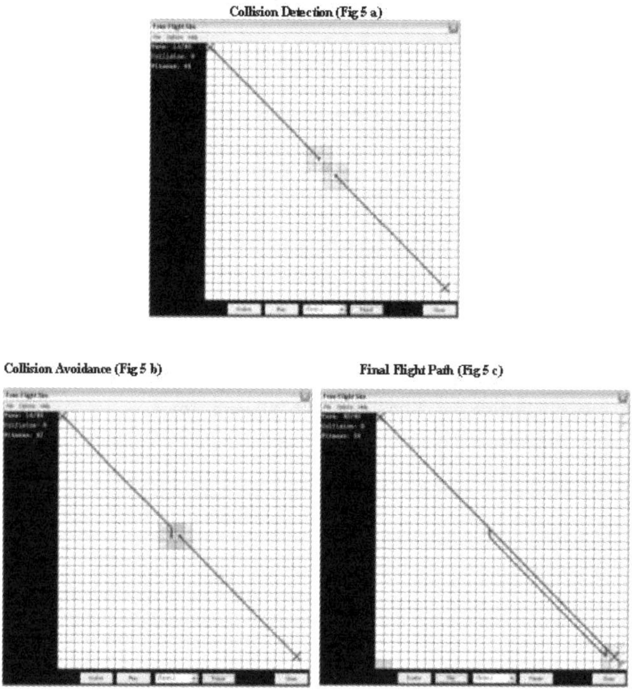

Fig. 2.5 5a: The agents moving in the environment towards their target with the shaded area representing their occupancy sensor and shows the two agents coming in close proximity of each other, detecting the conflict. 5b: The agent resolving the conflict by changing the direction heading. 5c: shows the final path taken by two agents at the end of simulation.

$$\frac{(x - x_0)^2}{a^2} + \frac{(y - y_0)^2}{b^2} \tag{2.4}$$

The center of the ellipse is at $x0$, $y0$, assuming $b > a$, b is called the minor axis and a is called the major axis. From the start position and end position for an Agent, the center of the ellipse is computed and the elliptical mission trajectory is generated. This is to ensure that the ANN can learn to follow an elliptical path rather then just moving the Agents in a straight line.

The ANN is trained to move the Agents towards their destination in desired time steps as computed by their elliptical mission trajectory. If the Agent reaches its destination early, there will be a penalty associated in terms of the extra distance which it will move away from its destination in the remaining time steps. To compensate for any collision avoidance manoeuvres extra time steps are allowed to an agent. If an Agent doesn't reach its destination in the desired time steps, then there will be a penalty cost based on the remaining dynamic distance to its destination.

2.6 Fitness Function

For the main experiments a powerful class of technique known as dynamic programming is used [37]. A particular sub-category known as dynamic time warping (DTW), that has been successfully utilized in automatic speech processing [320] is employed for computing the Fitness Function. DTW is a method for calculating the distance between two time-varying sets of values. The method seeks the best temporal alignment between the two samples under various constraints - the alignment can be visualized as 'stretching' (repeating) certain portions of each set at certain times. Given that alignment - which ensures start and finish alignment, and that all values from each set are used - a minimum distance between the two sets is found. This technique suits our test environments given the Agents mission path and dynamic nature of the actual path they take while exploring the environment. The agents may deviate from their mission path for a variety of reasons like avoiding a collision, implementing wrap around the environment etc. For computing the first fitness as the area between the two paths (mission path and actual path) this technique 'compensates' for relatively minor temporal differences, while still 'accentuating' significant temporal (and raw value) differences between the two paths. The first fitness is given by

$$f1 = D(M, N) \tag{2.5}$$

Table 2.3 Parameters used for the main experimental setup

Parameter	Value
World Dimensions	10x10
Number of Agents	2
Generations	1000
Time Steps	40
Population size	100
Hidden Nodes	10

M is the mission path $= b_1, b_2, b_m$ and N is the actual path $= a_1, a_2, ..a_n$. $D(1,1) = d(b1, a1)$ where d is the Euclidian distance $D(i,j) = min\{D(i-1,j), D(i-1,j-1), D(i,j-1)\} + d(bi, aj)$ Where $d(bs, at) = \sqrt{(bs_x - at_x)^2 + (bs_y - at_y)^2}$ at position x, y. The second fitness function remains the same as total number of collisions that occurred in the run.

$$f2 = C \tag{2.6}$$

where C is the total number of collisions detected in the run. The fitness function for the main experiment is

$$F = min(f1 + \alpha.P_n + f2) \tag{2.7}$$

where $\alpha.P_n$ is the wraparound penalty as is described in Equation 2 To ensure generalization, four different scenarios were taken for each run. Each scenario has different start and end position for the Agents, these scenarios ensure that the agents have colliding trajectories. For each run each scenario gives its own fitness which is then averaged out. This helps the ANN learn motion, avoidance and target acquisition beyond symmetrical paths.

2.7 Main Results and Analysis

The experiments and results with the new fitness function and generalization show that an ANN can be very efficiently trained for multi objective scenarios, viz. following an elliptical trajectory, Conflict Detection and Resolution, finding the target and avoiding the wraparound behavior. The results are recorded from the best performing ANN in fitness 1 with the lowest scoring fitness 2; we analyzed only those solutions where no collisions occurred. The best overall solution found in the experiment with Fitness Function Value 10. 82. This ANN keeps the off track error to its minimum, detects collision, avoids the collision and drives the agents towards their target. Hinton diagram display the output behavior of the hidden nodes with every time step of the run.

Table 2.4 Results of Experiment

	Fitness F1	Fitness F2	Fitness Value(F1+F2)
Best Case	10. 82	0. 0	10. 82
Worst Case	144. 70	0. 0	144. 70
Average Case			

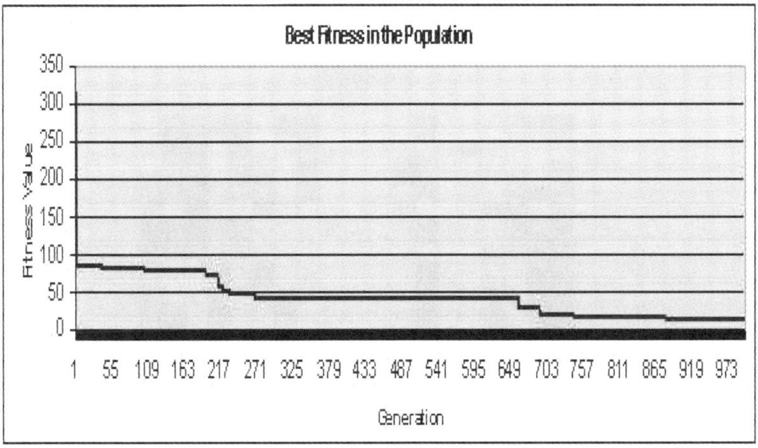

Fig. 2.6 Evolution Graph showing the best Fitness values in a population set of 100 for 1000 generations

It appears from Figure 2.7 that hidden node 2 activation state drives the agent towards its trajectory as well as collision detection. The Hidden node 1 is activated during collision resolution, hidden node 1 and 6 activates to bring the agent back to its trajectory following an off course resolution maneuver and hidden node 5 is activated when destination is reached. Hidden nodes 4 and 8 remain inactive during the run.

From Figure 2.7 it appears also that that hidden node 5 drives the agents to their trajectory path and node 3, 10 and 4 activates during conflict detection, resolution and resume own navigation respectively. Hidden Node 1and 7 remain inactive during the simulation run.

From Figure 2.7 it also appears that hidden nodes 7 and 3 initially drives the agent to its planned trajectory and then collision detection and resolution are regulated by hidden nodes 6 and 5 respectively. Node 3 drives the agent to resume its own navigation. Hidden nodes 10 remain inactive during the run.

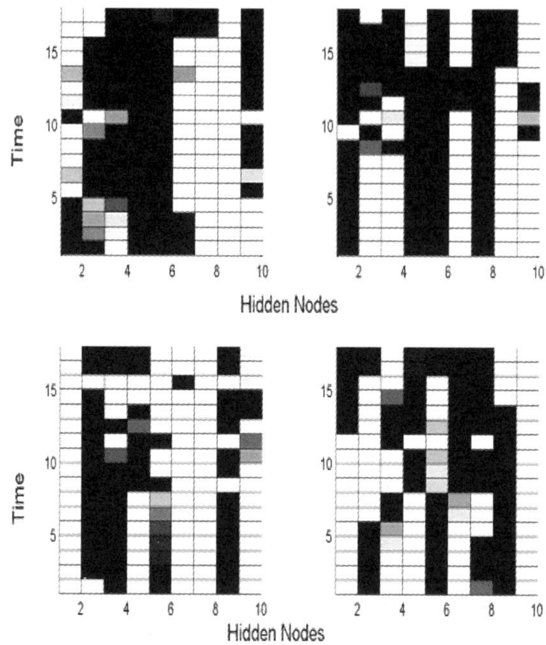

Fig. 2.7 Hinton diagrams showing the output behavior of the hidden nodes with every time step of the run

2.8 Conclusion & Future Work

Experiments and results shows that an ANN can be trained efficiently using Evolutionary techniques for collision detection and resolution in a 2-D environment using horizontal manoeuvre techniques. With the new fitness function and generalization after 1000 generations the Neural Network not only learns well to guide the Agents in a 2D environment to reach their desired destination while minimizing the cross track error (deviation from optimal trajectory) but also detects and resolves collisions with other agents in the environment.

On of the principle requirement of future Free Flight system will be robust CD&R mechanism. Since detecting conflicts with aircrafts on random routes is more difficult than if the air traffic were on structured-airways, the pilots/controller will have to rely on an automated system to detect problems and to provide solutions. Such a system can only be implemented by developing a robust and efficient CD&R algorithm.

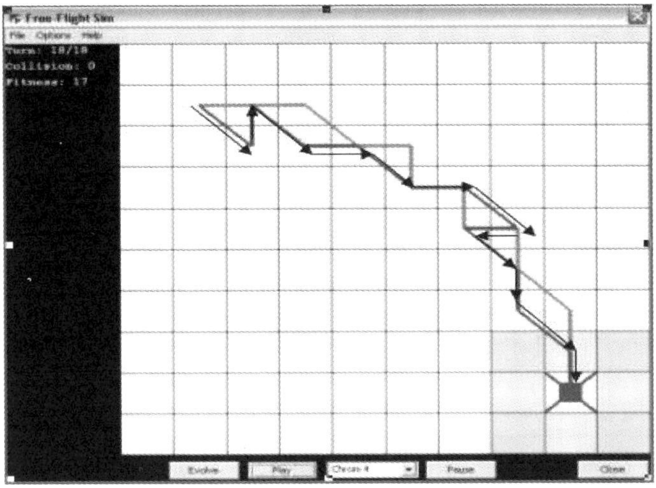

Fig. 2.8 A single agent movement in the environment showing that an ANN can be trained to follow an elliptical path, the light shade lines denotes the original mission trajectory and dark shade denotes the actual trajectory

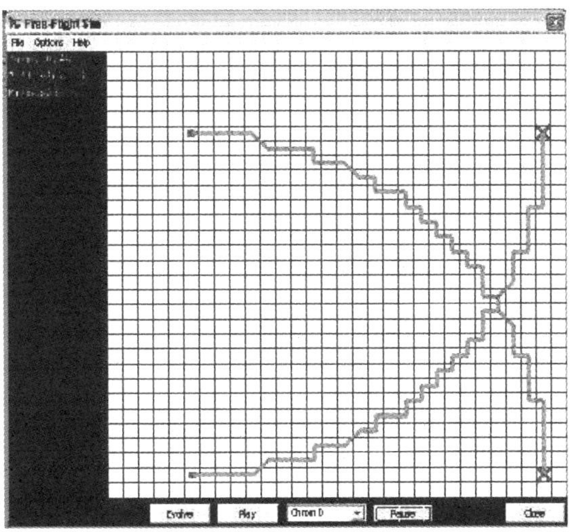

Fig. 2.9 The initial position of agents in the 2D environment for the main experiment setup with their destination marked as X. The elliptical trajectories displayed are optimal path to destination. The shaded rectangle shows a potential conflict zone

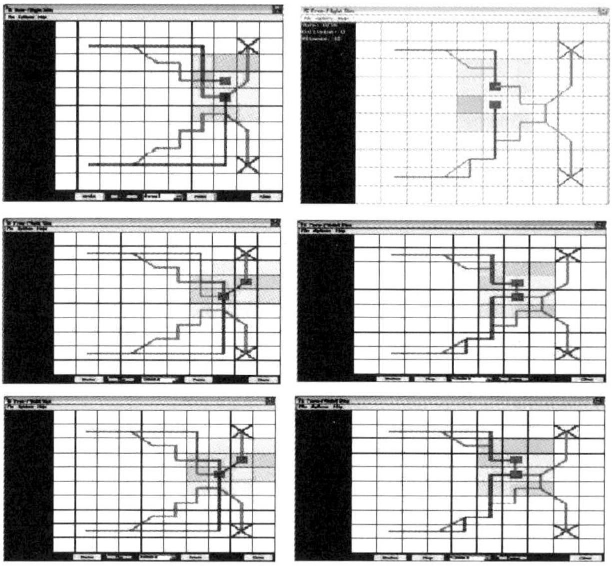

Fig. 2.10 A (top): Two scenarios showing the agents approaching each other and detecting a collision B(middle): Two scenarios showing the agents in collision resolution by change of heading. C(bottom): Two scenarios showing the agents reaching their destination without colliding with each other

Future work involves extending the model to three dimensions and to add other parameters for collision resolutions viz. speed, heading and vertical manoeuvres. Moving the environment from discrete to the continuous domain will bring new challenges in training and testing the ANN. Future developments on the 3-D environment with continuous domain will certainly affect the architecture of ANN and increase the complexity of the system,and may give a deeper insight in understanding its behavior.

Acknowledgements

This work is supported by the Australian Research Council (ARC) Centre for Complex Systems grant number CEO0348249.

Chapter 3

Cooperative Coevolution of Genotype–Phenotype Mappings to Solve Epistatic Optimization Problems

L. T. Bui, H. A. Abbass, and D. Essam

The ARC Centre for Complex Systems,
The Artificial Life and Adaptive Robotics Laboratory,
School of ITEE, UNSW@ADFA, Canberra, Australia
E-mail: {l.bui,h.abbass,d.essam}@adfa.edu.au

Genotype-phenotype mapping plays an important role in the evolutionary process. In this paper, we argue that an adaptive mapping could help to solve a special class of highly epistatic problems known as rotated problems. Our conjecture is that co–evolving the mapping represented by a population of matrices in parallel with the genotypes will overcome the problem of epistasis. We use the fast evolutionary programming (FEP) algorithm which is known to be unsuitable for rotated problems. We compare the results against the traditional FEP and a conventional co-evolutionary algorithm. The results show that, in tackling rotated problems, both FEP and the co-evolutionary FEP were inferior to the proposed model.

3.1 Introduction

The biological evolution can arguably and debatably be seen as an optimization process in which the fittest individuals of a species survive from the competition with others throughout the evolutionary chain. This feature attracts much attention, particularly in the field of optimization. To simulate evolution, a population of individuals are distributed randomly in a search space. They then evolve and compete overtime, and the population gradually approaches the area of optimality.

In biology, to clarify the structure of an individual (or organism), geneticists distinguish between the concepts of genotype and phenotype, or the genetic structure and the physical characteristics of the organism. The genotype is the genetic material inherited from the parents, while the phenotype is the expressed traits of an organism [65]. The mechanism by which genetic variations are mapped onto phenotypic variations has a strong impact on the evolvability of the organism [66].

In general, the genotype-phenotype mapping (GPM) has an important role in the interaction between the evolutionary process and the process whereby the organism interacts with the surrounding environment (learning process). Despite that GPM correlates the phenotype of an organism to its genotype, organisms with identical genotypes can express variations in their phenotypes. The Baldwin effect reveals that the behaviors learnt on the level of organisms have a high impact on the evolution of species [132].

The Baldwin effect works through phenotypic plasticity and genetic/environment canalization [293; 400; 404]. In phenotypic plasticity, the external environment can contribute to the formation of the phenotype; thus the phenotype does not depend solely on the instructions encoded in the genotype. Through learning, an individual may adapt to a mutation that would be useless to the individual without the extra learning step. Thus, if the fitness of the individual increases as a result of this mutation+learning component, the mutated gene may proliferate in the population. Evolution, however, cannot depend on this costly phenotypic plasticity alone. Evolution may then play a role in maintaining the learnt behavior through genetic/environment canalization. In genetic canalization, specific genes become resistent to genetic mutations; thus persist to exist in the population. In environment canalization, a stablizing natural selection process may favor genes which reduce environmental variance of a trait. The Baldwin effect differs from Lamarckian evolution in that there is no direct alteration in the genotype as a result of the learning occurring on the phenotypic level, while in Lamarckian evolution, a direct alteration in the genotype exists.

Given the importance of GPM in biology, we hypothesize that this mapping can play an important role in optimization as well. In particular, we have considered an adaptive GPM scheme to solve rotated optimization problems. In rotated optimization problems, a number of almost independent variables are rotated. The resultant variables thus become highly dependent and the optimization problem is defined on those rotated variables This problem is very difficult under the following two assumptions:

(1) We do not have access to the independent variables, otherwise, the optimization problem becomes trivial. Thus, due to the dependency

(interaction, epistasis) between the variables, the problem represents a major challenge to genetic algorithms.

(2) The optimization problem is black box, thus it is not possible to analyze the objective function analytically and explicit calculation of exact gradient information is not possible.

We assume that the GPM is neither unique or fixed; thus as natural selection favors genes, it also favors successful mappings. We co-evolve populations of genotypes and mappings (matrices). In this paper, we only look at linearly rotated problems, where the original space is linearly transformed to a new space. A genotype is mapped to it's phenotype by multiplying it with a selected matrix. Since we do not know the mapping, the genotypic space simulates the dependant variable space, and hence the population of matrices can simplify the problem by simulating the inverse of the mappings. Thus, a phenotype in this paper is the product of a chromosome in the genotype population and a matrix in the mapping population. As both populations co-evolve, the inverse effect of rotation emerges to relax the difficulty of the epistasis. To validate this model, we have compared its performance to that of conventional Fast Evolutionary Programming (FEP) [439] and a cooperative co-evolutionary version of FEP called CFEP [246]. We chose these two methods because the genetic operators of evolutionary programming are known not to work for rotated problems. Thus, if the proposed method converges, it will converge mainly because of the mappings, not because of the operators. In the rest of the paper, we first review work on co-evolutionary GPM, and then explain the proposed model, and the setup for the comparative study. The paper concludes with results, and discussions.

3.2 The use of co-evolution for GPM

There are several pieces of work on the integration of co-evolution with the process of GPM. In general, by allowing multiple populations that are interactively (cooperatively or competitively) co-evolved, GPM could be evolved in parallel with the evolution of genotypes. A typical example is the co-evolutionary model in the work of Potter and Dejong [307; 308] where the phenotype is built by collecting individuals from each sub population. Individuals could be either selected at random or from the best of each sub population. However, although this is a kind of of indirect encoding, the mapping structure here is fixed and non-evolved.

A thorough study about the co-evolutionary approach was carried out by Paredis [299]. The author proposed a symbiotic co-evolution system, called SYMBIOT, to solve 3-bit deceptive problems. It involved two cooperative populations, one is the population of solutions and the other is the population of permutations. The mapping between a permutation and a solution is implemented by using a permutation to define the order of genes in a solution. This ordered solution will be applied to the target function in order to get the fitness value of the solution. Each encounter between solutions and permutations involves two solutions and a permutation. These two solutions are used to generate a child. The permutation is used for both the parents and the child. For it, each encounter gives the permutation a pay-off value that is the average fitness of the parents divided by the fitness of the child. Each permutation has a history list of 20 most recent encounters. This list is updated continuously over generations. The sum of pay-off values in the history will be set to be the fitness of the corresponding permutation. In this way, the fitness becomes continuous and partial. This evaluating technique is called Life-Time Fitness Evaluation (LTFE).

Murao et al [285] proposed an idea to apply this approach to engineering design using genetic algorithms. The authors hypothesized that a GPM could help designers to explore the search space to find a possible solution candidate, but that the lack of prior knowledge of the problem poses a difficulty to the design of a proper GPM. In order to overcome this, it is possible to evolve the GPM along with the evolution of genotypes. This will therefore help to relax the aforementioned difficulty. Basically, the system is designed similarly to the SYMBIOT model described in Paredis's work [299]. In Murao's system, two populations are built, one is the population of sequences of genes, while the other is the population of sequences of permutations of positions in the genome. A genome is generated by the permutation of all the genes in the first population (the population of seeds) in the order defined by an individual (a rule) of the second population. However, instead of using LTFE, the fitness of a seed is determined by first playing the seed against all rules (permutations), and then the maximum value obtained is given to the fitness of the seed. The fitness of a rule is evaluated in the same manner. The empirical experiments showed that the co-evolutionary approach outperformed a conventional GA on a 3-bit deceptive problem. Although the system was limited to the binary domain, it gives a strong implication of the usefulness of the co-evolutionary approach in helping the GPM process.

As stated by Murao [285], an adaptive mapping would be convenient for engineering design problems where it is difficult or impossible to obtain the prior knowledge of the problem to determine a suitable mapping. In this situation, adaptive mapping can help to relax this difficulty. By adaptive mapping, we mean that a genotype is mapped indirectly to a phenotype. Although, to date, it is not clear that it is better to allow adaptive mapping in combination with or separated from the genotype, the application of co-evolution has great potential to adaptive mapping. The analysis of evolving populations of combined mappings and genotypes is outside the scope of this paper.

The above work shows that for a certain problem and with a suitable design, co-evolution can help search algorithms to progress better towards an optima than comparable conventional EAs. It would be ideal if we can find some way to effectively combine the adaptive mapping with co-evolutionary features. In previous work, a number of problems were considered by co-evolved adaptive mapping systems, such as a binary 3-bit deceptive problem where the mapping was the permutations of gene loci in the chromosome. Potter and De Jong [307; 308] proposed another scheme for adaptive mapping. For it, each gene was evolved cooperatively in differing populations, and combination of a single representative from each population defined a phenotype vector. However, this approach for real-parameter optimization is most suitable when variables do not interact. When high epistasis exists, the interaction between variables is much harder to capture.

3.3 The proposed algorithm

Undoubtedly GPM has an important role in both biological and computational evolution. It is the bridge for two parallel processes: evolution and learning. It is possible to use GPM as a tool to impose bias and to implement strategies to control the interaction between evolution and learning. One of these possibilities is the evolution of a GPM. Instead of a static representation, a mapping can be adapted to be better suited for a given problem. This corresponds to the situation where the mapping evolves under the effect of both the interaction of genetic materials, as well as with the surrounding environment.

For rotated problems, the genotype always has to undergo a transformation operation with a fixed rotated matrix before assessing its fitness against the target function. If the matrix is known in advance, it is easy to generate the inverse matrix and therefore the optimization process could more easily progress. However, in the case of the absence of information about the rotation matrix, the rotation becomes difficult. Therefore, if an adaptive mapping is applied, it should support this inverse process. Our conjecture is focussed on two aspects. Firstly, a mapping is defined in the way that a phenotype is generated by multiplying a genotype with an evolved matrix. Secondly, the mapping will be evolved in parallel with the evolution of the genotypes in order to adapt matrices towards the inverse matrix. This is

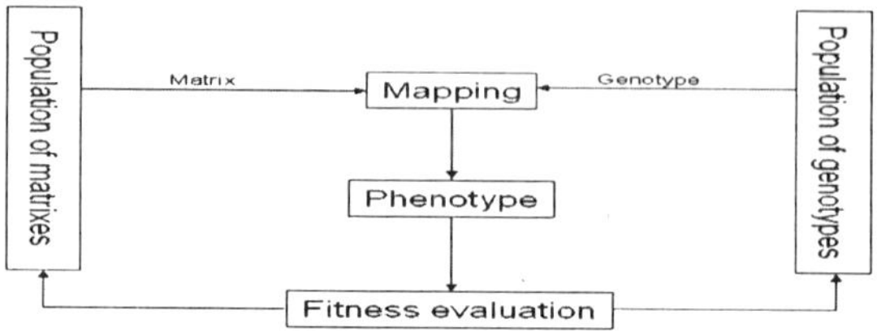

Fig. 3.1 The general framework of the proposed model.

an incremental development process in which good mappings are developed incrementally based on previous mappings.

In order to implement the evolution of mappings, we co-evolve the population of genotypes with a population of matrices, called the population of mappings, in a hope to evolve the inverse of the rotation matrix. The population of genotypes are implemented as normal except for the calculation of their fitness values. The evaluation of an individual's fitness is carried out by taking the phenotype resulting from mapping the genotype through a selected matrix, and then measuring how well it performs on the target function. The same applies for the population of matrices; the phenotype used to evaluate the fitness of a matrix is generated from mapping a selected genotype with the matrix. In this way, we integrate a cooperative approach into the proposed model (See Fig. 3.1). The fitness of an individual in the genotype population is partially dependent on the performance of individuals in the mapping population and vice verse. A good convergence to the inverse of the rotation matrix will help to reduce the distortion caused by the rotation matrix and a diverse population of genotypes will benefit the convergence of the mapping population.

There are a number of possible fitness assignment schemes, we adopt the "best" strategy to assign the fitness value for each individual in both populations. According to this strategy, an individual in a population will play with the best one in the other population. In other words, the mapping always takes into account a pair of a selected individual and the current best one in the other population. It is clear that the fitness of an individual is dependent on two aspects: the collaboration of the individual with the other and the convergence of the mapping population.

In order to run the model, we implemented the Fast Evolutionary Programming (FEP) as the underlying evolutionary algorithm, due to its successful performance on real-valued optimization problems [439]. FEP is known to perform badly on problems with high epistasis as our experiment below confirms. In general, FEP and EP are quite similar. The only difference is in the way to mutate the chromosome. In EP, the mutation is carried out by adding a Gaussian random value to each gene in the chromosome, while FEP uses a Cauchy distribution. As stated by Yao et al [439], the Cauchy distribution is better in exploring the search space than its Gaussian counterpart. A description of FEP is given in Algorithm 3.1.

Algorithm 3.1 The pseudo code for the Fast Evolutionary Programming

Require: Population P: μ individuals (pairs of real-valued vectors (x, η))
Ensure: Evolve P

repeat
  Create a population Q of μ children from P by modifying each individual in P in the following way
  for all individuals (each individual i has n members) **do**
    Generate R = N(0,1), a Gaussian random generated value
    Generate δ = a Cauchy random generated value
    Generate a child (x_i^c, η_i^c):
    $x_i^c = x_i + \eta_i\delta$ and $\eta_i^c = \eta_i exp(\tau' R + \tau N(0,1))$
    $\tau' = (\sqrt{2\sqrt{n}})^{-1}$ and $\tau = (\sqrt{2n})^{-1}$
    Evaluate the fitness values for the child
  end for
  Use tournament selection to select μ individuals from P+Q
  Replace all members in P by these new individuals
until Given conditions are satisfied
return New population P

In detail, the model can be described by its major steps as follows:

1. **Initialize**: both populations are initialized uniformly within prespecified ranges. For the genotype population, the ranges of genes are similar to the ranges of variables of a specific problem. Meanwhile, the values of the matrices are set initially in the range of [-1,1]. The phenotype that is used to evaluate the fitness of an individual is generated by mapping between the individual and a randomly selected one in the other population.

2. **Apply FEP to the genotype population**: The evolution of the genotype population is controlled directly by FEP. The fitness of a new individual is evaluated by using the current best mapping. The outcome of this step is a new population of genotypes.

3. **Apply FEP to the mapping population**: The Operation of FEP on the mapping population is the same as in Step 2. A new population of mappings are generated as a result.

4. **Examine the termination condition**: A pair of new populations from Steps 2 and 3 is denoted as a generation. If the error rate or the

number of evaluations reaches a certain level, the co-evolutionary process is terminated, otherwise the algorithm returns to Step 2.

3.4 A comparative study

3.4.1 *Testing scenario*

To validate the model, this system was tested on a rotated Rastrigin problem with 10 variables [377]. This is a minimized multi-modal optimization problem with a huge number of local optima. Further, it is a shifted, rotated, non-separable and scalable problem. Figure 3.2 visualizes the problem in both cases: with and without rotation. The problem is defined formally as in Equation 3.1.

$$f(x) = \sum_{i=1}^{D}(z_i^2 - 10cos(2\pi z_i) + 10) + fbias \tag{3.1}$$

in which z=M*(x-o), M is the rotation matrix, D is dimension, o is shifting vector, and $x \in [-5,5]^D$. Global optimum $x^\star$=o, and f($x^\star$)=fbias(10)=-330

We also tested this problem with two versions of FEP: a conventional FEP and a version of co-evolutionary FEP called CFEP. The CFEP is implemented similarly to CCGA1 in Potter and De Jong' work [308]. It uses FEP as the underlying EA instead of using conventional GA. The setting of all approaches are given in Table 3.1. In terms of the genotype-phenotype mapping, FEP uses a direct encoding scheme, while CFEP and the proposed model employ indirect schemes. CFEP's mapping structure is fixed because the way it converts the genotypes to phenotypes is kept unchanged over time, while the proposed model uses an adaptive mapping scheme. All algorithms were tested with 20 separate runs in which each run was associated with a different initial random seed.

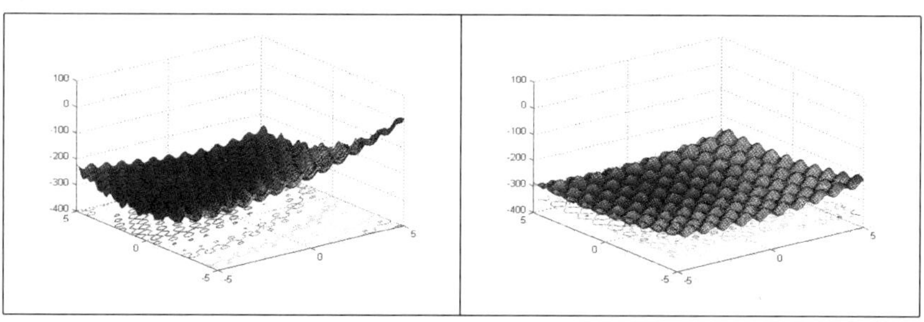

Fig. 3.2 Rastrigin function with and without rotation.

Table 3.1 The settings for the proposed model, FEP and CFEP

Parameters	The proposed model	CFEP	FEP
Population size	50	50	50
Initial η	0.1(for the population of mappings) – 3.0	3.0	3.0
Tournament size	10	10	10
Number of evaluations	100000	100000	100000

3.4.2 *Results*

From Table 3.2, it is clear that the proposed model is leading with a mean of -315.879, while the equivalent values of CEFP and FEP are just -271.588 and -310.618, respectively. This finding continues to be true when we examines the best, seventh, median, and worst (25th) runs. In all cases, including those not shown, the proposed model outperformed the others. The matrices seem to have helped the system to relax the difficulty of the rotation. Once again, the advantage of adaptive mapping is confirmed over fixed mappings. However, it is interesting to note that CFEP is inferior to FEP. This is another case in which an indirect scheme is outperformed by a direct one. To understand this, we examined the structure of CFEP in which each gene is evolved independently with a separate population. This kind of structure does not have any advantage in the case of rotation, as was pointed out by Potter and De Jong [308]. We visualized the fitness

Table 3.2 Fitness values achieved after 100000 evaluations

Parameters	The proposed model	CFEP	FEP
1st	-325.740	-320.050	-323.035
7th	-319.055	-288.650	-316.071
13th	-315.075	-274.283	-312.091
25th	-307.581	-194.001	-286.222
Mean	-315.879	-271.588	-310.618
STD	5.187	27.505	9.904

values over time on a convergence graph (Fig. 3.3). In the figure, we plot the fitness values recorded overtime at every cycle of 1000 evaluations. The graph shows that both CFEP and FEP converge quickly to their optima in the first 10000 evaluations, then stagnate. Meanwhile, the proposed model gradually approaches the optima. The fitness variance graph indicates that for the proposed model, the variance between the runs gets smaller over time, while for the others, this value quickly becomes unchanged. We note that the fitness is measured on the phenotype space. Thus, for CFEP, a fixed variance implies that the individual populations do not change. As for FEP, the mutation step somehow becomes low, thus solutions move around in a valley or a small neighborhood.

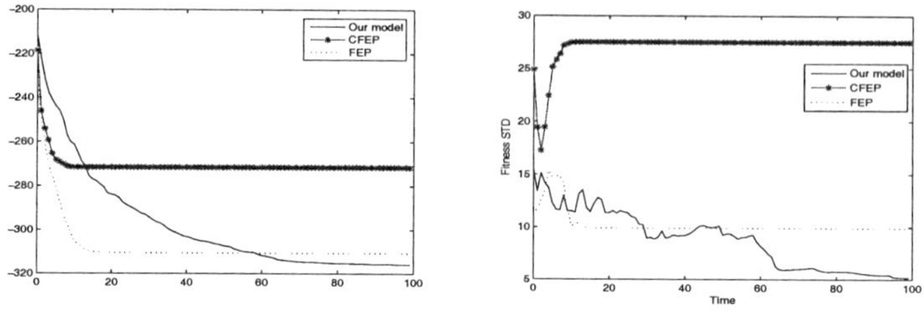

Fig. 3.3 Convergence graph: fitness and its variance (for 25 runs) over time.

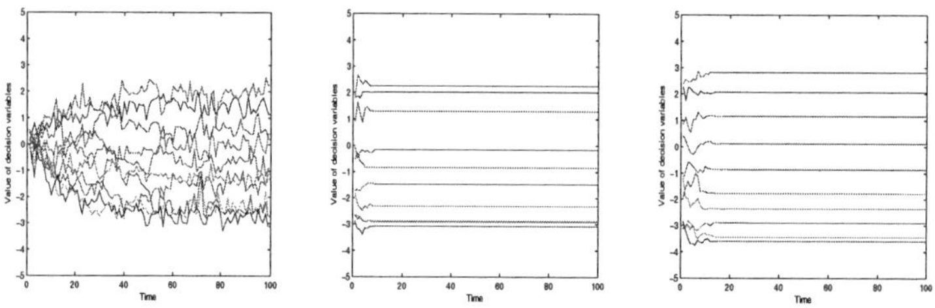

Fig. 3.4 The averaged phenotype of the best individual overtime for the proposed model (left), CFEP (middle) and FEP (right). Each phenotype contains 10 variables.

This is more clear when we examine the vector of decision variables for the best individual over time (Fig. 3.4). In the proposed model, this vector changes frequently; indicating high genetic activity. Meanwhile for FEP and CFEP, stagnation is strongly apparent.

The above phenomena can be explained by the way that the rotation directs the search to arbitrary local optima. CFEP and FEP rely only on the power of searching in the genotype space and have no control over the phenotype space. On the other hand, the proposed model can control the mappings by adapting them to the specific structures which facilitate the search by reducing the effect of rotation on the phenotypes. This seems to be the reason that the model gradually overcomes local optima to approach the global one. We hypothesize that the cooperation strategy also plays an important role in preserving the diversity in the populations over time. This diversity helps our system to get out of the local optima. For FEP, diversity is lost when it gets stuck at the local optima, as the mutation operator does not generate enough diversity to resist the effect of the rotation. Although, CFEP is supported by the cooperative co-evolutionary mechanism, it seems to be trapped in a local Nash equilibrium as stated in [308].

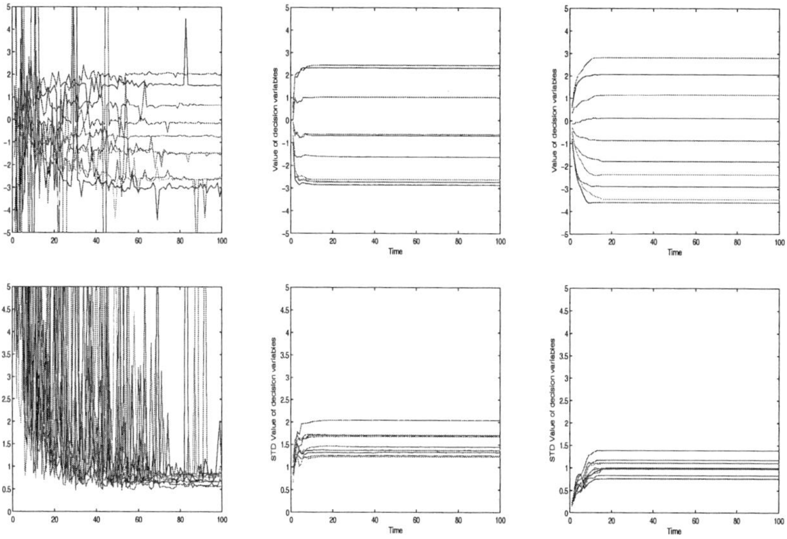

Fig. 3.5 The exploration of the phenotype space for the proposed model (left), CFEP (middle) and FEP (right): the 1st line is the averaged values of 10 variables and the second is the corresponding variances.

In order to understand the dynamics of the system and to consequently explain the convergence of each technique, we investigated how the techniques explored the phenotype space. The averaged values of each variable in the phenotype space are recorded overtime and they are plotted in Fig. 3.5. We find that CFEP and FEP quickly explore a large part of the space and then remain fixed in this area, while in the proposed model, variables keep changing over time; thus the adaptive mapping gives the proposed model the ability to continue exploring the phenotype space. Our technique also shows very good exploration in the genotype space (Fig. 3.6).

3.4.3 *Fitness landscape analysis*

In this section, we analyze the fitness landscape. This analysis helps us to understand the dynamics of the system [2]. The fitness landscape is constituted by the three components: genotype representation, fitness values and the operator to generate the neighborhood of related solutions. In our comparison, the three models have different techniques to build the fitness landscape; this is because they use different types of mapping: direct for FEP, fixed indirect for CFEP and adaptive mapping for our model.

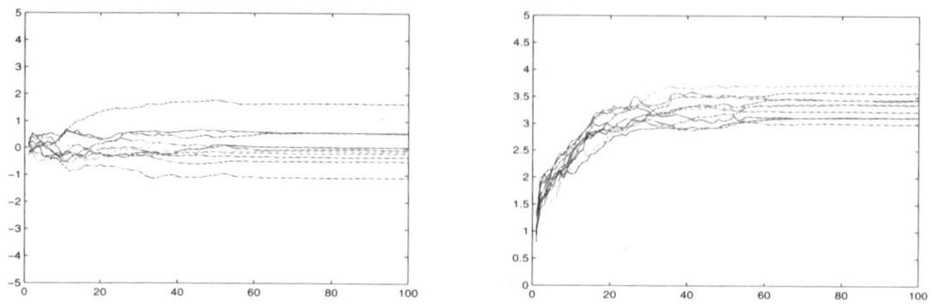

Fig. 3.6 The exploration of the genotype space for the proposed model: the averaged values of variables (left) and their variances (right) over time.

Figure 3.7 depicts the fitness histograms of the respective models. The proposed model clearly has a better distribution in the fitness space, while the others focus only on a small area. The distribution of the proposed model implies that there is a good collaboration between mapping and genotype populations. It also maintains diversity from generation to generation in the genotype space, and therefore in the fitness space as well.

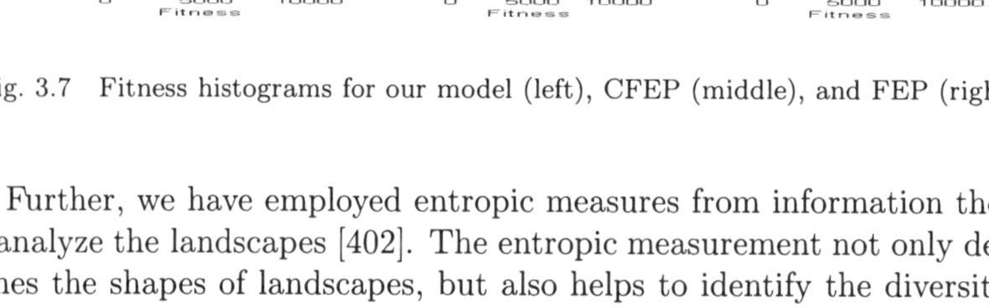

Fig. 3.7 Fitness histograms for our model (left), CFEP (middle), and FEP (right)

Further, we have employed entropic measures from information theory to analyze the landscapes [402]. The entropic measurement not only determines the shapes of landscapes, but also helps to identify the diversity of the populations overtime. The information contents (IC) (based on Shannon entropy) is used as a measure of the ruggedness of a landscape, and the partial information contents (PIC) measure is used to scrutinize the modality of the fitness paths (time series). All obtained information is given in Table 3.3. The parameter ϵ in the table is used to determine the measures on different levels of the flatness of the landscape. If ϵ =0, the calculation of measures is very sensitive to the difference between fitness values. If ϵ is maximal (the difference between the maximal and minimal fitness values in the path), the landscape is seen as flat and the measures become zero.

Table 3.3 Information contents (IC) and partial information contents (PIC) for the test problem using the proposed model, CFEP and FEP.

ϵ	**The proposed method**		**CFEP**		**FEP**	
	IC	**PIC**	**IC**	**PIC**	**IC**	**PIC**
0	0.476	0.619	0.425	0.177	0.294	0.026
100	0.379	0.143	0.119	0.015	0.098	0.005
400	0.247	0.088	0.001	0.00003	0.002	0.00006
1000	0.202	0.067	0	0	0	0

Obviously, for all ϵ values, our model always has a greater value for its IC. This means that the proposed model's landscape is more rugged than that of other models. It shows the strength of the proposed model's capability to explore the search space. By having this capability, the model must maintain a certain level of diversity in the phenotype space. That is why in the phenotype exploration graph, the proposed model's variables keep changing overtime. Meanwhile, CFEP and FEP quickly explore the phenotype space and remain unchanged. Therefore, their landscape's ruggedness are effected. As ϵ increases, their landscapes are quickly flattened out: with $\epsilon = 1000$, their information contents become zero, while the proposed model's value is 0.202. This fact also indicates the higher value of the information stability of the proposed model's landscape in comparison with the others. It is worth noting that FEP has the worst value of its information contents. This is because FEP does not have support from the co-evolutionary process in contrast to that our model and CFEP do.

The exploration capacity of the algorithms has also been assessed by the ability of each algorithm to discover local optima. This concept can be quantified by using the partial information contents (PIC) measure. In general, PIC helps to identify the modality of the landscape. The proposed model shows a strong ability to explore local optima.

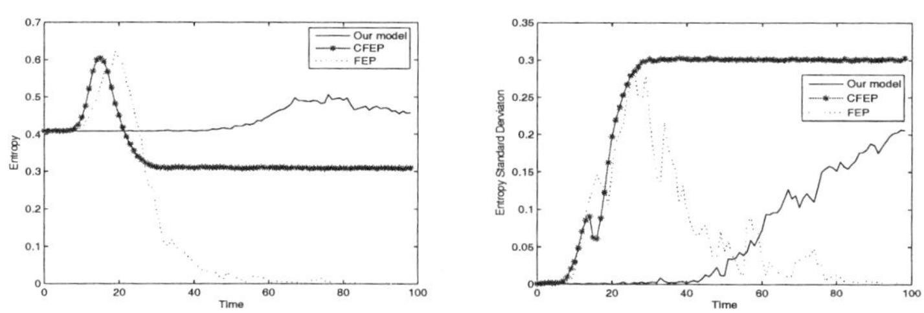

Fig. 3.8 The averaged entropy value (left) and its standard derivation (right) obtained over time for the proposed model, CFEP, and FEP.

Lastly, we investigated the values of IC over cycles of time using 1000 evaluations for each cycle (Fig. 3.8). All algorithms started with almost identical values of information content (≈ 0.4). In the first few time cycles, while the proposed model keeps this value quite stable, CFEP and FEP increased quickly. However, these values dropped sharply in the cases of CFEP and FEP (especially, FEP dropped to nearly zero) as time passed. Meanwhile, the proposed model continued to stay stable and even increased at the end. This fact once again confirms our above analysis that CFEP and FEP quickly become trapped in local optima and lose diversity. For the proposed model, the adaptive mapping drives the search mechanism to escape the local optima and approach the global one. The figure showing their variances is evidence for the above finding. For CFEP, the variance stays fixed after being trapped in a local optima. This indicates that the individual populations do not change. As for FEP, this figure reduces significantly over time. The reason is that the mutation step becomes low somehow, thus it cannot help FEP to escape from the local optima.

3.5 Conclusion

In this paper, we proposed a new model of adaptive mapping to solve rotated problems. GPM was implemented by using cooperative coevolution on two populations: one is the population of genotypes and the other is the population of matrices or mappings. The phenotype was generated by transforming selected genotypes by selected matrices. To validate the model, we compared its performance with that of the conventional FEP and a version of cooperative co-evolutionary FEP on the rotated Rastrigin problem. The results show that with the help of adaptive mapping, the proposed model clearly outperformed CFEP and FEP in all aspects such as the best achieved fitness value, convergence, and the diversity. We have also carried out an analysis on the fitness landscape to verify these findings.

In future work, we will continue to improve the representation and computational cost of the mapping.

Acknowledgement

This work is supported by the University of New South Wales grant PS04411 and the Australian Research Council (ARC) Centre for Complex Systems grant number CEO0348249.

Chapter 4

Approaching Perfect Mixing in a Simple Model of the Spread of an Infectious Disease

D. Chu and J. Rowe

School of Computer Science,
The University of Birmingham,
B15 2TT, Birmingham, UK
E-mail: {D.Chu, J.E.Rowe}@cs.bham.ac.uk

In this article we present an agent-based simulations of the spread of a vector borne disease in a population with limited mobility. The model assumes two types of agents, namely "vectors" and "people agents"; infections can only be transmitted between agents of different type. We discuss how the infection levels of the population depend on the mobility of agents.

4.1 Introduction

Recent outbreaks of infectious diseases, such as the SARS virus and the avian flu in south east Asia or recurrent outbreaks of Ebola in Africa underline the need to understand how diseases spread in a population. A major practical problem connected to those diseases is how to contain local outbreaks and prevent them to cause a global pandemic. A major problem here is of course the global mobility of people, especially infected people.

A commonality of the above mentioned diseases is that an infected person can directly infect another susceptible individual through some type of interaction (what type depends on the pathogen in question). The focus of interest of the present contribution is another type of infection that requires a mediating agent; so called *vector borne* diseases cannot directly be transmitted on from one infected individual to another, but require a vector (typically an insect of some sort) as an intermediate carrier of the pathogen. An example of such a vector borne disease is Malaria. Malaria parasites are passed on from one person to a mosquito if the mosquito feeds on the person's blood. Similarly, the next person to be bitten by the mosquito will then get infected. Malaria (and other vector borne diseases) are important killers far outstripping more widely publicized diseases such as Ebola or AIDS in the number of fatalities they cause.

There is a large body of mathematical theory modeling how infectious diseases (vector borne and otherwise) spread in a population. Mathematical approaches to model the spread of diseases usually assume that the population mixes perfectly. In the context of a discrete time model perfect mixing means the following: At every time step the probability that two members of the population meet is independent of who they encountered in the previous time steps. Clearly, real populations normally do not fulfill this criterion, but are spatially structured. In practice perfect mixing nevertheless turns out to be a very good approximation. This is so because a limited amount of local mixing might be sufficient to generate the global effect of perfect mixing.

While it is fairly well understood how diseases spread in a perfectly mixing population, there are relatively few attempts to study the spread of diseases in populations where this condition does not hold[211; 226; 295], that is in population where the local mixing is not sufficient for the population as a whole to approach global perfect mixing. In such cases the mathematics necessary to describe the systems tends to become very involved. In those situations, agent-based computer simulations[72; 197; 193] are a valuable tool, as they are very adept at modeling populations with limited mobility. In this article we will describe a simplified model of the spread of a vector borne disease in a population. The aim of this article is to investigate how the infection levels in the population depend on the mobility of the agents.

We describe our model in section 4.2; section 4.3 presents previous results describing a mathematical result of how the model behaves in the perfect mixing case. Section 4.4 thereafter describes the results of computational simulations for the case of limited movement. We provide a discussion and conclusion in sections 4.5 and 4.6 respectively.

4.2 Description of the Model

The model we describe in this article is not meant to be realistic with respect to the behavior of any real system. Instead we aim to study how infection levels depend on agent mobility in a bare-bone model of the spread of a vector borne disease. Once this basic understanding of the maximally simple case is reached, it will be possible to add further detail to the model.

ABMs are best described in order of their major components:

- Environment
- Agents
- Interactions

The environment of the model is a 2 dimensional continuous space of size $2L \times 2L$, where L is measured in some unit. The environment has no features other than that it provides a topology for the agents.

There are two types of agents in the model, "vectors" and "people". Those agents are mainly distinguished by their infection period, movement rules, and mode of infection transmission. An infection can only be transmitted between agents of different type. The number of agents of each type is held constant during a simulation run. Agents are thus characterized by their position in the environment and by their internal state, which can be either "infected" or "healthy".

At each time-step the agents take a random position within a square of linear size $2M$ centered around their current position. M is a parameter of the model and set independently for people and vectors; throughout this article we will refer to M as step-size of the agent. Movement is always subject to the constraint that the agents stay within the boundaries of the environment.

The only form of interaction between agents is transmission of an infection. At each time-step each vector interacts simultaneously with all people agents that are at most one unit away from it. If the vector is infected then all agents it interacts with will also be infected from the following time-step on. If the vector is not infected, but at least one of the people in its "bite area" is infected, then the vector will be infected from the next time-step on. In all runs presented here the bite area is a circle of radius 1 centered around the vector. Throughout all simulations presented here, vectors keep their infection for two time steps and people keep their infection for 40 time-steps. However, the model has re-infection, that is whenever an agent interacts with another infected agent, while it is already infected, then its remaining infection period is reset to its full value. So, for example, if a people agent is re-infected 39 time-steps after it has been infected the last time, then it will still have to wait for 40 more time-steps until it loses its infection again.

4.3 Behavior of the Model in the Perfect Mixing Case

Standard epidemiological models are mostly dealing with the case of what is called "perfect mixing." In the case of agent-based models perfect mixing

V	total number of vectors
P	total number of people
v	number of infected vectors
p	number of infected people
R_v	time for vector to recover
R_p	time for person to recover
b	biting area of a vector
W	world size (area)

is achieved if for each agent it is true that its position at time $t+1$ is independent of its position at time t. In the real world perfect mixing is hardly ever realized. Many systems are nevertheless well approximated by a perfect mixing approach: The present system seems to behave as though it perfectly mixes for relatively modest step sizes. If the agents' step size is 9 (or greater) the infection levels are negligible different from the perfect mixing case. This observation presumably generalizes to a large class of systems. Hence, examining models that assume perfect mixing is useful also for systems that are relatively far from the ideal of perfect mixing.

In [79] we showed that for the perfect mixing case the equilibrium of the present model is described by the following set of equations:

$$\begin{aligned} \frac{v}{V} &= 1 - \left(1 - s^2 q(p)\right)^{R_v} \\ \frac{p}{P} &= 1 - \left(1 - q(v)\right)^{R_p} \end{aligned} \tag{4.1}$$

Here the function $q(x)$ is defined as:

$$q(x) = 1 - \exp(-bx/(W))$$

Figure 4.1 shows some simulation results for the perfect mixing case; in this case we realized perfect mixing by assigning to each people agent at each time step a random position. Figure 4.1 shows the proportion of agents that are infected for various system and population sizes. As expected, the infection levels vary from zero (no infection) to one (all people agents infected). The infection levels of vectors (data not shown) are similar. A thorough discussion of the perfect mixing case including a discussion of how the infection spreads for vectors can be found in [79].

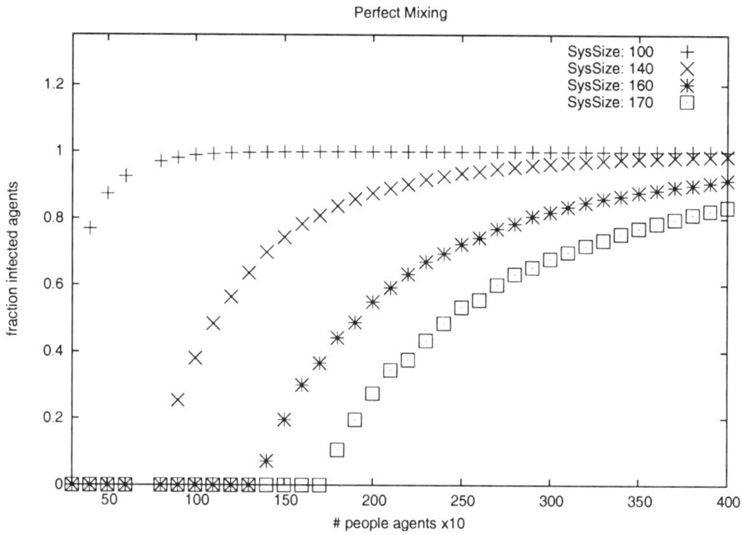

Fig. 4.1 The proportion of infected agents for various systems sizes. Along the x-axis the size of the people population is increased. The number of vectors has been kept constant at 10000 in all simulations. Every point marks the infection level in the population once the equilibrium has been reached; each data point has been obtained from the time average over 1000 time-steps of one simulation run of the system.

4.4 Beyond perfect Mixing

In this section we will describe the spread of a vector borne disease in a population that does not perfectly mix.

4.4.1 *No Movement: The Static Case*

The next simplest case to perfect mixing is the case where there is no movement at all. In the context of the present model it would make no sense to keep both agents and vectors fixed in space. Throughout this article, we will therefore only vary the mobility of the people agents while keeping the mobility of vectors constant at step size 1. In this subsection we will consider the case where the agents do not move at all; we will henceforth refer to this case as the static case.

Generally, the infection levels in this case are substantially lower than in the case of perfect mixing. For the population densities considered here we observe substantial infection levels only for the smallest system size ($L = 100$). When we increase the system size to $L = 140$ the infection level never surpasses 0.2. For even greater systems, the infection never

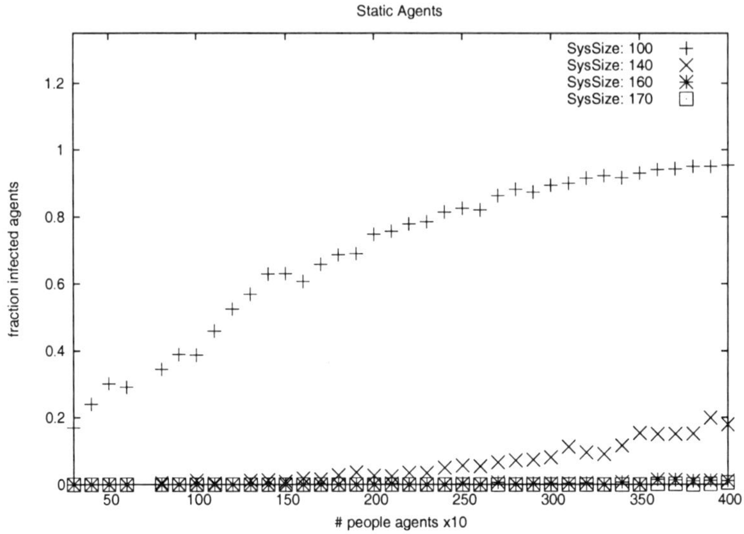

Fig. 4.2 Same as figure 4.1, but the agents are not allowed to move.

establishes itself.

4.4.2 *In Between*

The comparison between the perfect mixing case and the static case showed that there is a strong quantitative difference in the infection levels in the population between the perfect mixing case and the static case (i.e. no agent movement). In this section we will investigate the behavior of the system in between those two extremes. The question we will ask is how the system approaches perfect mixing.

Figure 4.3 illustrates the approach of the system to perfect mixing as the mobility of the agents increases. The first observation is that for small increases of the step size (up to a step size of 3) the infection levels increase as the step size increases. In the static model and a world size of 170 the infection cannot establish itself in the parameter range considered here. Once people agent mobility is introduced the infection levels increase up to a step size of 3; at this point the infection levels are clearly higher than in the perfect mixing case, particularly for low densities of people agents (fewer than 3000 people agents; see figure 4.3).

Increasing the step size beyond 3 will not lead to a further increase of the infection levels. Figure 4.3 shows that the infection levels in the perfect mixing case are *lower* than the infection levels in the case of a step size

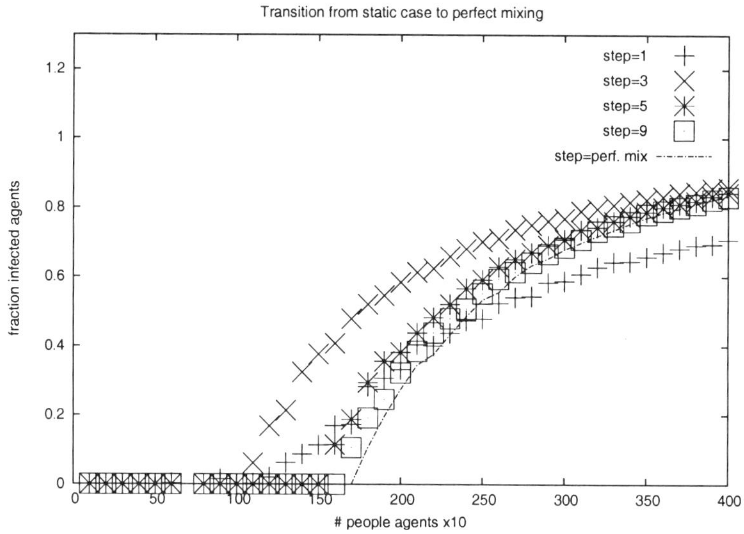

Fig. 4.3 The transition from the static model to the perfect mixing model in a system of size 170.

1. The same qualitative dependence of the infection levels on the step size was observed for a wide range of system sizes (data not shown). The step size at which the infection levels begin to decrease depend on the specific parameter settings.

4.5 Discussion

The dependence of the infection levels as shown by the present model is surprising. In this section we will explore possible explanations for the observed effect.

The low infection levels in the static case (if compared to the perfect mixing case) are readily explained by the following observation: In the case of no agent movement, a vector can only infect an agent, if the following two conditions are fulfilled:

(1) It picks up an infection from another people agent
(2) It can travel the distance from this infected agent to the people agent in question without losing its infection

Given the restrictions on the movements of vectors (step size 1), in the present model the second condition can only be fulfilled if the distance

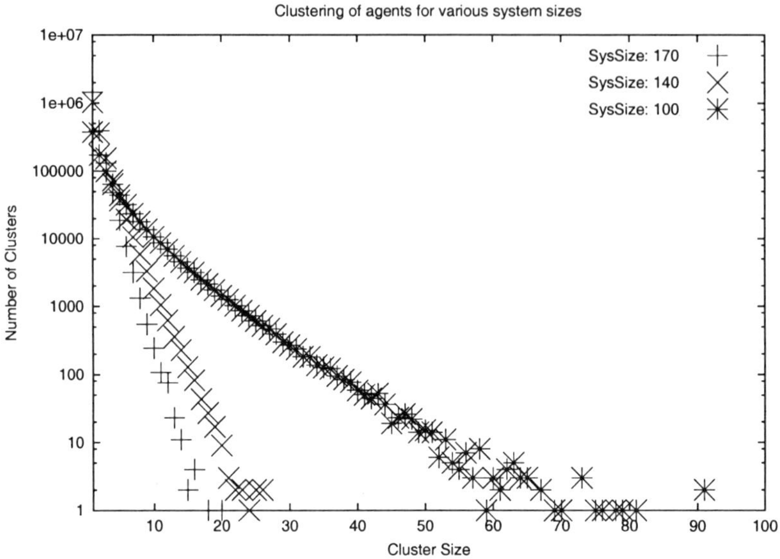

Fig. 4.4 The distribution of clusters for three system sizes. In all experiments there were 3000 agents in the system.

between the people agent to be infected and an already infected people agent is at most 3 units. This is so because the vector remains infected for only 2 time steps if no re-infection occurs; hence it could not carry an infection over distances larger than 3 units.

This then leads to a natural definition of clusters of neighboring agents: We say that two agents are neighbors if they are at most 3 units away from each other; two people agents are then defined to be in the same cluster if they are neighbors. In the case of no movement, the spread of the disease is restricted to clusters. If two agents are not in the same cluster, by definition one cannot cause the infection of the other, neither directly nor indirectly. Conversely, once there is neither an infected people agent nor an infected vector left in the cluster, the cluster will remain free of infection for all times.

The graphs in figure 4.4 show the distribution of cluster sizes for a system of 3000 agents. For these parameters the clustering is sub-critical (see [159; 25]) and the cluster size distribution is exponential. For the smallest system size ($L = 100$) the maximal cluster size of about 60 is considerably larger than in the case of the larger systems where clusters are never bigger than about 25. Also, by far the majority of all clusters in the larger system consist of only very few people agents.

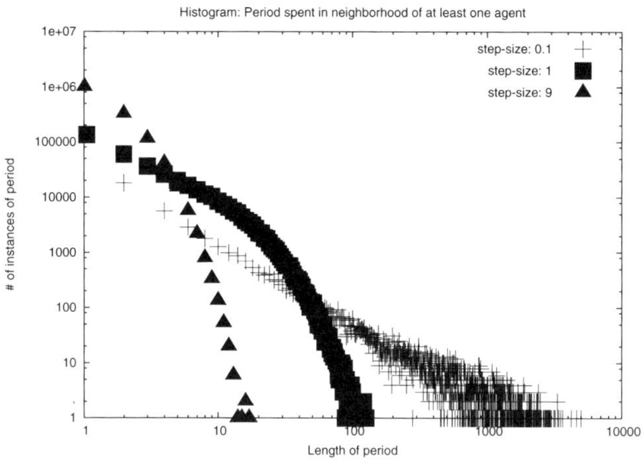

Fig. 4.5 The distribution of the contiguous periods spent in the neighborhood of at least one agent. The system size is $L = 170$.

In the case of no movement of the agents it can be expected that the infection can only be sustained in the larger clusters. Smaller clusters, particularly clusters of size one, can in principle sustain an infection, but will be much more vulnerable to random fluctuations; those will over time wipe out infections in smaller clusters. Hence, the main contribution to infections mainly comes from the large clusters. The larger the system size, the smaller the population density and maximal cluster size; furthermore, the proportion of the population contained in large clusters increases.

The change of the cluster size distribution is therefore the main reason for the rapid fall of the fraction of infected agents with increasing system size. The density of agents in the case of $L = 170$ is too low to sustain an infection if people agents are not mobile. This is in strong contrast to the perfect mixing case (fig 4.1) that shows considerable levels of infection in the population in the corresponding parameter range.

We will now discuss the transition from the static case to perfect mixing. There are two antagonistic effects that determine the infection in the case of a non-static population. Firstly, as discussed above, if agents do not move, then the infection cannot spread to larger parts of the population, thus essentially limiting the possibilities for the spread of the infection. As the step size increases, agents have more opportunity to find themselves in the neighborhood of other agents, and therefore also in the neighborhood of other *infected agents*. Overall, this greatly facilitates the spread of the disease. Hence, from this perspective one would expect that agent mobility is positively correlated with the infection level in the population.

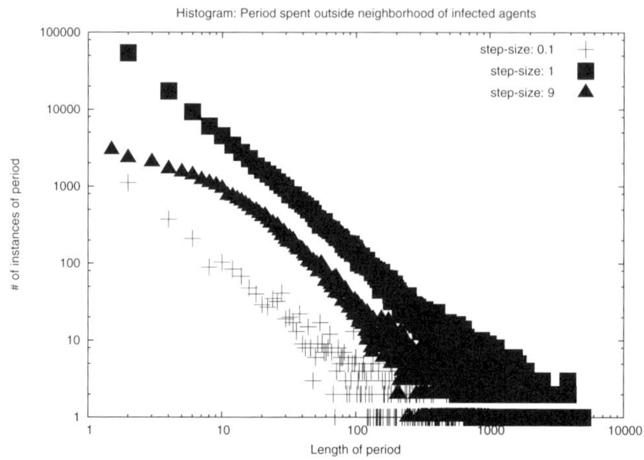

Fig. 4.6 Lengths of periods spent outside the neighborhood of at least one **infected** agent. The system size is $L = 170$.

On the other hand, increased movement also leads to a decrease of the typical number of contiguous time units spent in the neighborhood of a specific agent. More precisely, the overall time spent in the neighborhood of a specific agent is independent of the step size (for simulation times large enough), yet the contiguous time periods spent in the neighborhood of a specific agent strongly depend on the step-size. Figure 4.5 shows, for different step sizes, the distribution of the times agents have at least one neighbor.

In the case of very limited mobility (step size of 0.1) agents spend very long time periods in clusters; in the present simulations they spend up to the entire duration of the simulation in a cluster. At a step size of 1, agents are more mobile. Hence, correlations between successive locations decrease and clusters of agents are shorter lived. At this degree of mobility agents spend in a cluster is maximally of the order of magnitude of 100 time steps. At a step size of 9 the same number is under 20 (see figure 4.5).

It should be noted, however, that the overall time spent in the neighborhood of at least one other agent is independent of the step size. The effect of shorter periods in the neighborhood of agents is thus counteracted by more frequent visits with no net change (taken over sufficiently long simulation periods). The time spent in clusters is thus not relevant for the change of infection levels the step size varies.

A related measure is a better indicator of what is going on. Figure 4.6 shows the distribution of the time periods spent outside the neighborhood of *infected* agents. This is relevant because infection transmission is limited to areas in the neighborhood of infected agents. Figure 4.6 shows for the case of a system of size 170 and 1500 people agents that both the overall time spent outside a cluster and the distribution of the times spent outside the neighborhood of an infected agent increases with the step size. This explains the decrease of the infection levels as the step size increases. The explanation is as follows: Outside the neighborhood of infected agents, agents cannot get infected; on the other hand, agents remain infected only for a fixed amount of time steps. Extended periods outside the neighborhood of other infected agents will therefore necessarily lead to a loss of infection.

4.6 Conclusion & Future Work

In this article we reported some simulation results of how a vector borne disease spreads in a population of mobile agents. We found that for moderate levels of agent mobility the infection levels are highest, decreasing both if the mobility is increased and decreased. We also provided a qualitative explanation for this effect.

Future work will need to provide a quantitative explanation for this effect. A mathematical model of mobile agents that do not mix perfectly is needed in order to further elucidate the nature of the observed effect.

Chapter 5

The Formation of Hierarchical Structures in a Pseudo-Spatial Co-Evolutionary Artificial Life Environment

D. Cornforth[1], D. G. Green[2] and J. Awburn[3]

[1]*School of Information Technology and Electrical Engineering, UNSW@ADFA, Canberra ACT 2600, Australia. Email: d.cornforth@adfa.edu.au.*

[2]*Faculty of Information Technology, Monash University, Clayton VIC, 3800, Australia. Email: david.green@infotech.monash.edu.au.*

[3]*School of Environmental and Information Sciences, Charles Sturt University, PO Box 789, Albury, NSW 2601, Australia*

Enumeration of the factors underlying the formation of modules and hierarchical structures in evolution is a major current goal of artificial life research. Evolutionary algorithms are often pressed into service, but it is not clear how the various possible features of such models facilitate this goal. We address this question by using a model that allows experimentation with several candidate model features. We show how the notions of variable length genotype, variable genotype to phenotype mapping, pseudo-spatial environment, and memetic evolution can be combined. We quantify the effects of these features using measures of module size, and show that information shared between individuals allows them to build modules and combine them to form hierarchical structures. These results suggest an important role for phase changes in this process, and should inform current artificial life research.

5.1 Introduction

One of the key questions in the study of Artificial Life is to understand "open-ended complexity". That is, how do increasingly complex structures and behaviour arise in natural systems? In particular, is it possible to capture this phenomenon within a simulation model [34]?

Complex behavior arises from interactions between simple elements of a system [156], and includes clustering, modularity, and phase changes. Many models have successfully demonstrated emergent complex behavior from simple systems, for example, cellular automata [427] and Tierra [328]. These models have been able to demonstrate emergent effects such as the spontaneous appearance of parasitic organisms. However, they have generally not examined the interaction between open-ended evolution, local phenomena and global constraints. In particular, evolutionary models, which are often used in such investigations, are often limited to a fixed-length genotype. In contrast Harvey [180] suggested that a variable length genotype is essential for open-ended evolution, and Angeline and Pollack [15] have shown the efficacy of the same for the emergence of modules, and their combination to produce higher levels of problem abstraction, in conjunction with a co-evolving representation language.

Many evolutionary models employ a fixed genotype to phenotype mapping, while deJong [104] has shown the advantages of a variable mapping, and its role in building modularity. Furthermore, Channon and Damper [76] have suggested a co-evolving environment as a necessary feature for the discovery of truly novel behavior. Green has shown [157] that phase changes have an important role in explaining how more complex structures arise in complex systems.

In his book, *The Blind Watchmaker*, Richard Dawkins [92] illustrates the power of natural selection to produce complex results through the example of a child typing Shakespeare. Taking the line "*methinks it is like a weasel*" out of the play *Hamlet*, he points out that the chances of a child typing this line are virtually zero. However, if each time the child types a correct letter, and that letter becomes fixed, then the child's typing will quickly converge on the targeted line.

However, this model has selection directed towards a fixed target. Natural evolution does not proceed in such a simple linear way, but behaves more like a growing bush [153], with branching and pruning producing a host of different forms.

Modularity is one of the most prevalent ways of organizing complexity, both in natural and artificial systems [158]. Modules provide building blocks that simplify the construction of large systems. Plants for instance, are built of repeating modules including branches, buds and leaves. By restricting interactions, modules also simplify control of large complex systems.

In order to provide a test bench for the investigation of open ended evolution, we developed the Weasel World model [86]. In that work, we were able to demonstrate the development of phase changes, clustering and self-organization, as well as showing the importance of the interaction between local phenomena and global constraints. In this work, we go further and investigate how modules can be combined into hierarchical structures. In the next section, we describe the Weasel World model. Section 5.3 describes the experiments and section 5.4 the results. Section 5.5 summarizes our conclusions.

5.2 The model

Weasel World is an implementation of an evolutionary algorithm featuring a fixed size population P_g of individuals placed in a one-dimensional spatial environment. Many environments could be suggested, but because of our inspiration from the work of Richard [92], the environment consists of the entire text of Shakespeare's *Hamlet* [350], with upper case characters converted to lower case, and periodic boundary conditions to avoid edge effects. Individuals are spatially located within the text, and evolve to match their phenotype to the local text. An important feature of our model is that there is no fixed goal or "solution", but individuals are free to match as much text, and to grow as large, as they are able. In this (limited) sense our model is "open ended". The model consists of the following components:

- a genotype representation
- a genotype to phenotype mapping
- a selection mechanism
- reproduction and genetic operators
- a co-evolving memetic population
- a pseudo-spatial environment

We have adapted these features to suit the purposes of this work, as well as adding the concept of territory of individuals, and allowing the population P_g to share knowledge discovered about the environment by contributing to and accessing a shared information repository, similar to the notion of memetic evolution.

5.2.1 *Genotype to phenotype mapping*

As individuals evolve to match the text at their location, the phenotype is represented as a sequence of symbols p_i drawn from the same set H as those for the text of *Hamlet.* The set H includes all the letters of the English alphabet, plus punctuation and the space character,

$H = \{a \ldots z, .‘ - []; ?!() : ”\&\}$, where the ellipsis … indicates all the characters between a and z. For a phenotype of length n, we define the phenotype to be of the form $\langle p_1, p_2, \ldots, p_n \rangle : p_i \in H$.

The obvious representation for the genotype is also a string of symbols with the same definition, so that the genotype to phenotype mapping is the identity operator. However, in this new version of our model we introduce a variable mapping by using instead the set of integers K from zero up to some limit N_k. A symbol in the genotype $g_i \in H$ is mapped using the ASCII numeric to character conversion, while a symbol $g_i \notin H$ undergoes a variable mapping defined by an individual in a co-evolving memetic population, if a suitable individual exists. Otherwise, the symbol is regarded as "junk DNA", and is not represented in the phenotype.

For example, if the genotype consists of the integers $\{119, 2, 10, 115, 101, 108\}$, this is mapped as follows. The first, 199, is a member of H so the ASCII mapping is used and the phenotype becomes w. The next, 2, is not a member of H. However, assume that an individual having a code of 2 exists in the memetic population, and represents the mapping $2 \rightarrow ea$. The phenotype becomes wea. Assume that integer 10 has no individual in the memetic population, so is not expressed. The remaining integers in the genotype are ASCII codes for lower case letters s, e, l, and are simply copied into the phenotype to produce the complete string *weasel*. This mapping means that genotype length and phenotype length may be (and often are) different.

In our model, individuals have a fixed genotype length, but their offspring may acquire a longer genotype by adding a random gene during reproduction (within certain constraints described below). Additionally, offspring may have a shorter genotype length if the environment is unable to sustain them. The evolutionary process thus may lead to individuals of differing genotype lengths. In turn this means variation in phenotype length, with longer phenotypes able to achieve greater competitiveness by discovering potentially more building blocks.

5.2.2 *Selection mechanism*

Selection of individuals to reproduce is implemented using tournament selection, with the requirements of a fixed population size. An individual is chosen at random from P_g and examines its territory. If there are no other neighbors (individuals in its territory), it reproduces asexually. This means that there is no selection mechanism when individuals are small or located away from each other.

If the individual has one neighbor, they mate to produce two offspring. If there is more than one neighbor, two are selected at random and compared using two measures. The first is the length of the largest block of contiguous matches at any position in the rival's local environment or territory. For example, a phenotype "*metweel*" with local environment "*methinks it is like a weasel*" will score only 3 matches for "*met*", as the longest contiguous block. The second measure is the number of matches divided by the phenotype length. Either of these measure may be normalised as follows:

$$\varphi_{norm} = 2\frac{\varphi_i - \varphi_2}{(\varphi_1 + \varphi_2)}, \tag{5.1}$$

where φ_1 is the measure of the first individual.

Competition proceeds using a modified Pareto technique as follows:

(1) If one rival is dominant in both measures it wins.
(2) If one measure is equal in both rivals, comparison is based on the other.
(3) If neither rival is clearly dominant, both measures are normalised, using equation 5.1. The measure with the largest normalised value is selected, and rivals are compared on that measure alone.
(4) If there is still no winner after the preceding comparisons, one of the rivals is randomly selected.

The first individual selected mates with the winner of the tournament, producing two children. This process is repeated until the new generation reaches the fixed size.

In the case of sparsely distributed individuals with small, non-overlapping territories, reproduction will be completely asexual. However, as the phenotype length of individuals increases, their territory size will increase so that they gradually come into contact with others. The reproduction will become a mixture of asexual/sexual and then finally entirely sexual as sub-populations merge.

5.2.3 *Reproduction and genetic operators*

Mutation occurs by replacing a randomly selected gene with a random integer selected from the set K. Rather than a fixed mutation rate, the rate is specified relative to genotype length, which Harvey [5] suggests is more appropriate for variable length genotypes. In our model, the mutation rate is the number of mutations per genotype, rather than a fixed rate. For example with a genotype length 5 and mutation parameter 1, each gene will have a 1 in 5 chance of being mutated. A standard single point crossover was used for sexual reproduction.

The increase-length operator, suggested by Harvey [5], increases genotype length by exactly one character. A random gene is generated and inserted into the offspring at a random position in the genotype. The decrease-length operator removes a random gene from the offspring at a random location.

5.2.4 *Memetic evolution*

We wish to study the formation of modules and their combination, so here we define a structure for modules. It should be noted that we do not define what the modules will contain or how they combine to form larger structures, but leave that up to the evolution of the model. We define a module as any group of two or more adjacent symbols in the environment, and facilitate the discovery of frequently occurring modules by maintaining a list separate to the population. This is similar to the co-evolving representation described in [6].

Modules are stored in a fixed size population of memes P_m, that evolves over time, where individuals are subject to competition and selection. This is distinct from the genetic population Pg. Memes consist of a module plus a unique integer key. When a gene from an individual in P_g matches a key in P_m, the module is added to its phenotype.

At each generation, individuals in P_g identify the largest contiguous text fragment matched by their phenotype. This becomes a new candidate individual for the memetic population P_m, along with a fitness value for insertion given by the length of the fragment multiplied by the number of times it was encountered within the territory of the individual. The memetic population Pm reproduces a new generation by selecting memes from P_g with the highest insertion fitness value. This is illustrated in Figure 5.1.

Memes from P_m are used during the genotype to phenotype mapping, if a match is found between the gene and the key associated with the meme. The memetic population also loses memes by deletion, according to a deletion fitness value, defined as the length of the fragment times the number of times it has been used for mapping.

Genetic population P_g	Memetic population P_m
Genotype: [9][10][32][119][2][115][101][108]	key module
Phenotype: "they weasel"	[2] ⟶ [ea]
Territory: "of other nations; they clip us"	
Longest contiguous match: "the", frequency 2	[9] ⟶ [the]

Fig. 5.1 Memes are selected from the genetic population according to their fitness.

5.2.5 *Global parameters*

Two global variables affect the evolutionary process: nutrient and territory ratio. Nutrient represents a global common resource in the environment, and is implemented in the model as a single variable that is increased by a fixed amount at each generation, and reduced according to the total of the phenotype length of all individuals in the population P_g. This represents individuals from P_g consuming resources. If the nutrient amount falls below zero, reproduction entails a reduction in size of the genotype at a rate controlled by a parameter. Otherwise, genotype lengthening takes place at reproduction.

Territory ratio controls the size of the territory relative to the individuals' phenotype length. An individual exists in the centre of the territory, i.e. if the phenotype length is 5 and territory ratio 100, the territory will consist of 250 characters to the left of the position, the 5 characters it currently occupies, and 250 characters to its right.

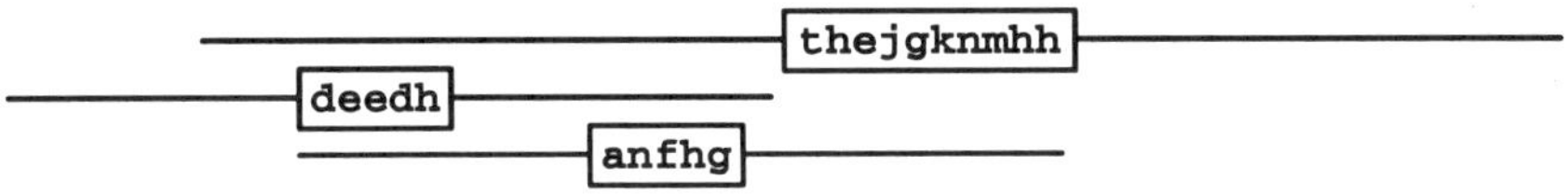

Fig. 5.2 Illustrating the territory of individuals of varying phenotype length.

Given the different length phenotypes, territory size may be different for each individual. When the territories of two individuals grow large enough for them to meet, they are able to commence sexual reproduction, and competition for mates. This is illustrated in Figure 5.2. The individual with phenotype "deedh" is able to find one neighbor with "anfhg", so reproduces with it. The individual with phenotype "anfhg", although having the same size neighborhood as "deedh", has two neighbors, due to the juxtaposition of individuals in the environment. In this case, the two neighbors would compete for the right to reproduce. The individual with phenotype "thejgknmhh" has more neighbors due to its larger territory size.

5.3 Experiments

Our research question is to determine the effects of various features of the model upon the size of modules formed, and the types of higher order structures. The features to be assessed are variable length genotype, variable genotype to phenotype mapping, memetic population and interaction with global parameters. To quantify the effect of these on module formation,

Table 5.1 A summary of the parameter values used in the four experiments.

	Experiment			
Parameter	**1**	**2**	**3**	**4**
Lengthening rate	0	1.0	1.0	1.0
Initial no. of characters	100	5	5	5
Mutation rate (per genotype)	1.0	1.0	1.0	1.0
Territory ratio	10	10	10	10
Size of memetic population	0	0	100	100
Nutrient	0	0	0	10,000

a measure is needed that can be applied to compare the operation of the model with or without these features. A direct measure of the size of and distribution of modules is inappropriate, for example, since it cannot be measured when the memetic evolution is disabled, as no modules will be produced.

In this work we have chosen to measure the size of modules formed using phenotype length per gene, and the usefulness of those modules using the number of matches per gene. These measures are proxies for the average module size and frequency of use. All experiments were performed with a population Pg of 100 individuals placed at a random location within the text. Periodic boundary conditions were used, so that the end of the Hamlet text was considered to be adjacent to the beginning for the purposes of calculating proximity. All experiments consisted of 500 iterations, and were repeated 10 times.

The first experiment is to establish a baseline, and uses a fixed length genotype, fixed genotype to phenotype mapping, no memetic population P_m and no nutrient effects. In order for the results to be comparable with other experiments with variable length genotype, the genotype length was set to 100. In the second experiment, we introduced the variable length genotype. In the third experiment, we introduced the variable genotype to phenotype mapping, and a memetic population. In the fourth experiment, we introduced a nutrient limit. Parameters for these experiments are shown in Table 5.1.

5.4 Results

The results of all four experiments are summarised in Table 5.2, which shows mean and standard error of selected measures at the conclusion of experimental runs.

Experiment 1 is a baseline test with no variable length genotype, no variable mapping and no nutrient. Figure 5.3 shows phenotype length per

Table 5.2 A summary of results, showing mean (standard error) for each measure. Experiment 4 clearly shows increased values for the size of modules formed (phenotype length per gene), and the usefulness of those modules (matches pre gene).

	Experiment			
Measure	**1**	**2**	**3**	**4**
genotype length	100 (0)	505 (0)	505 (0)	3.2 (0.53)
phenotype length	34.0 (0.8)	170.0 (3.16)	773.5 (30)	24.5 (4.18)
pheno. length / gene	0.34 (0.008)	0.337 (0.006)	1.53 (0.059)	7.86 (0.62)
matches	1.906 (0.036)	2.0 (0)	2.0 (0.0034)	7.64 (0.81)
matches per gene	0.019 (0.00036)	0.00396 (0)	0.00397 (0)	3.46 (0.95)

gene and matches per gene, averaged over 10 repeated runs of the model. There are no observable trends: these values reflect the underlying probability of 41 valid letters selected from 121(0.34), and of finding 2 contiguous letters in the individual's neighborhood with 100 genes. Any increase in either measure due to selection is balanced by mutation. In this experiment there was no opportunity for building modules or forming higher structures.

In the second experiment, the variable length genotype was introduced, and the genotype grew at the rate of one gene per iteration, reaching a final value of 505 (Table 5.2). The phenotype length reached 170, as the phenotype length per gene was roughly constant at around 0.34 (Figure 5.4). The number of matches quickly increased from near zero to 2 (Table 5.2), as favorable amendments suggested by the lengthening operator were selected. Subsequently, the mutation, selection and lengthening operators balanced to keep this constant, so the matches per gene then diminished (Figure 5.4). Adding a variable length genotype by itself does not provide any opportunity for building modules or higher structures.

In the third experiment we introduced a co-evolving representation in a memetic population. The genotype length grew as before, but the phenotype length was able to grow larger than in experiment 2, because of the presence of modules adapted to the environment (Table 5.2). This is reflected in a higher phenotype length per gene, shown in Figure 5.5. The number of matches, although rising more rapidly than experiment 2, did not achieve a higher value at the end of the run (Table 5.2). In fact the matches per gene had a behavior similar to experiment 2. Although the variable genotype to phenotype mapping succeeded in allowing the formation of modules, a limit on their size was very quickly reached, so this does not represent a convincing mechanism for building hierarchical structures.

In the fourth experiment, we introduced a nutrient limit that had the effect of producing oscillations in the genotype length of individuals. Because of this environmental pressure, the genotype length at the end of the run was only 3.2 (Table 5.2). However, the phenotype length was 24.5,

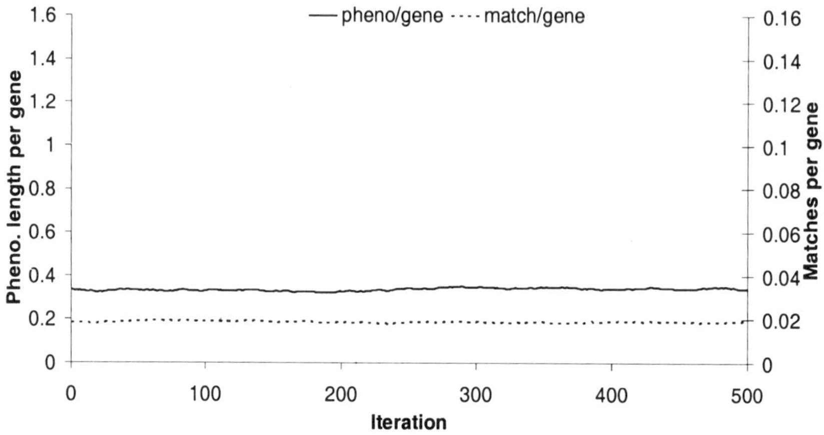

Fig. 5.3 Results from experiment 1, using a fixed length genotype, fixed genotype to phenotype mapping, no memetic population Pm and no nutrient effects.

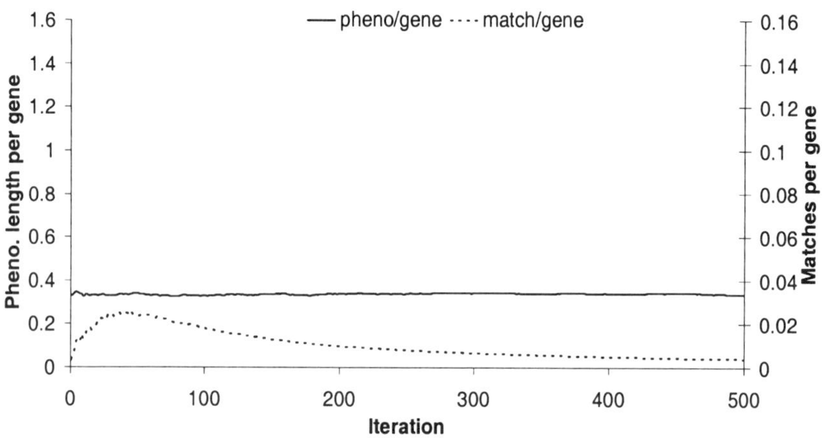

Fig. 5.4 Results from experiment 2, after introducing the variable length genotype.

and the phenotype length per gene continued to rise during the run (Figure 5.6). Modules formed represent increasingly longer segments of the text, and provided the opportunity for higher order structures to form. The average number of matches reached 7.64 (Table 5.2), as matches per gene continued to grow during the run up to 4.

Table 5.3 provides some examples of the five most used modules at

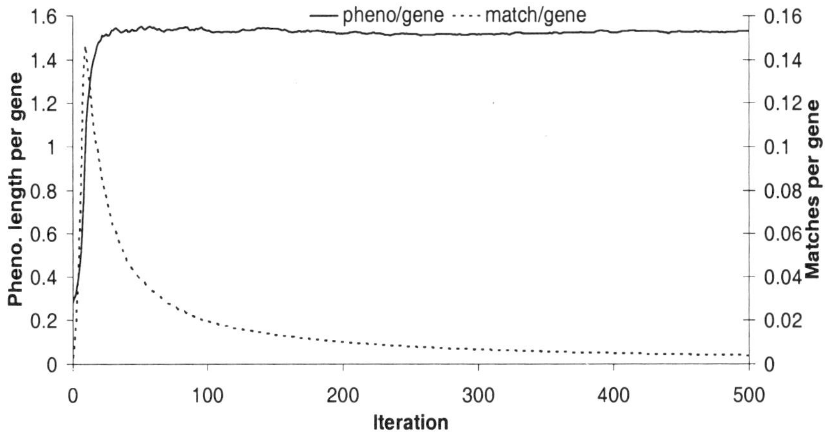

Fig. 5.5 Results from experiment 3, using a variable genotype to phenotype mapping and a memetic population..

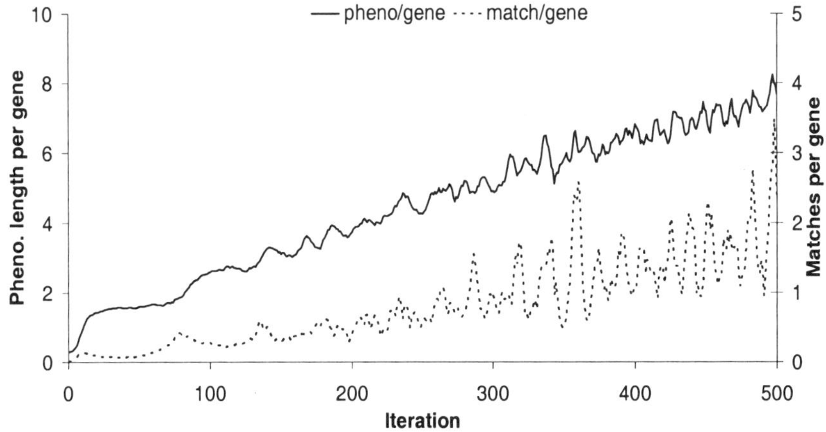

Fig. 5.6 Results from experiment 4, after introducing the nutrient limitation.

the end of a typical result of experiments 3 and 4. The module includes the symbols inside square brackets, separated by commas. Numbers refer to other modules, so the "end result" column provides a full decoding of the module (what would be added to the phenotype). Notice that the modules are generally larger for the results from experiment 4, suggesting that the phase changes experienced have enabled the formation of larger

Table 5.3 A comparison of some modules formed during experiments 3 and 4.

Experiment 3		Experiment 4	
Module	**End result**	**Module**	**End result**
[35, g]	". g"	[49, h]	"l work on h"
[r.]	"r."	[82, 21]	"ient as"
[p]	" p"	[135, 64]	" work on hi"
[62, k]	"nd! k"	[75, 15, 54]	" work on "
[g, 35]	"g. "	[, 67]	" madnes"

modules. Notice also the prevalence of symbols $g_i \in H$, which require further decoding, indicating hierarchical structures.

5.5 Discussion

One of the aims of artificial life research is to construct computer models that reproduce in some sense the complexity of biological systems [6]. Although evolutionary algorithms are often used, it is not clear which features are appropriate. In this work we have utilized a novel artificial life model to incorporate features that have not previously appeared together, allowing a comparative study of how they affect the formation of modules and hierarchical structures. This model incorporates a variable length genotype, a variable genotype to phenotype mapping, a separate memetic population encoding a collection of modules, and situates individuals in a pseudo-spatial environment to examine local effects between individuals as well as the interaction of global parameters on individuals.

The results indicate that although the variable length genotype, variable genotype to phenotype mapping, and the co-evolving population of memes increase the size of modules formed, the most dramatic effect upon module size is caused by phase changes, where the model is forced to oscillate between growth and consolidation phases.

These results have implications for the understanding of the formation of hierarchical structures in evolution. They imply a role in evolution of phase changes, although at present further work is required to identify exactly what systems this would apply to, and how it could be exploited. However, this should alert researchers to the consequences of choosing model features that can either restrict or enable desired model behavior. Finally, it has not escaped our notice that this model provides a potential research tool for investigating patterns within data of various kinds, including text and protein sequences.

Acknowledgements

This work was supported in part by a grant from Charles Sturt University under the Early Career Researcher scheme. The simulations were carried out on CSU's Cluster Computing Centre. David Green's work was supported by the Australian Research Council and the Australian Centre for Complex Systems.

Chapter 6

Perturbation Analysis: A Complex Systems Pattern

N. Geard, K. Willadsen and J. Wiles

ARC Centre for Complex Systems,
School of Information Technology and Electrical Engineering,
The University of Queensland, QLD 4072 Australia.
E-mail: {nic,kaiw,janetw}@itee.uq.edu.au

Patterns are a tool that enables the collective knowledge of a particular community to be recorded and transmitted in an efficient manner. Initially developed in the field of architecture and later developed by software engineers [138], they have now been adopted by the complex systems modelling community [417]. It can be argued that, while most complex systems models are idiosyncratic and highly specific to the task for which they are constructed, certain tools and methodologies may be abstracted to a level at which they are more generally applicable. This paper presents one such pattern, Perturbation Analysis, which describes the underlying framework used by several analytical and visualisation tools to quantify and explore the stability of dynamic systems. The format of this paper follows the outline specified in [417].

Pattern name Perturbation Analysis
Classification Dynamics, State Space
Intent The Perturbation Analysis pattern provides a quantifiable measure of the stability of a dynamic system.
Also known as None

6.1 Motivation

A complex dynamic system is one consisting of multiple elements, where the future state of the system is determined by a function f of its current state,

$$\mathbf{s}(t+1) = f(\mathbf{s}(t))$$

where $\mathbf{s}(t)$ is the state of the system at time t.

The typical feature of interest of complex dynamic systems is their asymptotic behaviour as $t \to \infty$. The set of states towards which a system converges under these conditions is known as an *attractor*. Attractors may be fixed points, limit cycles, or non-repeating 'chaotic' attractors. Systems may contain single or multiple attractors. The set of initial states of a system that converge to a given attractor forms the *basin of attraction* of that attractor.

Complex dynamic systems are widespread: genetic networks, economies and ecosystems are all examples. One of the emergent features of these types of distributed systems is their robustness or stability. Systems in the real world operate in noisy environments and are subject to perturbations from a wide variety of internal and external sources. In many cases, despite short term fluctuations, the long term behaviour of these systems is remarkably stable.

When modelling such systems, it is useful to be able to quantify this level of stability. For example, in a genetic regulatory system, where basins of attraction have been equated to cell behaviours [218], the stability of a system may reflect the phenomena of cell differentiation during development. Early in the developmental process, cells are sensitive to signals from their environment: transplantation experiments have demonstrated how embryonic cells can adopt the fate of their new neighbours rather than their original fate. As development progresses, the stability of cell types increases, and almost all fully differentiated cells will retain their original fate when transplanted [428]. The differentiation process itself is robust to fluctuations in external factors, such as temperature variation and nutrient levels, as well as internal factors, such as the stochastic nature of many genetic and cellular processes [225]

The Perturbation Analysis pattern provides a general framework for measuring the effect of changes to a system's current state on its long-term behaviour. These measurements may then be used as the basis for calculating more specific quantities, such as the rate of convergence or divergence of two nearby trajectories, or the probability of a perturbation causing a system to switch between different attractors.

6.2 Applicability

The Perturbation Analysis pattern requires a dynamic system, consisting of:

- a finite set of elements, each of which may take a discrete or continuous value; and
- a deterministic updating function.

The Perturbation Analysis pattern is useful in the following situations:

(1) A dynamic system is subject to some intrinsic or extrinsic perturbation, and it is desirable to stochastically or systematically explore and quantify the effects of these perturbations.
(2) A dynamic system can settle down into one of several possible behaviours and it is desirable to know either the likelihood of a system reaching a specific stable behaviour, or the probability of a system switching from one stable behaviour to another.
(3) A dynamic system is being used for prediction and it is desirable to know how far into the future its behaviour can be confidently predicted if there is some uncertainty as to its initial state.

6.3 Structure

The relationships between the classes involved in the Perturbation Analysis pattern are detailed in Figure 6.1.

6.4 Participants

State stores a state of the system $\mathbf{s}$, represented as a vector the values of each of the n elements,

$$\mathbf{s} = (s_0, \ldots, s_{n-1})$$

System applies an update function f to update the values of each element of a state,

$$\mathbf{s}(t+1) = f(\mathbf{s}(t))$$

Perturber applies a perturbation function p to create a new state from an old state in a systematic fashion,

$$\mathbf{s}' = p(\mathbf{s})$$

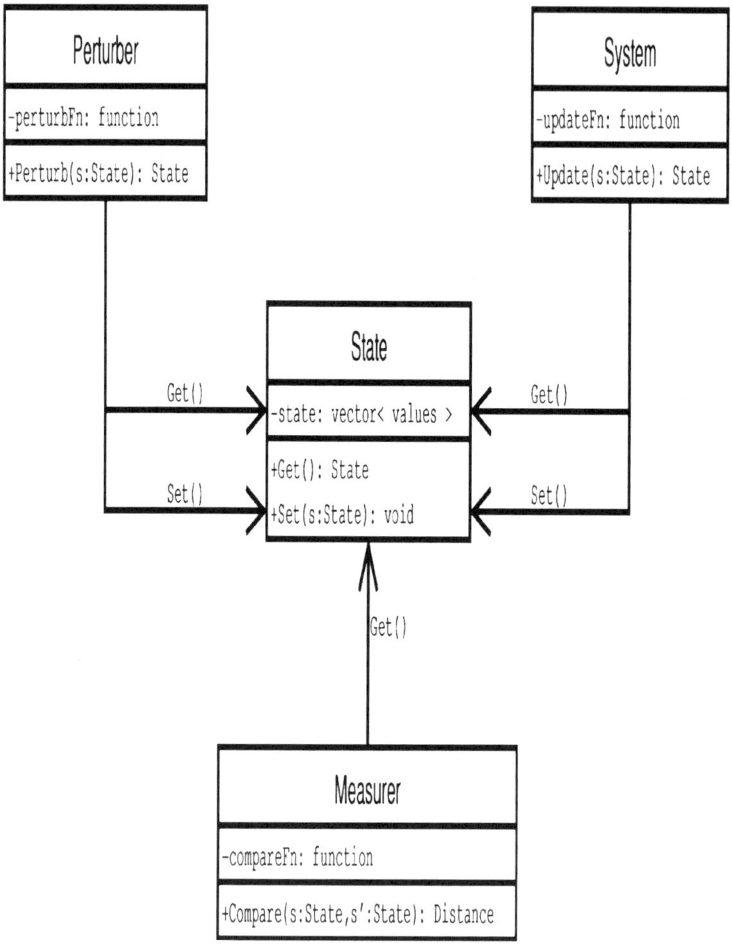

Fig. 6.1 A class diagram describing the types of objects involved in the Perturbation Analysis pattern and the relationships that exist between them. Each object lists the private data variables it contains (indicated by a minus), and the public functions it provides (indicated by a plus), together with their arguments and return values.

Measurer quantifies the distance d between two states according to some metric m,

$$d = m(\mathbf{s}, \mathbf{s}')$$

Concrete examples of the update and perturbation functions, and of the distance metric, are provided below, in the Implementation Section.

6.5 Collaborations

A dynamic view of the interactions between objects in the Perturbation Analysis pattern is shown in Figure 6.2.

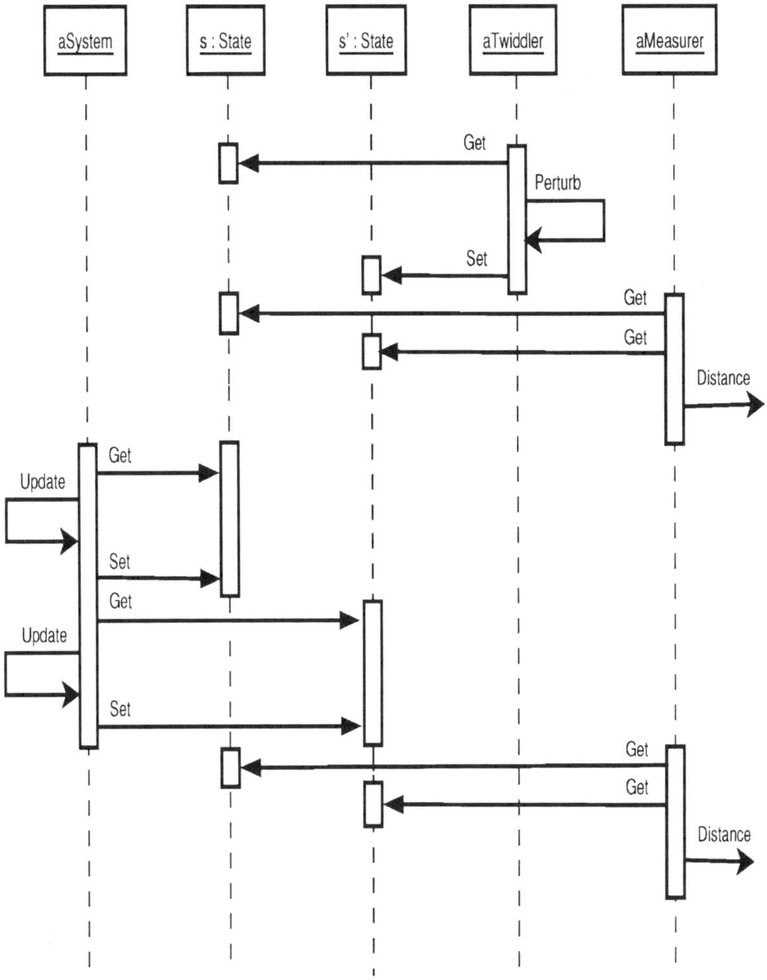

Fig. 6.2 A sequence diagram of the interactions between objects in the Perturbation Analysis pattern. Time runs vertically from top to bottom. The activation bars in the lifeline of each object indicate when that object is active in the interaction. Horizontal arrows indicate interactions – in this case function calls.

(1) Note that two States are maintained at all times: one (**s**) corresponding to the original system trajectory and another (**s**′) to the perturbed trajectory.
(2) Perturber sets the value of the perturbed State according to the application of the Perturb function to the original State.
(3) Measurer uses the Distance metric to calculate the distance between the original and perturbed trajectories.
(4) System uses the Update function to advance each of the states by one iteration (or time step).

6.6 Consequences

The Perturbation Analysis pattern has the following benefits and limitations:

(1) The pattern facilitates perturbation of a system and collation of distance measurements which can then be analysed using other methods. Two examples of the type of context in which the Perturbation Analysis pattern can be applied are provided below in the Sample Code Section.
(2) The pattern allows for a range of perturbation functions, distance metrics and system updating functions, each of which can be varied independently. Examples of these functions and metrics are provided below in the Implementation Section.
(3) Because the pattern only specifies a single iteration of the perturb and measure cycle, it supports the investigation of both annealed and quenched systems. In a quenched system, the structure of the system and the update function are static through time: measurements of a quenched system are specific to that particular instance of the system. In an annealed system, the basic parameters of the system (level of connectivity and type of updating function) are static, but the specific pattern of connectivity and set of updating functions are generated anew at each time step: measurements of an annealed system reflect basic properties of an entire class of systems.
(4) One limitation of the pattern as described here is that it requires a deterministic system updating function. While there is no reason that the pattern could not be applied to a stochastic system, doing so raises several issues that have not been addressed here relating to the structure of state spaces and the nature of attractors (see, e.g., [181]).

6.7 Implementation

The Perturbation Analysis pattern is generally applied as an iterative procedure. That is, a large number of perturbations and measurements are carried out in order to provide an estimation of the stability of a particular system or class of systems. A single iteration of perturbation and measurement may be described (in pseudocode) as follows:

```
# Set the initial state of the system.
State s = initialState

# Perturb the current state.
State s' = Perturber.Perturb (s)

# Measure the distance between the original and perturbed states.
startDist = Measurer.Distance (s, s')

# Update both the original and perturbed states.
s = System.Update (s)
s' = System.Update (s')

# Measure the distance between the updated states.
endDist = Measurer.Distance (s, s')
```

The main variables in this procedure are the nature of the update and perturbation functions and the distance metric. Each of these aspects may be varied independently.

(1) *Defining an update function.* The update function is defined by the dynamic system to which the Perturbation Analysis pattern is being applied. An example of an updating function in a discrete dynamic system is provided by Kauffman's Random Boolean Network model [218]. In this model, the value of a state element, σ_n, at time $t+1$ is some random Boolean function, f_n, of its K inputs at time t,

$$f_n(t+1) = f_n(\sigma_{n_1}(t), \dots, \sigma_{n_K}(t))$$

An example of a continuous update function is the sigmoid function used in many neural network applications,

$$f(x) = \frac{1}{1+e^{-x}}$$

where x is a weighted sum of the inputs to a particular element.

(2) *Defining a perturbation function.* Perturbation functions on discrete states can be either systematic or stochastic. A systematic perturbation function varies state elements systematically (e.g., by incrementing or decrementing an integer value, or by negating a Boolean value). A stochastic perturbation function varies state elements without regard for the sequential nature of the element values (e.g., randomly assigning a new integer value within the allowable range). A single application of either type of perturbation function will involve the alteration of one or more elements. The properties of the perturbation function are therefore:

- the number of elements being varied; and
- the mechanism (i.e., systematic or stochastic) used in their variation.

Several possibilities exist for perturbing continuous state element values. Unlike the discrete case, it is not possible to systematically explore the set of *all* possible perturbations. Therefore perturbations are generally applied in a stochastic fashion, by the addition of noise generated according to some distribution (e.g., Uniform or Gaussian) to some or all of the elements. The properties of the perturbation function that can be modified are:

- the number of elements modified by the addition of noise; and
- the parameters of the distribution used to generate the noise, for example, the mean and standard deviation of a Gaussian distribution.

Another possibility for perturbing continuous state elements is to define a discrete-valued structure embedded within the continuous state space and then systematically perturb the system within the bounds defined by this structure. For example, consider a system with three elements, in which the values of each element are constrained to the range $[0, 1]$. It is possible to define a three-dimensional cube within this space and constrain the initial states and perturbations to the vertices of the cube. If greater resolution is desired, the cube may be subdivided to introduce the midpoints of the edges and the centre of the cube. This method enables a continuous space to be explored and perturbed in a systematic fashion.

(3) *Defining a distance function.* In the case of discrete states, a distance function applies some transformation to the set of distances between individual elements of the state. The most common transformation performed in discrete systems is summation (e.g., Hamming distance), though other transformations such as the average or the sum of squares may be used.

When the values of the state elements are continuous, the standard distance metric is the Euclidean distance between the original and perturbed states. For a system with N elements, the Euclidean distance m between states $\mathbf{s}$ and $\mathbf{s}'$ is given by,

$$m(\mathbf{s}, \mathbf{s}') = \left[\sum_{i=1}^{N} (\mathbf{s}'_i - \mathbf{s}_i)^2\right]^{\frac{1}{2}}$$

where $\mathbf{s}_i$ is the value of the ith element of state $\mathbf{s}$.

(4) *Alternative distance measures.* Some methods for measuring the effects of perturbations do not require the initial distance measurement indicated above. An example method is the standard basin of attraction stability measurement commonly used in Random Boolean Network models [330; 14] – in this analysis the distance comparison is based solely on the final basins of attraction of the perturbed and unperturbed states.

6.8 Sample code

Boolean network attractor stability

A common application of the Perturbation Analysis pattern is to estimate the stability of an attractor in a Boolean network through either stochastic or systematic perturbation of attractor states. The typical unit of measurement in this usage case is whether or not the perturbed state reaches the same basin of attraction as the unperturbed state. Repeated trials are used to provide an estimation of the stability of a given basin of attraction, where stability is defined as the probability that a perturbation to a state does not change the basin of attraction.

One standard approach to obtaining such a stability measurement is to look only at the states in the attractor [181; 330; 14]. In this situation, the resulting measurement is the probability that the perturbation of an attractor state will move the system to a different basin of attraction.

(1) Choose a state $\mathbf{s}$ in attractor a.
(2) Perturb $\mathbf{s}$ to obtain $\mathbf{s}'$. This step is generally performed by flipping n elements of the Boolean state, where n is a small integer (frequently one).
(3) Iterate the trajectory starting at $\mathbf{s}'$ until it reaches an attractor a'.
(4) Store the value $a = a'$.
(5) Repeat steps 1 to 4 some number of times and calculate the average of the values stored in step 4.

An interesting characteristic of Boolean networks is the change in their behaviour as the degree of connectivity of the network (K) varies.

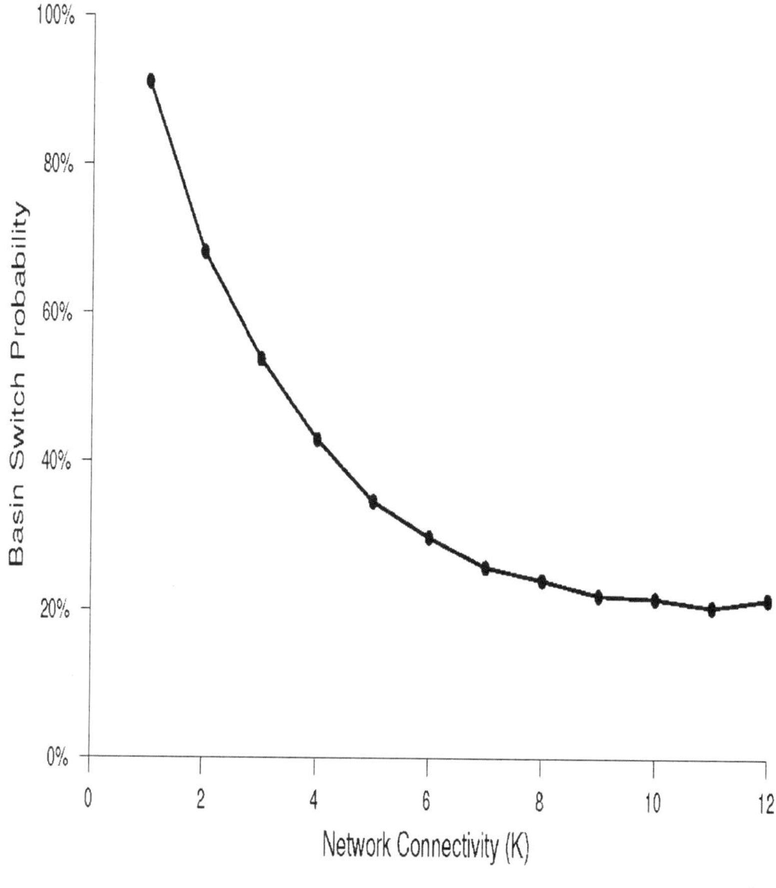

Fig. 6.3 Variation of attractor stability with increasing degree of connectivity in an $N = 12$ random Boolean network.

By repeated application of the above procedure, an approximation of the stability of the attractor states can be obtained for a range of connectivity values. Figure 6.3 shows the results of measuring stability in the above manner on a Random Boolean Network model with $N = 12$, by iterating through N perturbations of all 2^N system states and recording the probability of the target attractors of the original and perturbed states being different. This observed decrease in system stability is consistent with general expectations of the behaviour of the Random Boolean Network model.

Lyapunov characteristic exponents

One purpose for which the Perturbation Analysis pattern may be applied is estimating the largest Lyapunov exponent in order to determine the stability of an attractor. The Lyapunov exponents of a system measure the exponential rate of convergence or divergence of two nearby trajectories. If the largest Lyapunov is negative, the attractor is stable. If the largest Lyapunov is positive, the attractor is chaotic, and the magnitude of the exponent gives an indication of the time scale on which the future behaviour of the system becomes unpredictable.

The Lyapunov exponent λ is given by,

$$\lambda = \lim_{t \to \infty, \delta x_0 \to 0} \frac{1}{t} log \frac{|\delta x_t|}{|\delta x_0|}$$

where δx_t is the separation of the original and perturbed trajectories at time t. While methods do exist for determining the Lyapunov exponent directly from the equations describing a system's dynamics, it is also possible to approximate the value from a series of data points. The procedure for estimating the largest Lyapunov exponent is as follows [361]:

(1) Choose an initial system state.
(2) Iterate the state $\mathbf{s}$ until it is located on an attractor.
(3) Perturb $\mathbf{s}$ to obtain $\mathbf{s}'$. This step is generally performed by adding a small amount of Gaussian noise (mean 0, standard deviation 1×10^{-9}) to each of the state elements.
(4) Calculate the Euclidean distance d_0 between $\mathbf{s}$ and $\mathbf{s}'$.
(5) Iterate both trajectories.
(6) Calculate the new Euclidean distance d_1.
(7) Calculate and store the value $log|\frac{d_1}{d_0}|$.
(8) Perturb $\mathbf{s}$ to obtain $\mathbf{s}'$ such that distance between them is d_0 in the direction of d_1. This step can be carried out by adjusting each element i of state $\mathbf{s}'$ such that,

$$\mathbf{s}'_i = \mathbf{s}_i + \frac{d_0(\mathbf{s}'_i - \mathbf{s}_i)}{d_1}$$

(9) Repeat steps 5 to 8 some number of times and calculate the average of the values stored in step 7.

The number of iterations required to reach an attractor in step 2 and the number of iterations of steps 5 to 8 required for the value of λ to converge may vary. Similarly, it can be useful to repeat the calculation process using different initial states (step 1) and different initial perturbations (step 3). It is important to note that if a system contains more than one attractor, then the value of λ will be specific to the particular basin of attraction that contains the initial state.

One way to analyse the behaviour of a dynamical system is to explore the behaviour of a family of parameterised functions. For example a family of linear systems may be described by the function $f_m(x) = mx$ where m is varied over the real numbers. It is then possible to observe how the dynamics of the system change as the function is changed. The same technique may be applied to investigate the behaviour of more complex systems, such as neural networks, by the inclusion of a gain parameter g that scales the net input into the update function f,

$$f_g(x) = \frac{1}{1 + e^{-gx}}$$

As g affects the slope of the sigmoid function, modifying g from very small to very large results in a sweep from the linear range, through the nonlinear range to the Boolean range when the function is saturated.

By calculating the value of the Lyapunov exponent (using the same initial state each time) for each value of g, the range of dynamic behaviours of a particular system can be visualised. Figure 6.4 shows how perturbation analysis may be used to visualise the dynamics of a recurrent neural network [120]. The network used in this example consisted of 20 fully-connected nodes, with weights drawn from a Gaussian distribution with mean 0 and standard deviation 1. The gain parameter g was varied from 0.2 and 40 with increments of 0.2. An initial system state I was generated by setting the activation of each node to a value in the range [0, 1]. For each value of g the system was initialised to I and the procedure described above was used to estimate the Lyapunov exponent (Figure 6.4, top). In addition, the average activation of the network was recorded for each iteration of the calculation, providing an alternative visualisation of network dynamics (Figure 6.4, bottom). These two complementary views provide a comprehensive picture of the dyanmics of a system across a range of weight scales, revealing such features as bifurcations, fixed point and cyclic attractors, and chaotic behaviour.

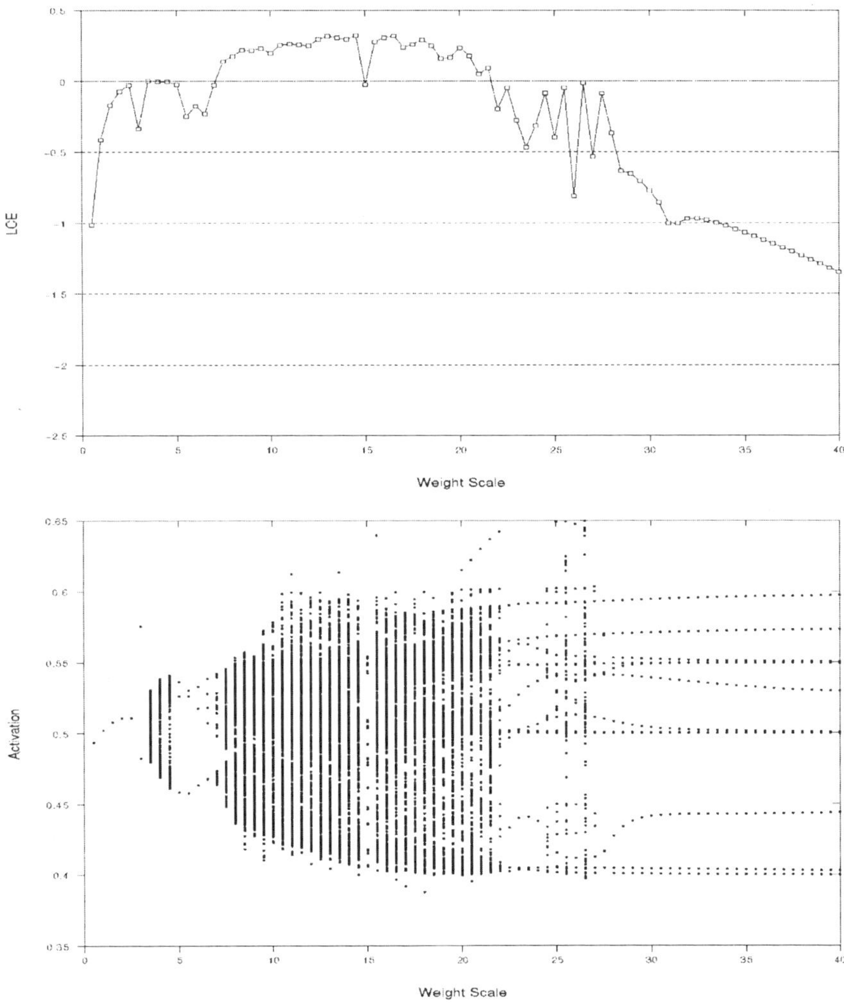

Fig. 6.4 Lyapunov exponent (top) and activation diagram (bottom) for a fully connected 20 node network as g is scaled from 0 to 40. Note the correlation between fixed point and cyclic attractors, indicated by single or multiple discrete points on the bottom chart, with negative Lyapunov values. In contrast chaotic attractors, with positive Lyapunov, values appear as as 'smears' of points.

6.9 Known uses

The concept of perturbation analysis as an exploratory tool was first formalised in the realm of Discrete Event Dynamic Systems, where it was developed to estimate the gradient of performance measures with respect to variation in system control parameters (see [191] for a history and overview of perturbation analysis in this context).

Within the field of complex systems, perturbation analysis has been used on an *ad hoc* basis by numerous researchers as a means of exploring the stability of genetic regulatory systems (e.g., [330]). Perturbation analysis has also been employed in a more principled fashion, to generate theoretical results about system stability: Derrida's annealed approximation method [105] illustrates the use of the Perturbation Analysis pattern on an annealed version of the Random Boolean Network model. This analytic tool uses an annealed random Boolean updating function, a stochastic perturbation process involving all N state elements and a state distance metric based on the normalised overlap of the states' values. The annealed approximation method was used to show that $K = 2$ connectivity in the Random Boolean Network model described a phase transition between the ordered and chaotic behaviour of the system. The annealed approximation method has since been used in different situations to identify phase transitions in the behaviour of networks of multi-state automata [357], and Boolean networks with scale-free topologies [14].

Lyapunov exponents have been used by mathematicians as an indicator of chaotic systems for some time. During the 1980s, several approaches were developed to allow the Lyapunov exponent to be determined from time series data [426], allowing the recognition of chaos in systems whose generating equations were unknown. Subsequent studies introduced the use of neural networks as general models of dynamic systems, typically for econometric and financial time series prediction tasks (e.g., [102]). More recently, simulations of high dimensional neural networks and systematic measurement of Lyapunov exponents has been used to investigate routes to chaos in high dimensional nonlinear systems [11]. Finally, the techniques described here have been extended and used to develop intuitions about the formation and stability of attractors in network models of gene regulation [141].

6.10 Summary

This paper has used the formal framework of patterns to describe a standard technique for analysing the stability of complex dynamical systems. The Perturbation Analysis pattern can be applied to a variety of discrete and continuous systems, as demonstrated by the random Boolean network and neural network examples detailed above. This form of stability analysis allows the effects of intrinsic and extrinsic perturbations on the dynamics of a system to be quantified. This paper also serves as an example of how the software engineering concept of patterns can be used to formalise modelling techniques and strategies for effective communication within a research community.

6.11 Acknowledgements

This pattern was developed during a Patterns Workshop held by the ARC Centre for Complex Systems on 6-7 July 2005.

Chapter 7

A Simple Genetic Algorithm for Studies of Mendelian Populations

C. Gondro and J.C.M. Magalhaes

CRC for Cattle and Beef Quality,
Animal Science,
University of New England,
Armidale, NSW 2351, Australia
cgondro@une.edu.au

Departament of Genetics,
Universidade Federal do Parana,
Curitiba, PR 81540-970, Brazil - C.P. 19071,
jcmm@ufpr.br

Evolutionary processes and the dynamics of Mendelian populations result from the complex interactions of organisms with other organisms and with their environment. Through simulations of virtual organisms the basic dynamics of these populations can be emulated. A conceptual model is used to define the universe, the hierarchical structures and a small set of rules that govern the basic behavior of these virtual populations. At the organism level a simple genetic algorithm is used to model the genotype of the entities and the Mendelian genetic processes. The model is implemented in an educational simulator called Sigex. From a small set of low level rules at the organism level, higher-order population and environmental interactions emerge that are in accordance to those postulated by the theory of population genetics.

7.1 Introduction

The two fundamental principles of *Evolutionary Theory* are that hereditary variability is the result of biological processes commonly referred to as factors of evolution; and, individuals who are more successful in survival and reproduction are differentially selected. These principles can only be explained taking into account the underlying genetic processes that operate on populations [334]. Genetic processes have classically been studied in *Mendelian populations* which consist of communities of potentially interbreeding, bisexually reproducing organisms. The main implication of the definition is that a Mendelian population consists of a single discrete species reproductively isolated from other species; thus genetic material can flow within the population whilst preserving its genetic identity from other species. Mendelian populations are important because higher animals, including humans, fall under this category and an understanding of population structures has immediate practical applications in medicine, animal breeding, ecology and wildlife management [179]. Further, the dynamics of allelic and genotypic frequencies in these populations can be studied in biological time scales (months or years) and help understand how changes can occur in evolutionary time scales (millions of years).

Evolutionary processes and the dynamics of Mendelian populations result from the complex interactions of organisms with other organisms and with their environment. Population genetics is the field of science that tries to explain these relationships or, more formally, it is the study of the effect of genetic processes on entire populations and how evolution factors modify species through time [179]. As a classic numeric branch of genetics, it has usually used a reductionist approach [269] operating mainly through mathematical models. These models are necessarily simplifications of reality, abstracted of the phenomena's complexity while trying to emphasize some of their aspects; for instance models of selection acting on the frequency of alleles at a single locus do not consider the effects of genes of other loci on fitness neither linkage or epistasis among genes. This is necessary since models can become very complex, as more parameters are included. Simpler and more tractable models can be overly distant from reality, whilst more realistic models can become too complex to be tested experimentally. Further, not infrequently population data is sparse and costly to gather, when not unobtainable; as for example the fossil record which is usually very fragmented. Within this context computer simulations can be an important tool to link theoretical abstractions with the complexity found in nature [72].

A sound scientific theory relies on a small number of hypotheses or axioms to generate complex models [269]. Analogously, computational models based on a small number of rules can exhibit complex behaviors. Phenomena that seem to be operating in nature, given the underlying constraints, can be emulated through computer simulations allowing data generation, model testing, experimental planning, emergence and investigation of complex phenomena not easily derived from theoretical postulates [72]. A particularly interesting approach to address complexity issues at workable levels is through Artificial Life [3] where computational simulations have been used to test new approaches into theoretical biology topics with promising results [238]. Holland [196] [197] introduced Echo models as a means of understanding complex adaptive systems which evolve by natural selection through the interaction of agents among themselves and with the environment. An Echo model is an empirical model in the sense that it encapsulates the mechanisms deemed the most relevant of the system. Even if holding similarities to classifier systems, Echo uses less abstract rule-activating messages making interpretation of the system easier [196]. The best known implementation of Echo was developed by Hraber *et al.* [200]. Other computational systems for studying complex systems include the seminal Tierra [327], the more generic Swarm system (www.swarm.org) and Sugarscape [121].

Evolutionary factors such as drift, mutation, migration and selection seldom act in isolation and studying the individual effect of these factors in natural populations can be a daunting task. In this paper we present a model along the general lines of Echo to simulate Mendelian populations using virtual organisms which allows studying different genetic processes and how the allelic and genotypic frequencies are affected over time. We use conceptual modeling [77] to define the universe, the hierarchical structures and a small set of rules that govern the basic behavior of these virtual populations. At the organism level a simple genetic algorithm [195] is used to model the genotype of the virtual entities and the Mendelian genetic processes. Our model is implemented in a freely available simulation package called Sigex.

The remainder of the paper is structured as follows. Section 2 briefly reviews the basics of genetic algorithms. In section 3 we overview the conceptual model of the virtual organisms, introduce the simulator Sigex and describe with some detail the genetic algorithm used to constructed the genetic structure of the agents. In section 4 a simple simulation example of the Hardy-Weinberg principle using Sigex is discussed. In section 5 we present some conclusions and future work.

7.2 Genetic Algorithms

The most widely disseminated Evolutionary Computation (EC) branch, Genetic algorithms (GAs) date back to Hollands [195] seminal work. GAs are the EC class of optimization heuristics which most closely mimic evolutionary processes at a genetic level; traditionally organisms are represented as linear bitstrings which are referred to as chromosomes; this is the canonical GA [195]). The value in each position of the bitstring is an allele (0 or 1) and the position itself is a gene or locus. The combination of values (alleles) in the bitstring (chromosome) maps to a phenotypic expression, such as a parameter to be optimized. From the above it is clear that GAs operate at two structural levels: a genotypic and a phenotypic one. Selection operators are carried out based on the overall chromosome value (phenotype) while search operators act on the genotype, modifying the chromosome which may or may not change the phenotypic expression. The main search operators are recombination and mutation.

7.2.1 *Search operators*

Recombination is a search operator that does not generate new sources of variability in the populations albeit introducing new variation. It operates by combining parts from two or more parents to generate one or more offspring. The drive behind recombination is to generate new variability in the population by manipulating the component sources of variation to explore new combinations.

Figure 7.1 illustrates a one-point crossover in a binary GA. Briefly, two parents are selected for recombination, a breakpoint in the chromosome is randomly determined, and from the breakpoint onwards the two chromosomes swap the remainder of their bitstrings.

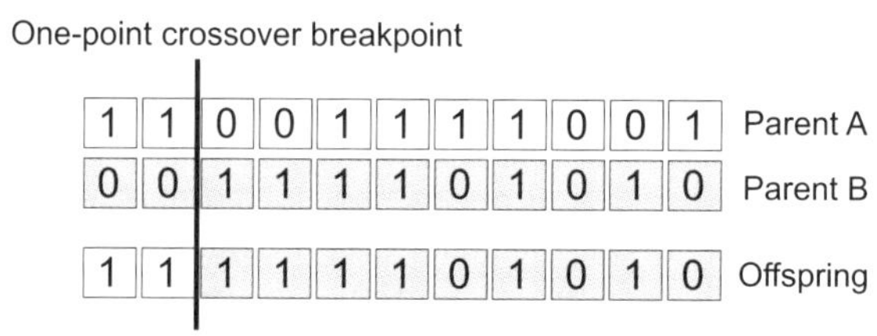

Fig. 7.1 One-point crossover in a binary genetic algorithm. A breakpoint is randomly selected and the two chromosomes swap bitstrings after the breakpoint. Recombination is a search operator which explores available population variability by testing new combinations. No new allelic variability is generated through recombination but it does generate new variation in fitness values.

In contrast to recombination, mutation generates new allelic variability in the population. The general principle is that new offspring are created by a stochastic change to a single parent. Figure 7.2 shows a point-mutation bit-flip in which an allele of a parent is randomly selected to be flipped. The most common approach is to assign a small uniform probability that mutation will occur and test each position of the bitstring; if the mutation operator returns true the bit at the position is flipped.

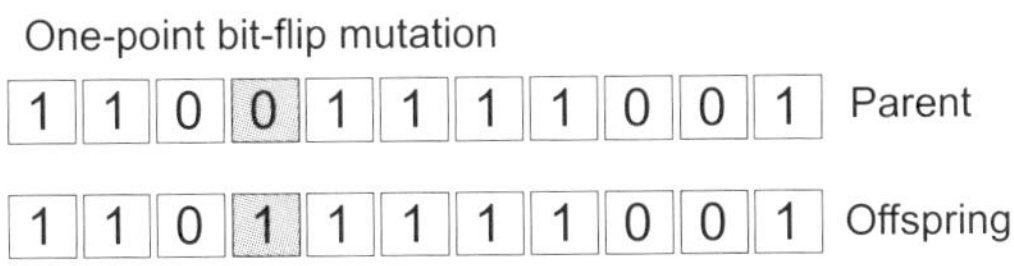

Fig. 7.2 Point-mutation bit-flip in a binary genetic algorithm. New offspring are produced by a random change to the parent. In the example allele at position 4 was selected for mutation and flipped from zero to one. Mutation is a source of new variability in a population.

7.3 Conceptual Model of Mendelian Populations

Our universe of virtual organisms was designed as a conceptual model, meaning that the key components and low-level interactions of the system were empirically determined. The key aspect of the model was to abstract the main mechanistic properties of the biological population. In order to do this the model was structured according to the following steps:

(1) Description of the components of the system, properties and low-level interactions.
(2) Definition of analogies between the biological system and the virtual environment.
(3) Definition of the hierarchical structures of organization with each level establishing a set of elements and relations.
(4) Implementation from lower to higher order levels. Once the elements and relations of the model were defined these were grouped in hierarchical levels and implementation was restricted to the lower level elements and relations. Higher level interactions emerge from lower level orders.

These steps are clearer if we look at the model itself. Our model consists of a much reduced abstraction of reality formed by the universe (biosphere) which is the uppermost hierarchical level and two subsets forming the environment and the population. The environment has a single level set of three elements, namely *food*, *barriers* and *space* each with its own set of properties. The population has three tiers: *organisms* (tier 3), *chromosomes* (tier 2) and *genes* (tier 1), each with its own set of elements and relations. Figure 7.3 depicts an abstract three level hierarchical structure with elements and the interactions between levels. At the intersections of the layers the properties of the higher level tiers emerge from the interactions of the lower level elements.

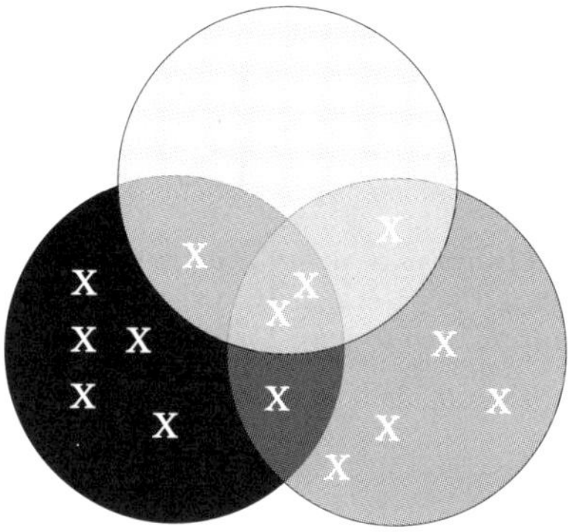

Fig. 7.3 Elements and relations in a three layer structure. At the intersection the properties of the higher level tiers emerge form the interactions of the lower level elements.

Differently from Echo [200] which uses haploid agents with a single chromosome and asexual reproduction, our virtual organisms are an abstraction of Mendelian populations, meaning that they are a single species of freely interbreeding diploid organisms with two sexes on an XY system. There are two genes in the sex chromosomes and seven genes distributed in a variable number of autosomes (between 1 and 7). Each gene has between two and four allelic variants with user-defined phenotypic expressions within a certain interval limit. The genes through their phenotypic expressions express characteristics that intimately relate to the universe ensuring a rapid evolution of the population. For example the gene for *vision* determines the line of sight of the organism which is an important trait for searching for food in the environment and finding a partner for reproduction. The genes not only relate to the environment but they also relate to other organisms, as for instance the *fight* gene which defines the level of aggressiveness of an organism. An exception to this aspect is a *neutral* gene that has no phenotypic expression in any allelic combination. This gene is important in drift and migration studies. The only genetic processes included in the model were mutation, segregation, recombination and reproduction. This model was implemented in an educational software package – Sigex – that is used for teaching population genetics and evolution subjects at undergraduate and graduate levels (figure 7.4). Sigex allows simulation studies using virtual organisms and the generation of population data files that can be studied with the *Analysis* module that covers the main topics of population genetics. A full description of the model and the simulation tool are available from www.sigex.com.br/genetics.

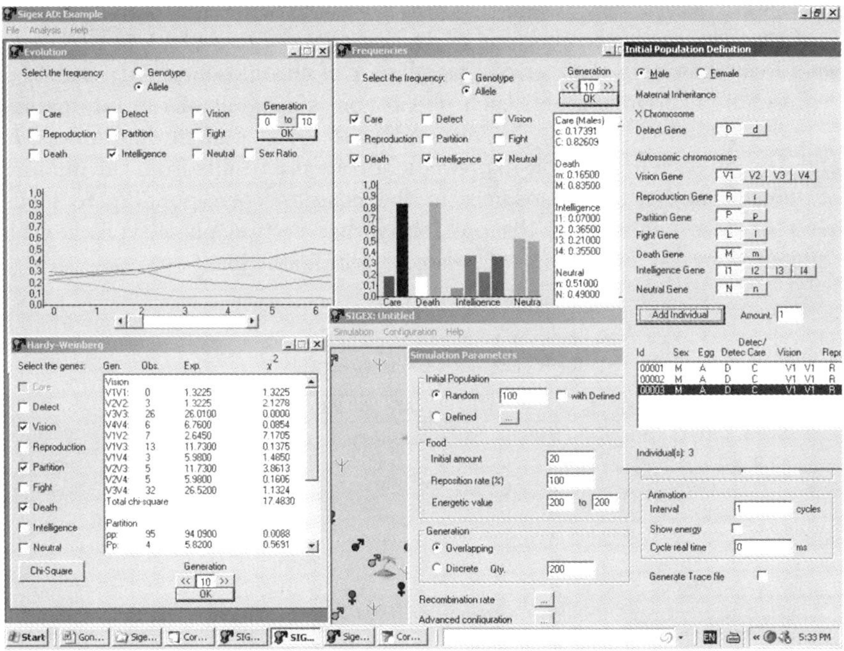

Fig. 7.4 Sigex – an educational package for studies of population genetics and evolution. The program consists of four modules: a simulator of virtual organisms, a genotype editor, a data analysis tool and a manual/tutorial of population genetics and evolution.

7.3.1 *Virtual organisms as a simple genetic algorithm*

Each organism is formed by two structures: an identification structure and a genetic structure. The first structure simply stores the organisms information with a unique identifier that allows tracking its activities in the environment and retrieving parental information.

The genetic structure defines the organism itself. It is based on a simple genetic algorithm that represents the organisms genotype. It consists of two homologous bitstrings, in an analogy to DNA molecules. Each bitstring is defined as a chromosome.

Here we define that if M is a chromosome; M_1 and M_2 form the homologous chromosome pair and m_x is a gene and the phenotypic expression resulting from the genes action, where x is an integer that stands for the position (locus) of the gene in the chromosome and n_{xy} are the alleles of gene m_x established in such way that y is the number of bits in locus x and n indicates the number of zeros found in this particular section of the bitstring.

Consider that a chromosome M codes traits m_j ($x = 1, 2, 3, .., j$) and each locus has its y number of bits. A pair of chromosomes (M_1 and M_2) represents an organism and each phenotype is a result of the interaction between the two alleles n_{xy} corresponding to locus x of chromosomes M_1 and M_2. So each phenotypic expression of trait m_x results from the number of zeros from both chromosomes in the position corresponding to locus x. This does not exclude the possibility that certain phenotypes can be originated by the interaction of genes from different loci.

Table 7.1 Possible coding levels for a gene with 2 bits ($y = 2$).

	M_1		M_2		**Zeros**
0	0	0	0	0	4
1	0	0	0	1	3
2	0	0	1	0	3
3	0	0	1	1	2
4	0	1	0	0	3
5	0	1	0	1	2
6	0	1	1	0	2
7	0	1	1	1	1
8	1	0	0	0	3
9	1	0	0	1	2
A	1	0	1	0	2
B	1	0	1	1	1
C	1	1	0	0	2
D	1	1	0	1	1
E	1	1	1	0	1
F	1	1	1	1	0

The different combinations of values for each locus x can be classified according to the number of zeros as shown in table 7.1, representing a locus x for $y = 2$. From table 7.1 it is evident that in this case there are five coding levels. For $y = 1$ there would be three coding levels. More bits in y imply in more available coding levels.

The higher the value of y used the lower is the probability of certain combinations appearing. Table 7.2 presents the associated probabilities for $y = 2$ *(1:4:6:4:1)*.

Table 7.2 Levels and probabilities for $y = 2$.

Zeros	**Probability**
2	$6/16 = 0.3750$
1	$4/16 = 0.2500$
3	$4/16 = 0.2500$
0	$1/16 = 0.0625$
4	$1/16 = 0.0625$

We can thus define that the set of possible combinations of zeros and ones form the set of alleles of trait m_x. The set of m_x genes forms chromosome M_j, the haplotype. For every gene m_x in M_1 there is a homologous of the same size and in the same position in M_2. Chromosomes M_1 and M_2 form the genotype.

In this manner binary structures with any number of genes and each gene with any number of alleles for diploid organisms can be very easily created. Table 7.3 shows a binary structure with two chromosomes M_1 and M_2 and six genes.

Table 7.3 Binary structure of an organism.

Genes m_x	**Chromosome** M_1	**Chromosome** M_2
1	00	10
2	01	11
3	11	01
4	01	11
5	0	1
6	1	0

The main genetic operators: segregation, recombination, mutation and reproduction operate on the binary structures. These operators are executed before a new organism is formed. Each operator is discussed below.

We have so far mentioned only a pair of chromosomes (M_1 and M_2). Obviously natural organisms may have more than a single pair of chromosomes. To obtain independent segregation between loci belonging to different chromosomes it is not necessary to use more than a single binary string for each haplotype. Independent segregation can be obtained specifying a recombination ratio of 0.5 between the last locus of a chromosome and the first locus of the following chromosome. In this manner it is possible to emulate the existence of several binary strings (chromosomes) in a single string.

Recombination is processed in the same manner as independent segregation, with the difference that the recombination ratio between the loci is below 0.5. To obtain absolute linkage between loci a 0.0 recombination ratio can be determined.

Operationally, recombination and independent segregation are obtained by defining breakpoints between loci, testing the probability that a recombination will or not occur against a computer generated random number. If the recombination should occur the string is swapped with its pair after the breakpoint (figure 7.1). The main differences to a conventional GA are that (1) recombination occurs only within an organism and not between organisms and (2) since organisms are diploid, allelic interactions (such as *dominance*) are possible. Recombination within a locus follows the same procedure described above, thus generating new alleles.

Mutation is also a probabilistic event in the binary string of the organism. If the event returns true, the bit(s) will be inverted ($1 \rightarrow 0$ or $0 \rightarrow 1$). Mutation rates can be modeled for each allele, locus, chromosome or organism. It is important to establish appropriate mutation rates ensuring that these occur within the desired frequencies.

The use of two homologous binary strings is a natural approach to emulating diploid organisms. It also makes reproduction a simple procedure, being necessary only to randomly select one string (chromosome) from the *father* and one from the *mother* to form a new organism, after the previous operators have been executed.

Offspring are generated applying the genetic operators to the structure of the parents according to the following steps:

(1) Define temporary structures to store the bitstrings of the parents modified by the operators.
(2) Segregation and recombination between the binary strings of both selected parents.
(3) Selection of one of the strings from the parents.
(4) Mutation.
(5) Creation of the new organism.
(6) Discard of the temporary structures

7.4 Hardy-Weinberg Equilibrium in a Virtual Population

To illustrate the use of our model, the Hardy-Weinberg law was selected, which is one of the basic principles of population genetics. This principle essentially states that in a large (theoretically infinite) random mating population, in the absence of evolution factors – mutation, migration, genetic drift and selection – the relation between allelic and genotypic frequencies remains constant from generation to generation and the genotypic frequencies are determined by the allelic frequencies [179]. Mathematically for a single locus with two alleles this relationship is expressed by the simple equation:

$$(p + q)^2 = p^2 + 2pq + q^2 = 1$$

where p and q are alleles. The immediate implication of the principle is that it allows estimation of genotypic frequencies in cases where not all genotypes are phenotypically distinguishable. It also allows checking the equilibrium of the population which is the first step to identify if evolution factors are altering the population structure.

To test for Hardy-Weinberg equilibrium the assumptions of the model have to be taken into account. In Sigex the natural choice is the *neutral* gene which has no phenotypic expression and is thus not subject to selective pressures. We ran a simulation using discrete generations over 20 generations. In Sigex discrete generations are simulated by having the parents lay *eggs* instead of juveniles; when a user-defined number of eggs has been generated, the parental population dies out and the eggs hatch to form the new generation. Each generation consisted of 1000 organisms. The initial population was randomly generated, mutation rate was set at zero and the *neutral* gene was placed on an individual chromosome not to be affected by selection on linked genes. The resource *food* was supplied at a level that it did not exert a high selective pressure on inferior phenotypes (500 *food units*, 200 *calories* per unit, replacement ratio 1 : 1).

Figure 7.5 shows the allelic and genotypic frequencies for the *neutral* gene over 20 generations. The dashed lines correspond to the allelic frequencies of the two allelic variants *n(p)* and *N(q)* and the solid lines are the genotypic frequencies $nn(p^2)$, $Nn(2pq)$ and $NN(q^2)$. Since there is no selection, mutation or migration on this gene, the frequency changes over the generations are essentially due to genetic drift. The effect of drift on frequencies is evident observing the frequencies of the homozygotes in comparison to the frequencies of the heterozygotes, since the frequencies of the former, in populations in Hardy-Weinberg equilibrium, are defined by p^2 and q^2 of the allelic frequencies p and q making them more susceptible to changes in the allelic frequencies than the heterozygotes which change by $2pq$.

Table 7.4 summarizes the allelic and genotypic frequencies. The traditional method for testing Hardy-Weinberg equilibrium is the χ^2 test. The

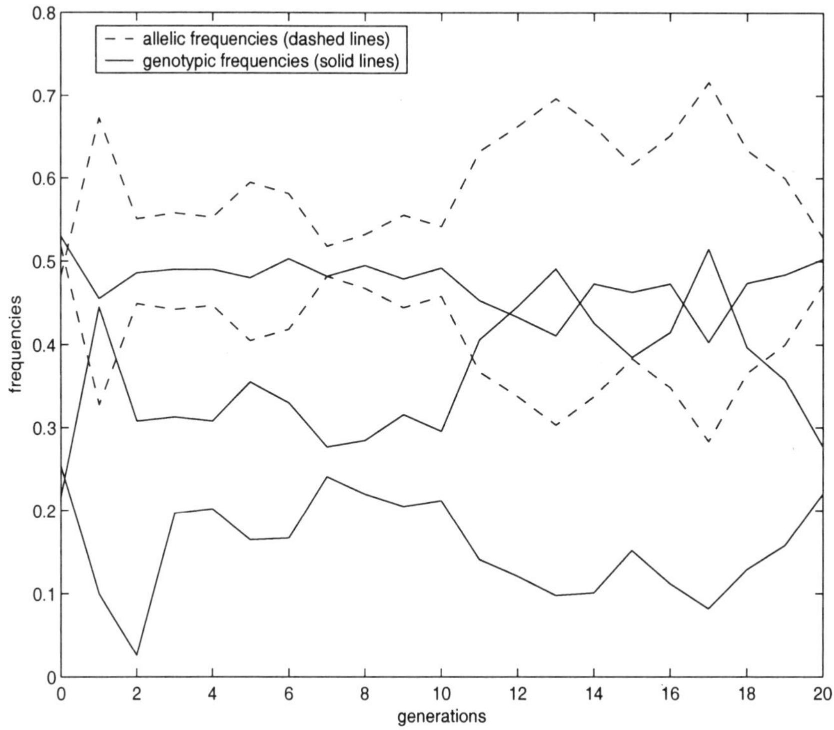

Fig. 7.5 Allelic and genotypic frequencies for the *neutral* gene over 20 generations in a population of 1000 organisms.

probability values associated with the χ^2 are shown in bold in table 7.4. From these it can easily be observed that even though the frequencies vary across the generations, the population remains in Hardy-Weinberg equilibrium over time. The frequency fluctuations can largely be attributed to drift. Even though population sizes of 500 and above are considered sufficiently large to minimize the effect of drift over a small number of generations (over many generations even populations of 10^6 organisms show drift effects), the actual breeding population is much smaller (Ne), increasing the effect of drift on the frequencies [179].

This simple example illustrates the use of our model for population genetics studies. For validation studies that we carried out to test Sigex the reader is referred to [150]. A study comparing the adequacy of different methods for estimating effective population size (Ne) using population data derived from the simulator was presented in [379].

Table 7.4 Allelic and genotypic frequencies for a population of 1000 organisms over 20 generations. P-values are the probabilities associated to the chi-square test for Hardy-Weinberg equilibrium.

	Allelic frequencies		Genotypic frequencies			
Generation	**n(p)**	**N(q)**	**nn(p^2)**	**Nn($2pq$)**	**NN(q^2)**	**p-values**
Parental	0.5190	0.4810	0.2540	0.5300	0.2160	**0.0516**
1	0.3275	0.6725	0.1000	0.4550	0.4450	**0.2974**
2	0.4490	0.5510	0.0260	0.4860	0.3080	**0.5739**
3	0.4420	0.5580	0.1970	0.4900	0.3130	**0.8338**
4	0.4470	0.5530	0.2020	0.4900	0.3080	**0.7792**
5	0.4050	0.5950	0.1650	0.4800	0.3550	**0.8980**
6	0.4185	0.5815	0.1670	0.5030	0.3300	**0.2900**
7	0.4820	0.5180	0.2410	0.4820	0.2770	**0.2718**
8	0.4675	0.5325	0.2200	0.4950	0.2850	**0.8545**
9	0.4445	0.5555	0.2050	0.4790	0.3160	**0.3419**
10	0.4580	0.5420	0.2120	0.4920	0.2960	**0.7758**
11	0.3675	0.6325	0.1410	0.4530	0.4060	**0.4187**
12	0.3375	0.6625	0.1210	0.4330	0.4460	**0.3157**
13	0.3035	0.6965	0.0980	0.4110	0.4910	**0.3784**
14	0.3375	0.6625	0.1010	0.4730	0.4260	**0.0679**
15	0.3835	0.6165	0.1520	0.4630	0.3850	**0.5098**
16	0.3485	0.6515	0.1120	0.4730	0.4150	**0.1880**
17	0.2836	0.7164	0.0820	0.4031	0.5148	**0.4033**
18	0.3660	0.6340	0.1290	0.4740	0.3970	**0.4994**
19	0.4000	0.6000	0.1580	0.4840	0.3580	**0.7922**
20	0.4717	0.5282	0.2203	0.5028	0.2768	**0.8134**

7.5 Conclusions and Future Work

In this paper a simple genetic algorithm was presented for simulating Mendelian populations of virtual organisms. Using homologous bitstrings and adapting conventional GA search operators, the basic Mendelian genetic processes were implemented in virtual entities at low hierarchical levels. This structure at the organism level is sufficient to manifest at a higher level, the dynamics of populations.

The simple conceptual model we designed and implemented in Sigex was extensively tested through custom developed simulations evidencing a good fit of the resulting population data to the current theory of population genetics. Within the limits of the model several topics in population genetics were addressed: Hardy-Weinberg equilibrium, selection, genetic drift, mutation, migration, linkage disequilibrium, fitness and adaptation [150]. Sigex was primarily designed for educational purposes to allow students to design, run and interpret population studies, whilst at the same time emphasizing that evolution factors do not operate in isolation but together [151].

It is clearly noticeable that as for real biological systems, lower hierarchical level components interact to originate complex higher-level properties. Even though the model is constructed informally it is designed to be a model of the evolutionary theory. This means that it is a set of objects, properties and relations that must satisfy an abstract structure (axiomatics) as the one presented by Magalhaes and Krause [256], in the same way that a real biological system must, in principle, satisfy. This informal approach was adopted due to its approximation with the methods of experimental science; thus the model was built based on the intuitive perception of the biological system from which it was abstracted. The virtual populations had to reflect the same collective properties of natural Mendelian populations and this was the main validation criterion. Nevertheless, once it has been extensively tested, it can be used in new situations as a prospective method.

We are currently interested in classic experiments in population genetics, applying the original analysis methods on data obtained from simulations. The results of these experiments should help determine the suitability of virtual populations as surrogates of natural populations for population genetics studies. An example of a classic experiment with *Drosophila* is found in Buri's [60] drift experiment. Frequency variations were used to estimate the effective population size, *Ne*. This parameter, which allows estimation of the occurrence of genetic drift and inbreeding in finite populations, can be estimated through other methods using different sources of data [62]. All these methods can be applied since full datasets of the population dynamics are generated, including the genealogical history of each organism and its genotype.

Virtual populations can be a useful tool for research in population genetics. Even though computational simulations cannot accurately depict the dynamics of natural populations, for theoretical studies and as an initial approach to test a new model *artificial* data can be used prior to obtaining experimental data. This approach not only allows testing on rapidly obtainable, controlled data but can also help in determining which experimental data is relevant, assisting in the design of the experiment.

Acknowledgement

We thank Mr. Omar Achraf for the C++ modules in Sigex and Professor Brian Kinghorn for his comments on this manuscript.

Chapter 8

Roles of Rule-Priority Evolution in Animat Models

K.A. Hawick, H.A. James and C.J. Scogings

Computer Science, IIMS, Massey University - Albany
North Shore 102-904, Auckland, New Zealand
Email: {k.a.hawick, h.a.james, c.scogings}@massey.ac.nz
Tel: +64 9 414 0800 Fax: +64 9 441 8181

Evolutionary behaviour in "animat" or physical-agent models has been explored by several researchers, using a number of variations of the genetic algorithmic approach. Most have used a bio-inspired mutation/evolution of low-level behaviours or model properties and this leads to large and mostly "uninteresting" model phase-spaces or fitness landscapes. Instead we consider individual animats that evolve their priorities amongst short-lists of high-level behavioural rules rather than of lower-level individual instructions. This dramatically shrinks the combinatorial size of the fitness landscape and focuses on variations within the "interesting" regime. We describe a simple evolutionary survival experiment, which showed that some rule-priorities are drastically more successful than others. We report on the success of the rule-priority evolutionary approach for our predator-prey animat model and consider how it would apply to more general agent-based models.

8.1 Introduction

Evolutionary behaviour in physical agent or animat models [424] is both a philosophically intriguing problem and also a computationally demanding one. Other workers have demonstrated exciting emergent [116] properties of animat models, starting from a set of very low-level instructions or microscopic behaviours [326; 4; 194]. It now seems indeed reasonable to describe collections of such animats as artificial life [241; 64] systems. We have previously reported on emergent macroscopic "life-forms" or large scale patterns in a predator-prey model with some carefully chosen microscopic behaviours [209; 210; 186]. Introducing evolutionary behaviour into our model to evolve a better predator or a "longer surviving prey" is interesting but involves what is essentially a brute force exploration of the model's phase space. In this paper we report on what happens when we do **not** evolve **microscopic behaviour** or instructions which are all assumed to have a priori probabilities, but rather what happens when we assume the micro-animat behaviour to be made up of a **pre-selected set** of **high-level rules**. We apply evolutionary algorithms to exploring combinations of pre-evolved rule priorities. We believe this is an important mechanism in real life, in that presumably many of the high level features in real life-forms, once established, can be reused in interesting ways. We consider the analogy with bio-programming and real programming is that we are experimenting with different combinations of relatively small numbers of sub-programs rather than arbitrary large combinations of micro-instructions.

In this paper we briefly summarise and review our predator-prey animat model in section 8.2. We show some pictorial results of different rule combinations in section 8.3 and discuss some parametric and statistical metrics to characterise the overall results of simulations and model behaviour in section 8.4. We report on an evolutionary survival experiment in section 8.5 and explore how the formulation and application of this hierarchical rule prioritisation approach might be generalised for other animat simulations in section 8.6.

8.2 Rule-Based Model

Our model [209; 210; 186] is based around a set of rule-controlled individual animats which can move, breed, eat and die in a "flatland" of discrete x,y integer coordinates. More than one animat can exist on a single physical cell and the boundaries are not fixed: the size of our model world simply expands as animats explore it. In all our experiments we initialise the model with a block of animats near the origin and time-step the rules so each animat has the opportunity to take one action at each time step. Two well-known systems with similar objectives to ours are Tierra [326] and Avida [4].

Within both Tierra and Avida each animat (cell) is represented by a set of integers that represent a command string, consisting of low-level instructions for the individual cell. Each cell has an internal maximum command string length but not all instruction locations may be filled or valid. Instructions are commands such as "nop", "if-not-0", "inc", "dec", "push" and "pop" for the simple accumulator-based cells. They are a kind of simple cell programming language and include instructions to divide cells and attempt to inject instruction sequences into other cells. Cells are contained within a structure called a GeneBank.

One of the fundamental differences between Tierra and Avida is that cells are considered to be 'in a soup' in Tierra, whereas in Avida cells interact only with their (nearest or next-nearest) neighbours on a periodic 2-dimensional grid structure.

Depending on the characteristics of the system under consideration in Tierra and Avida, cells can either undergo reproduction or random mutations. In reproduction the user has control over how the candidate that will reproduce is chosen: either the oldest or largest can be chosen, the one with empty space near it, or a random cell.

Users can choose to enable point mutations of the entire population and mutations on certain operations such as copying cell instruction sets, dividing instruction sets between parent and daughter cells and deletion of instructions from cells. If point mutation is enabled in the system, then at the end of each evolutionary step inside both systems the total population of cells inside the GeneBank is considered as targets for mutation.

Using point mutations, the number of mutations that are to occur is given as the product of the number of active cells inside the GeneBank multiplied by the maximum creature size multiplied by the probability of a point mutation within the system. Once the total number of mutations has been calculated, then random cells are chosen from within the GeneBank, and inside those cells a random int (defining the cell's instruction sequence) is changed, thus introducing a new instruction. Note that it is possible that a single cell may be hit with repeated mutations, making it evolve more quickly than other cells in the system.

Thus, the rules in both these systems are very low level. Furthermore, after the evolution process of a cell has completed, there are no guarantees that its instruction set will produce any meaningful set of operations, let alone a "better" individual. There are no simple ways, in Avida or Tierra, to "pre-package" a group of low-level instructions into a high-level "routine" for easily manipulation. This is precisely the approach that we take in our exploration of how we can better evolve a predator-prey system: we have defined a small number of operations that we believe characterise the behaviours of our predators and their prey.

Our model has some interesting spatial properties and we are presently working to compare it to other spatial models and iterated games such as the Prisonner's dilemma [297].

8.2.1 *Our Predator-Prey Model*

We first developed our model with only two very definite animat types. Predators (known as "foxes") and prey (known as "rabbits") coexist in the "flatland" simulated by the model. Within our predator-prey model there are several redundant parameters that control microscopic details of the animats. These rules fall into two broad categories: (a) rules affecting the environment such as how far a rabbit can "see" or how long a fox takes to get hungry; and (b) the set of rules that govern the behaviour of an individual animat. We describe choices for category-a parameters in previous work, but generally most of the model behaviours are insensitive to these. In this paper we focus on permutations of category-b rules that govern behaviour of individual animats. These rules are typical of the type of rule that is changed in a system with animats that are evolving by following a genetic algorithm such as Tierra or Avida.

We are exploring how the model varies by changing the rule priority permutation that animats use, rather than changing the basis set of possible rules themselves. In our previous work, we have used the same set of rules for every animat. These rules were chosen by us as a good base set providing a rich set of pattern formation and are stable against small variations. An outline of these rules is:

A fox will:		A rabbit will:	
0	eat rabbit if adjacent;	0	move away from adjacent fox;
1	move towards rabbit if hungry;	1	breed if adjacent to rabbit;
2	breed if adjacent to fox;	2	move towards another rabbit;
3	move towards fox if not hungry;	3	move randomly.
4	move randomly.		

Note that foxes and rabbits can only move towards each other if they are within the *radius of perception*. Outside of this distance, animals cannot be seen. The order of these rules is important. Each rule has a condition and if that condition is true, the rule is executed and any later rules will be ignored. Thus rule 0 has a higher priority than rule 1 and so forth. For example, if a fox is not hungry and is adjacent to another fox but not adjacent to a rabbit, than rule 2 (breeding) will be executed and rules 3 and 4 will be ignored. Note that the conditions for rules 0 and 1 will be false in this case. The "move randomly" rule is a catch-all and does not have a condition. Thus if all else fails the animat will move randomly.

We can therefore denote the behaviour of a particular animat i at time t by $\mathcal{B}_i(t)$ which will result in the action expressed by the rule priorities in effect. Suppose the animat i has a current rule list $\mathcal{R}_i = [r_0, r_1, r_2, ..., r_j]; j = 0, 1, ..., N_R - 1$ Rules are evaluated in strict order and the first that can be applied is actioned. In this paper we restrict the list length N_L to being identical to the number of possible rules N_R and so each of our rules appears precisely once in an animat's list. Our rules are formulated as a "first matching" so that it is not useful to duplicate rules in the list, but it would be possible to omit some possible rules randomly or otherwise, so that $N_L <= N_R$.

A permutation $\mathcal{P}_k; k = 0, 1, ..., N_P - 1$ such as $\mathcal{P}_k = [1, 0, 2, 3, 4]$ can be applied to the rule priority lists. It is the space of possible permutations $\mathcal{P}_k$ that we are exploring in this present paper. The number of possible permutations is drastically smaller than the number of combinations of an arbitrary string of N_R instructions $N_R^{N_R}$. For operational reasons the last rule is always the "move randomly" rule in our present experiments, so we can only in fact permute only $4 = N_R - 1$ rules for the fox model described. The number of permutations is therefore $4! = 24$.

In our earlier work we explored the sensitivity of our model to starting conditions and geometric arrangements. We found that it is remarkably stable against different geometric starting arrangements of rabbits and foxes, and that the medium- to long-term behaviour is only affected in terms of absolute population sizes by the initial conditions. In this present paper we report only on medium- to long-term effects that emerge from the same simple initial starting arrangement of interleaved blocks of rabbits and foxes.

8.3 Resultant Behaviours from Prioritisation

We have investigated the macroscopic pattern formations that result from using different permutations of rules. In this paper we focus on the fox-species and the resulting behaviours for the 24 different subspecies of fox in the presence of "normal" or base species rabbits. We present some configuration snapshots showing the macroscopic pattern variations and summarise our findings in table 8.1 for each of the 24 fox subspecies.

The following figures show the situation in the model at time step 800. We chose this time regime, as our previous work suggests that all system memory of the initial animat configuration is erased by step 400. Each run of the model uses the same random number seed so the outcome would have been identical each time except that the order of the fox rules has been changed to one of the 24 possible permutations.

A number of experiments were run where the order (and thus the priorities) of the rules were changed. In order to limit the number of experiments, one species (either rabbit or fox) retained its usual rules while the order of the rules for the other species were changed. Each rule set is referred

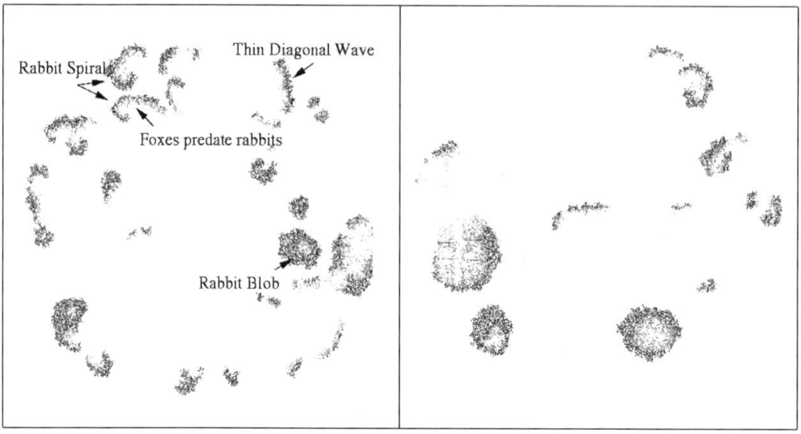

Fig. 8.1 Situation at step 800 for (a) the original Fox rule set 0-1-2-3-4 and (b) Fox rule set 2-0-3-1-4

to by a sequence of digits which show the new order for the rules. For example: rule set 1-3-0-2-4 means that the rabbit rules are as usual and the fox rules are in the order 1, 3, 0, 2 and 4. In this paper we focus on fox rule combinations, but where a rule set is given if it has 5 digits it refers to a fox rule, if it has 4 digits it is a rabbit rule.

In previous work we analysed the macroscopic patterns that emerge from the microscopic model [185]. We identified shapes we refer to as: rabbit clumps; defensive lines; horizontal, vertical and diagonal wave-fronts and left or right spirals. We use this vocabulary in discussing and analysing the configuration snapshots shown below.

Figure 8.1 (a) shows step 800 for the predator-prey model with the original (base) rule set and (b) shows time step 800 for the fox rule set 2-0-3-1-4. These two diagrams illustrate the two main visual effects caused by changes in rule priorities. Certain rule sets (including the original set 0-1-2-3-4) produce an overall "circular" pattern with clusters of animats spread more or less evenly across the area. These clusters comprise the usual spirals, wave fronts and dense "blobs" previously analysed in [186]. Each wave front consists of a leading edge of rabbits fleeing from a following line of foxes. It is noticeable that the wave fronts in this type of diagram are predominantly "compact", that is the line of foxes is keeping up with the rabbits.

The second diagram in Figure 8.1 (rule set 2-0-3-1-4) illustrates the second type of effect where the overall circle pattern is destroyed and larger distinct clumps form. In addition many clumps show that foxes have scattered away from the line of rabbits. This effect is caused by rule 1 (move towards rabbit) appearing low down in the list of priorities.

Both the diagrams in Figure 8.1 contain relatively low numbers of animats - 12,609 rabbits and 2,683 foxes with the original rule set and 8,004 rabbits and 3,648 foxes in rule set 2-0-3-1-4. Note that the number of animats in a "good circle" pattern is always significantly higher than in the "degraded circle" where the lower numbers of animats is caused by rule 3 (move towards another fox) appearing low down in the priority list. This shows how rule 3 has more of an effect on the population than rule 2 (breed) which can only be used if another fox is adjacent.

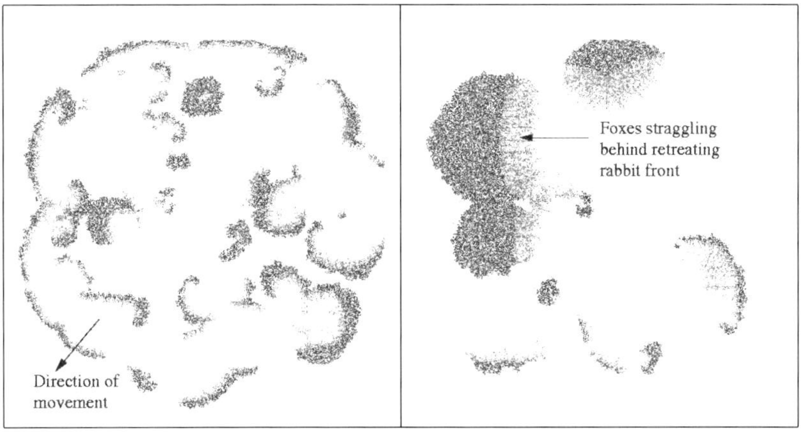

Fig. 8.2 Situation at step 800 for (a) Fox rule set 3-0-1-2-4 and (b) Fox rule set 0-3-2-1-4. In contrast to figure (b), figure (a) has a good circular shape

Figure 8.2 shows two rules sets which illustrate an "in-between" situation. Rule set 3-0-1-2-4 contains a reasonable circle but several wave fronts show foxes falling slightly back due to rule 1 being in the centre of the list of priorities. Whereas rule set 0-3-2-1-4 shows a poor circle with distinct foxes trailing behind the rabbit lines due to rule 1 being at the end of the list. Both rule sets contain relatively high levels of animats due to the high priority given to rule 3 – 21,277 rabbits and 6,139 foxes in 3-0-1-2-4 and 22,129 rabbits and 8,021 foxes in 0-3-2-1-4.

In Figure 8.3 a) and b), rule set 3-1-2-0-4 contains a high number of animats – 23,735 rabbits and 8,818 foxes. Likewise, rule set 3-0-2-1-4 also contains high numbers of animats (for a diagram with a degraded circle) with 17,429 rabbits and 8,974 foxes. The degraded circle and scattered foxes in 3-0-2-1-4 is due, as usual, to rule 1 being low in the list. The

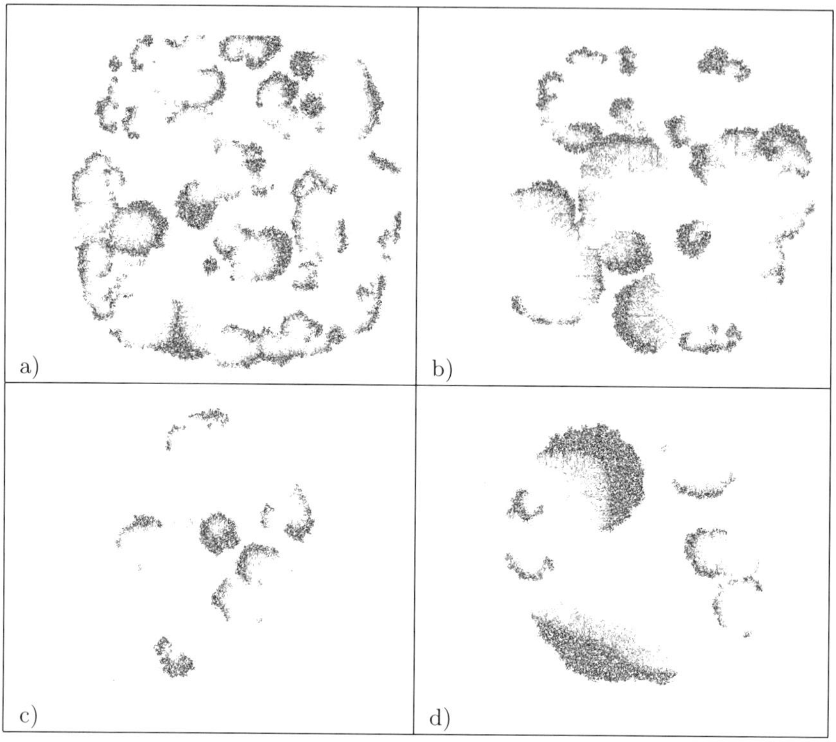

Fig. 8.3 Situation at step 800 for (a) Fox rule set 3-1-2-0-4 and (b) Fox rule set 3-0-2-1-4, (c) Rabbit rule set 2-0-1-3 and (d) Rabbit rule set 0-2-1-3

relatively high number of animats in both these rule sets is due to rule 3 being the first rule in the list.

Figure 8.3 c) and d) show results for foxes following their original rules while the rabbit rule order has been changed. Rabbit rule set 2-0-1-3 shows low numbers of rabbits (5,790) because rule 1 (breed) has moved to the end of the list just before rule 3. Thus less rabbits breed and the fox population is also low (1,329) because of a lack of prey. Rule set 0-2-1-3 shows rabbits "herding" together because of the combination of rule 0 (move away from fox) and rule 2 (move towards other rabbit) as the first two rules in the list. This causes many rabbits to have a "double movement" and thus large clusters of rabbits form in front of pursuing foxes.

In general the rule list permutations we have examined are remarkably robust and all give rise to macroscopic patterns without extreme population explosions or extinctions. Generally the length scales of the model are fairly stable too – all the configurations shown in figures 8.1 to 8.3 are all on the same length scale.

Table 8.1 contains all tested rule permutations for foxes and table 8.2 contains all tested rule permutations for rabbits. Each entry in these tables indicates the number of foxes and rabbits at step 800.

Table 8.1 4! = 24 permutations of rules controlling fox behaviour showing numbers of rabbits and foxes at step 800 and some description of the observed behavioural effects.

Permute code	Priority List	Rabbits at step 800	Foxes at step 800	Description
0	0-1-2-3-4	12,609	2,683	good circle, no straggling
1	0-1-3-2-4	21,598	8,113	good circle, no straggling
2	0-2-1-3-4	3,800	2,557	no circle, some straggling
3	0-2-3-1-4	3,800	2,557	identical to permute code 2
4	0-3-1-2-4	21,598	8,113	identical to permute code 1
5	0-3-2-1-4	22,129	8,021	no circle, heavy straggling
6	1-0-2-3-4	16,217	5,163	good circle, no straggling
7	1-0-3-2-4	23,715	7,644	good circle, no straggling
8	1-2-0-3-4	15,052	6,784	weak circle, slight straggling
9	1-2-3-0-4	10,884	5,534	good circle, no straggling
10	1-3-0-2-4	24,180	6,881	good circle, slight straggling
11	1-3-2-0-4	23,735	8,818	good circle, no straggling
12	2-0-1-3-4	8,004	3,648	no circle, heavy straggling
13	2-0-3-1-4	8,004	3,648	identical to permute code 12
14	2-1-0-3-4	15,471	7,574	no circle, blobs, heavy straggling
15	2-1-3-0-4	11,131	10,961	weak circle, some straggling
16	2-3-0-1-4	6,172	7,192	weak circle, heavy straggling
17	2-3-1-0-4	11,131	10,961	identical to permute code 15
18	3-0-1-2-4	21,277	6,139	good circle, no straggling
19	3-0-2-1-4	17,429	8,974	weak circle, heavy straggling
20	3-1-0-2-4	24,180	6,881	identical to permute code 10
21	3-1-2-0-4	23,735	8,818	identical to permute code 11
22	3-2-0-1-4	10,723	8,100	weak circle, some straggling
23	3-2-1-0-4	13,669	5,774	weak circle, some straggling

Several of the rule sets were found to be degenerate. In particular, if rule 4 (random move) for foxes precedes the rules for breeding or eating, then every fox moves randomly and never breeds or eats which results in the fox population becoming extinct within a few time steps. Thus the rule sets of interest always place rule 4 as the last rule in the list. Similarly, rule sets involving the changing of the order of the rules for rabbits always retain rule 3 as the last in the list.

Of the useful rule sets for foxes, six were found to be identical to other rule sets. This occurs when rule 3 and rule 1 follow each other since rule

1 is only used if the fox is hungry and rule 3 is only used if the fox is not hungry thus only one of these two rules will ever be used. If they appear in sequence they have the same effect as if they appeared in the opposite order – thus, for example, rule set 0-1-3-2-4 has an identical effect to rule set 0-3-1-2-4.

Table 8.2 3! = 6 permutations of rules controlling rabbit behaviour showing numbers of rabbits and foxes at step 800 and some description of the observed behavioural effects

Perm. code	Priority List	Rabbits at step 800	Foxes at step 800	Description
0	0-1-2-3	12,609	2,683	good circle, no straggling (original rule set)
1	0-2-1-3	14,831	4,522	no circle, huge rabbit herds, heavy straggling
2	1-0-2-3	18,474	9,115	good circle, slight straggling
3	1-2-0-3	14,461	4,938	good circle, no straggling
4	2-0-1-3	5,790	1,329	no circle, no straggling
5	2-1-0-3	9,018	3,500	weak circle, no straggling

The frequency that animats (of a particular type) will enact a particular rule can be measured for a particular time-slice and averaged over all animats of that type to obtain what we refer to as action-probabilities where $P_j(t)$ is the probability that action or rule j is enacted at time t. In fact, we expect the dependency to be more complicated as our system is known to have a periodically varying population envelope function for animats of a particular type.

The measured action-probabilities are both time dependent and also depend on the circumstances (local environment) of individual animats. However, averaging over all similar animats in the population reveals that the mean probability of rule j is a good differentiator between the different behaviours that arise from different rule permutations.

Table 8.3 Rule priority action-probabilities for various rule sets from step 1 to step 1000

Rule	original set	set 0-2-1-3-4	set 2-1-0-3-4	set 3-0-2-1-4
0	0.081	0.052	0.042	0.048
1	0.432	0.495	0.527	0.483
2	0.029	0.035	0.039	0.034
3	0.074	0.071	0.067	0.085
4	0.385	0.346	0.326	0.350

The rule action-probabilities were measured during several runs of the model with different rules sets and appear in table 8.3. Table 8.3 clearly shows that the two seemingly most important rules, namely rule 0 (eat) and rule 2 (breed) are actually used the least. This is because they depend heavily on their accompanying "precondition" rules 1 (move towards a rabbit) and 3 (move towards another fox). In fact, foxes spend roughly half their time moving towards food (rule 1) and this is consistent even when rule 1 is given a low priority (see 3-0-2-1-4). Similarly, rule 3 is also used in a consistent manner although there is a slight increase in use when it is given a high priority (see 3-0-2-1-4).

8.4 Behavioural Metrics and Analysis

As table 8.1 shows, the different rule priority lists give rise to significant behaviour variations. Some interesting metrics that can be used to characterise these differences are discussed below. One metric is the size (and density) of typical animat clusters such as wave fronts. It has already been shown how the placement of fox rule 1 (move towards rabbit) can cause foxes to "straggle" behind wave fronts. This appears to be independent of the direction of travel of wavefronts.

Another metric is the relative animat populations. Table 8.4 shows how changes to the priority of the rules for foxes affect the rabbit population. Although the grouping criteria are vague, it is clear that rule 0 (eat) has little or no effect. The rabbit population size is governed by the placement of fox rule 2 (breed). If rule 2 is placed near the top of the list, then more foxes will be produced to prey on the rabbit population which will consequently be lower. Similarly, if rule 2 is placed near the end of the list then a higher rabbit population results. The rule priority also affects the general dispersal of the population. If rule 1 (move towards rabbit) is placed near the top of the list then a good circular pattern will result. This is because foxes are often moving towards rabbits which, in turn, flee from the approaching foxes, forming more wave fronts in an expanding circular formation.

Table 8.4 Rule sets grouped according to number of rabbits at step 800

	low number of rabbits	medium number of rabbits	high number of rabbits
good circle		0-1-2-3-4, 1-0-2-3-4, 1-2-0-3-4, 1-2-3-0-4	0-1-3-2-4, 1-0-3-2-4, 1-3-0-2-4, 1-3-2-0-4, 3-0-1-2-4
no/weak circle	0-2-3-1-4, 2-0-3-1-4, 2-3-0-1-4	2-1-0-3-4, 2-3-1-0-4, 3-2-0-1-4, 3-2-1-0-4	0-3-2-1-4, 3-0-2-1-4

Table 8.5 shows how changes to the priority of the rules for foxes affect the fox population. The grouping is not as clear as that for the rabbit population but does show certain trends. Once again the priority given to rule 0 (eat) seems to have no effect.

Table 8.5 Rule sets grouped according to number of foxes at step 800

	low number of foxes	medium number of foxes	high number of foxes
good circle	0-1-2-3-4, 1-0-2-3-4, 1-2-3-0-4	0-1-3-2-4, 1-0-3-2-4, 1-3-0-2-4, 3-0-1-2-4	1-3-2-0-4
no/weak circle	0-2-1-3-4, 2-0-1-3-4	0-3-2-1-4, 1-2-0-3-4, 2-1-0-3-4, 2-3-0-1-4, 3-2-0-1-4, 3-2-1-0-4	2-1-3-0-4, 3-0-2-1-4

Another metric is an analysis of fox lifestyle and breeding habits which change with each rule set. For example the mean lifetime of foxes using the original rule set 0-1-2-3-4 is 32 (time steps) whereas the mean lifetime of foxes using rule set 2-0-1-3-4 is 26. In the original rule set 40% of foxes die of old age (at 50 timesteps) and 60% starve to death whereas in rule set 2-0-1-3-4, 20% die of old age and 80% starve to death. These differences occur because the original rule set leads with rule 0 (eat) and 2-0-1-3-4 leads with rule 2 (breed).

8.5 An Evolutionary Survival Experiment

We believe we have identified the gross behaviour classes of the possible rule set permutations in our model. We can therefore introduce a very simple genetic evolution process into the model and observe which subspecies is most fit. Fitness of course can have several interpretations. In our present discussion we simply take it to mean survivability. We set up a model run whereby all subspecies of the model configuration are substituted randomly by a particular subspecies. In the case of foxes, the relative populations of each subspecies rapidly diverge.

8.5.1 *Evolution Procedure*

In this simplified evolutionary process we simply inject the different subspecies into the fox population after the initial set up. Individual pairs of animats breed producing offspring that are randomly (50:50) the same type as one or either parent. We have not yet introduced any mutation. Unsuccessful species therefore die out and are lost from our population. That is the aim of this particular experiment – simply to determine which are the most likely subspecies to survive.

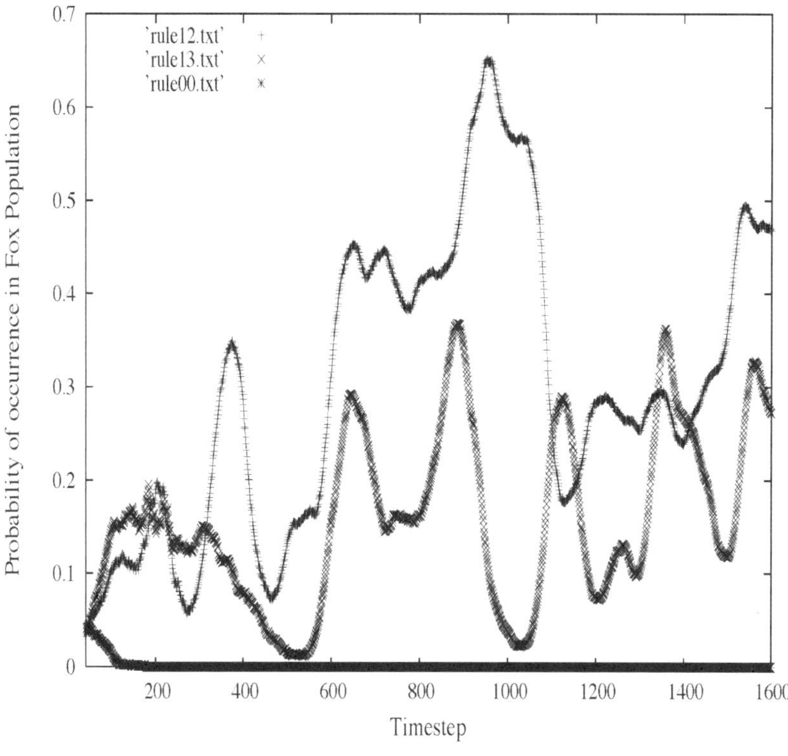

Fig. 8.4 Survival frequencies for the different species of Foxes

We have tried several random seed combinations for this experiment and the emergent pattern is similar in all cases. We conclude that there are enough random effects in this experiment to separate out pathological spatial effects. We anticipate that certain conditions that arise for some pockets of the population such as being at the end of a spiral or being at the heart of a rabbit clump might lead to special behaviour and affect survivability. In this paper we only report on behaviour averaged over the whole system and hence without analysis of potential anomalies.

We believe this is a useful exercise to explore survivability in our model, prior to the more open ended studies involving genetic cross over and mutation that will considerably expand the size of the fitness landscape.

8.5.2 *Survivability*

Figure 8.4 shows how the relative frequencies of the different subspecies of fox dramatically diverge from an initial equality. Foxes with a rule set starting with rule 2 (breed) have a very much stronger tendency to survive.

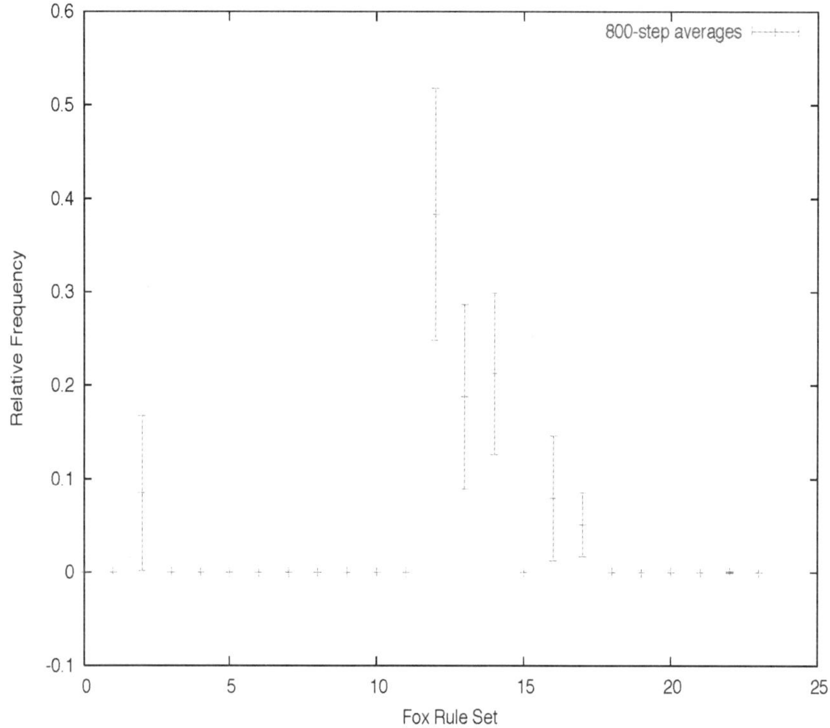

Fig. 8.5 Survival frequencies and variances for all fox types averaged over last 800 steps in a 1600 step run where multiple species were injected at step 40

Figure 8.4 shows the measured survival frequencies for fox subspecies, normalised over the entire remaining population. The curves for rules 0, 12 and 13 are shown for illustrative purposes. Generally as expected the population follows periodic boom-bust variations. The interesting feature that arises from the different fox subspecies is due to their different relative propensities for survival. Generally rules fall into one of two survival categories - long term survival or short term extinction. In this particular experiment there is no evolutionary mechanism for an extinct subspecies of fox to be later re-introduced into the population.

Figure 8.5 shows the relative survival frequencies of the sub type populations averaged over the last 800 steps of a 1600 step run where the different subspecies were injected at step 40. Prior to step 40 the run used the original rule set for all foxes just to build up a large and uncorrelated base population. The graph shows the value and variance of the relative survival frequencies. Note that most subspecies are effectively extinct and the subspecies that have survived are rules 12,13,14,16,17 which all have rule 2 (breed) as their top priority. However anomalously foxes of rule set 2 also survive and rule set 15 which also has (breed) as its top priority does not. This latter may be explained by the low priority rule 15 (2-1-3-0-4) foxes give to eating – therefore they easily starve. Rule set 2 (0-2-1-3-4) survives because of lucky chance combinations of eating and breeding (and their precondition rules for moving towards food and breeding partners respectively) in the priority list.

8.6 Generalising the Approach

By considering high-level "common sense" ideas for the behaviour of individual animats, we have arrived at a permuted-rules model that is amenable to evolution. In this paper we have only considered very simple probabilistic inheritance. It is relatively straightforward to apply point mutation into the animat rule lists, but it is not entirely clear yet what the best way to provide genetic crossover is.

It is interesting to ask what makes a successful animat anyway? Survival has meaning in the context of particular subtypes. Our model is simple enough that we can control and limit some of the interactions that would otherwise lead to an irreducible set of species-species reactions. It is likely however that we can engineer animats with particular traits that will be especially "fit" for certain circumstances that may occur in the model.

We have managed to introduce a very limited sort of evolutionary operation into our predator prey model. The evolution operator is not very satisfactory albeit adequate for these initial investigations of relative survivability. We believe it is straightforward to introduce point mutations into the process but are still investigating good ways to define a genetic cross over representation and operator that will allow a controlled exploration of new sorts of animat.

There are of course other phenomena to explore. We are currently exploring what might make a particular animat more or less attractive to a potential breeding mate to encourage promotion of particular traits in the animats.

We are also considering the particular local circumstances that arise in the macroscopic patterns that emerge in our model. It is not clear how to link microscopic and macroscopic behaviours, but it is likely that for example foxes at the horns of a wavefront will behave differently from those gorging in the midst of a rabbit clump. We would like to investigate particular survival niches that might arise from such macroscopic effects.

We believe we have considerably reduced the search space for evolution in our model using the permutation approach and fixed rule sets. We believe the rule sets have some relevance to the exploration of gene functions in real life systems.

8.7 Conclusions

We have codified the important behaviours of our predator prey animats and shown how they can be represented by permutable gene sequences. We have categorised the major properties of fox and rabbit sub-species in our model using just 5 and 4 gene sequences respectively. Our model is remarkably robust and we have been able to identify some key behaviour regimes of stability.

We have a number of developments planned for the future [184]. Firstly, a "do nothing" rule could be added to our basis set and this could be used as the last rule, giving an N_L of 5 and a permutation space of $5! = 120$. We believe such a "nop" rule will only change the time constants in our model but we have yet to investigate this.

Secondly, we have focused on foxes in this paper since they have a richer rule permutation possibilities. We have yet to explore the survival propensities of different types of foxes in the presence of different rabbits. We hope to map out this model space semi-automatically based on the categorisation ideas presented in the present paper.

Finally, we believe our approach makes an exploration of a model phase space much more manageable than the common "shot gun" approach to genetic algorithms. Our approach of combining a carefully controlled evolutionary exploration of model phase space with a microscopic model known to exhibit emergent macroscopic properties may be of relevance to studying other life systems – artificial or otherwise.

Acknowledgements

This work has benefited from the use of Massey University's "Monte" compute cluster and thanks also to the Allan Wilson centre for Molecular Biology for the use of the "Helix" Supercomputer.

Chapter 9

Gauging ALife: Emerging Complex Systems

K. Kitto

School of Chemistry, Physics and Earth Sciences,
The Flinders University of South Australia,
Bedford Park 5042, Australia
Email: kirsty.kitto@flinders.edu.au

Despite its early successes, ALife has not tended to live up to its original promise, with any emergent behavior very rarely manifesting itself in such a way that new higher level emergence can occur. This problem has been recognised in two related concepts; the failure of ALife simulations to display Open Ended Evolution, and their inability to dynamically generate more than two hierarchical levels of behavior. This paper will suggest that these problems of ALife stem from a missing sense of *contextuality* in the models of ALife. A number of theories which exhibit some form of contextual dependence will be discussed, in particular, the gauge theories of quantum field theory.

9.1 Life and ALife

Artificial Life is a diverse and sometimes disparate field, with a number of differing goals, theories, models and proposed outcomes. There is a drive to understand concepts such as complexity and emergence as they relate to the field. There are the attempts to generate emergent, cell-like behavior invitro, and insilico. Some models, such as Avida, in addition to their primary role as a platform for the investigation of digital life, are being used to investigate current theoretical and evolutionary problems. Then, there is the attempt to achieve a living organism, what might be called *Strong ALife.* This diversity is indicative of a new vital field, with a number of promising avenues of research, however, it also indicates what might be perceived as a lack of direction; there are no obvious answers in the field, no well accepted theories that might be used to guide research. These two perceptions are in essence compatible, a newly established field will lack theories, which can acquire the status of dogma in a more developed field of knowledge. One of the few mature theories relevant to artificial life is the theory of evolution, but, although this concept is relied upon extensively in both fields, it does not have one universal interpretation or even definition. In particular, the way in which evolution couples with *development* is still very poorly understood.

9.1.1 *Development*

This process, whereby a phenotype emerges from its specifying genotype, poses some of the most important, as well as the least understood problems in the field of biology. In particular, there is a lack of robust models, and a general questioning over whether such models are even feasible. As such, this is a field which could benefit from artificial models, and yet few complex artificial models exist. There are a number of simple morphogenic models such as Dawkin's tree growing program [92], but these models do not seem to capture the true complexity of the coupling between the phenotype and the selective process. For example the tree growing program requires that an observer choose the most 'pleasing' configuration at each time step.

Development involves some very interesting phenomena. Consider for example *phenotypic plasticity*; "the ability of a single genotype to produce more than one alternative form of morphology, physiological state, and/or behavior in response to environmental conditions." [413] This phenomenon amounts to a situation where evolutionary important traits do not have to be 'genetic' (immune to environmental effects). Thus, this phenomenon reinforces an often cited (but perhaps not truly recognized) fact that the phenotype is a product of *interaction* of the genotype *and* the environment. That is, the phenotype that develops in biological systems depends in a strong way upon the environment that surrounds it; a different environment can result in a vastly different organism. Phenomena such of this are not generally possible in ALife simulations, a genotype results in only one phenotype. The *contextuality* of genotypic behavior is not generally recognised.

Even the complexity of genotype-genotype reactions is rarely acknowledged. Most organisms are diploid (*i.e.* having more than one copy of each bit of information or strand of DNA), and yet artificial organisms are generally haploid (possessing only a single strand of DNA). It is often claimed that the phenomena of evolution can be captured without this added complexity, but, in the next section, we shall examine two problems indicating that this may not be the case.

It is my contention that the contextuality of actual biological systems is missing from artificial life, and that this manifests itself in their consequent lack of complexity. This article will propose a manner in which we might start to incorporate this missing contextuality.

9.1.2 *ALife and Emergence*

In order to achieve some form of artificial life, it will be necessary that our artificial models display some sort of 'strongly' emergent behavior. However, two results in the field suggest that this is not happening.

Firstly, it is widely accepted that ALife models do not display Open Ended Evolution (OEE) [35; 36], but very little research has been conducted into the reasons why this might be the case. Rather, the tendency has been to construct ever more complicated environments, or to throw out existing environments and to almost arbitrarily develop new ones. Although there is sometimes an explanation of the reasons behind this substitution [239], there is very little general analysis of the source of this failure.

A second problem with ALife models their general failure to exhibit more than two levels of hierarchical development [44]. Without the generation of new, more complex levels there will be a highest level of complexity attainable, a point emphasised by the Ansatz for Dynamical Hierarchies (ADH):

> *Given an appropriate simulation framework, an appropriate increase of the object complexity of the primitives is necessary and sufficient for the generation of successively higher-order emergent properties through aggregation.* [442]

This statement has some rather profound consequences. It limits the overall complexity that might be obtained from any system formed as part of a dynamical hierarchy and thus violates what Rasmussen *et. al.* call the "complex systems dogma" which claims that we can use simple rules and states to generate complex behavior. If the ansatz is true then our current modelling methodology may not be able to produce dynamical structures of unlimited complexity. Before we can consider the truth or falsity of the ansatz, we must consider the core concepts of simulation framework and object complexity that it utilises.

Object Complexity: This is the key term of the ansatz, specifying the complexity of the individual objects in the hierarchy. It is important that we are able to define the complexity of the objects contained in a hierarchy in order to make use of, or even to test the ansatz. Unfortunately there is no formalised definition of object complexity.

Simulation Framework: This rather *ad hoc* definition of object complexity in the ansatz is made possible by the requirement that the simulation framework be given. A simulation framework is provided by the specification of the rule set of the system. Thus the requirement that a simulation framework be given forces the rule set of the system to be fully specified, and to remain constant, throughout an application of the ansatz to some system or set of systems.

Perhaps the most interesting aspect of the ADH comes from the observation that the rule set defining the simulation framework is precisely the rule set used to define object complexity in all analyses performed to date. With this point in mind, we might rephrase the ansatz as:

> *Given an appropriate* rule set, *an appropriate increase of the* rule set *of the primitives is necessary and sufficient for the generation of successively higher-order emergent properties through aggregation.*

This makes the ansatz appear to be rather trivial, a point independently raised by Rasmussen *et. al.*:

> *The* ansatz *is in some sense trivial: Assume that we have a minimal rule set that generates a particular dynamical hierarchy but only up to order N. If we stay within this simulation framework, it is necessary to add new rules to generate an additional order* $(N+1)$ *of emergence. How can the system generate a new, higher level of behavior unless something new is added to the elements?* [443]

This triviality leads to a strange outcome that arises if we attempt to consider it in the light of our current understanding of object complexity as obtained from examining the rule set of agent based models. It appears there is some sense in which the complexity of any system satisfying the ansatz is built into the lowest level objects [160]. In an agent based model, this means that the most complex objects are those defined as primitive objects, the available interactions are effectively 'used up' as the zeroth order objects interact and form bonds with each other to form new higher order objects. This appears to be rather counter-intuitive. In the natural world, higher order objects often seem to have a higher complexity.

The ansatz has been subject to some debate [161; 443; 113]. One key point of disagreement concerns the possibility of comparing the object complexity of two different hierarchies. The simulation framework restriction reduces the applicability of the ansatz rather dramatically, limiting the result to individual systems. No comparison between different systems is allowed as they are automatically considered to be part of a different simulation framework.

These problems suggest that the ansatz may be part of some sort of fundamental misconception; one that leaves the ADH appearing reasonable until a more indepth analysis points to inconsistencies in its implications. It is my belief that the ADH, actually *points to* the missing contextuality of current ALife models and that without this contextuality the models will not be capable of displaying 'strong' emergence and hence OEE. For example, the interactions of the standard agent based methodology, we can immediately identify the problem; interactions are defined in the primitive objects, there is no room in which new interactions or behavior *can* emerge.

This problem, while less obvious in some of the other ALife environments, still exists. Consider for example Avida [239], one of the most comprehensively developed environments. Avidans evolve within an environment that rewards them for their ability to solve a set of defined problems, but this set of problems does not generally change, or increase during this process. Certainly, the environment of the Avidans is also affected by the (biotic) interactions between the Avidans themselves, however, the (abiotic) environment of problems is not likely to be sufficiently complex for either 'strong' emergence, OEE behavior, or levels upon levels of hierarchical structure to form. Attempts to increase the complexity of the simulation, such as the current drive to introduce diploidity will certainly increase the amount of behavior displayed by Avida, but until a much higher amount of both biotic and abiotic interaction becomes possible it is unlikely that Avida will not display the truly complex behavior of the natural world.

We are now in a position to, at least partially recognize what the term 'strong' emergence implies in the field of ALife; strongly emergent phenomena can in turn be used in the creation of higher levels of new emergent phenomena. If such behavior was manifesting in ALife simulations, then it is probable that they would display more open ended behavior. However, the emergence exhibited by artificial simulations to date appears not to be amenable to interacting in any higher level interactions.

9.1.3 *Complexity and Contextuality*

Although there is an ongoing debate surrounding the definition of complexity which shall not be covered here,[1] there is a general agreement that

[1] See for example, Bruce Edmonds' *Hypertext Bibliography of Measures of Complexity*, which, while no longer maintained, contains a vast list of articles that cover the many different ideas surrounding complex systems. This, and a number of other resources are available at http://bruce.edmonds.name, including Edmonds' PhD thesis [117], which

emergent behavior, such as that sought by ALife researchers, results from the dynamics of complex systems.

The above discussion, suggests that a sense of contextuality should be apparent in, at the least, our understanding of complexity as it relates to the developmental process, and possibly ALife modelling in general. However, there are very few theories of complexity that achieve this.

There is one class of theories available that incorporate contextuality; what might be termed *observer driven* theories and definitions of complexity. However, such theories are often ignored, as they offer no easy answers in our attempts to understand complexity. Some examples of theories that fall into this class include:

Pattee's Epistemic Cut which separates our knowledge from the thing that it describes; description from construction, the observer from the system, the genotype from the phenotype [301; 172; 171]. This concept has been used to examine a number of different systems, all generally recognised as complex.

Rosen's definition, where a system is defined by Rosen as complex if it can only be described by a combination of inequivalent descriptions, each defined with respect to his *Modelling Relation* [338].

Casti's conception of complexity as the number of nonequivalent descriptions that an observer can generate for a system of interest [71].

An epistemic cut places the observer in a very different class from the system that they are observing, but there is often a certain flexibility about where the cut can be placed. Thus it recognises that there is a sense in which how the observer is specified will change the system under examination. However the epistemic cut does not give much indication of how we might incorporate this dependency into our modelling. The Rosen and Casti definitions start to give us a greater understanding of this problem; if a system is contextual, then different observers will describe it in different ways, and it is likely that at least some of them will be inequivalent, thus the system will be complex according to Rosen and Casti. Therefore we can claim that contextual systems *are* complex, and may therefore display more of the behavior sought by the ALife community, such as 'strong' emergence.

9.2 Incorporating Context into our Models

This section will explore some of the ways in which context has, to some extent, been incorporated into our modelling techniques.

---

examines many of the varying definitions of complexity, before Edmonds settles on a definition that is highly contextual.

9.2.1 *The Baas Definition of Emergence*

Baas has developed a theory of hyperstructures, hierarchies and emergence [21] which incorporates observer dependence explicitly. First, Baas defines a set of primitive objects or entities, which are termed *first-order structures*, denoted $\{S_i^1\}$, where $i \in J$ is some index, finite or infinite, denoting the structures that are first-order in nature. If these structures exhibit one or more *first-order properties*, then these are denoted $P = O^1(S_i^1)$. Properties are dependent upon the observational mechanisms O^1 identified as relevant to the first-order structures. Note that these observational mechanisms may or may not be dynamical in nature. Subjecting the first-order structures to a family of interactions, Int, as allowed by their first-order properties we may obtain new, *second-order structures* $S^2 = R(S_i^1, O^1(S_i^1), Int^1)_{i \in J}$ where R stands for the result of the construction process. These second-order structures may themselves be observed by a set of mechanisms, O^2 (which may be equal, overlapping or disjoint from O^1), resulting in new second-order properties. This set of definitions allowed Baas to propose a *definition of first-order emergence*, where P is an emergent property of S^2 iff $P \in O^2(S^2)$, but $P \notin O^2(S_{i_1}^1)$ for all i_1.

These ideas can be naturally extended, with a family of second-order structures forming $\{S_i^2\}$ which may then form (perhaps with the assistance of first-order structures) third-order structures $S^3 = R(S_{i^2}^2, O^2, Int^2)$, and so on, with an *N-th order structure* being represented generically by $S^N = R(S_{i^{N-1}}^{N-1}, O^{N-1}, Int^{N-1}), i_{N-1} \in J_{N-1}$, and the associated emergent behavior defined with respect to the lower levels as occurred above.

This definition does provide for a sense of observer dependence, and therefore incorporates a sense of contextuality. However, Baas does not show a general way in which this dependency interacts with the system of interest in any dynamical sense. This leads to the somewhat unsatisfactory nature of the definition, we have learned very little through its application. This problem was highlighted during ALife VIII with the presentation of a system that clearly pushed the boundary of what should be considered a dynamical hierarchy [113] and yet was rightly claimed to be hierarchical in the Baas sense.

The question of how to incorporate context in a more systematic manner arises. There is another class of contextual models which have not been recognised as such in ALife, although they are starting to be incorporated (see for example the Quantum Coreworld website http://non.fiction.org/~await/qcw/); the quantum theories of physics are intrinsically contextual [276; 7].

9.2.2 *Quantum Mechanics*

Quantum theory has a well developed formalism. It is also an inherently contextual theory, as has been shown by its violation of the by now infamous Bell inequalities [276].

There are indications that the formalism of quantum theory can be used more generally than has been appreciated, specifically that the need to apply a quantum rather than a classical theory arises whenever contextual systems are under examination [7]. Aerts and coworkers have developed a generalisation of the quantum formalism that explicitly takes into account the contextuality of quantum mechanical systems, the SCOP (or State COntext Property system).[2] SCOP-type systems are now being used to describe a much more generic set of systems than those that occur in the quantum world, providing for example a new theory of concepts able to deal with the 'pet-fish' problem [134]. We shall not discuss this work, the interested reader is referred to the references. Instead, we shall turn our attention to an alternative formalism also utilising the contextuality of quantum mechanics; quantum field theory.

9.2.3 *Gauge Theories*

Gauged Quantum Field Theories such as Quantum Electrodyanamics (QED), and Quantum Chromodynamics (QCD) are some of the most successful theories of physics. Obviously this paper shall not discuss gauge theories in any detail whatsoever[3] rather, this section shall sketch out the form of gauge field theories. Some of the more complicated mathematical expressions are kept to the footnotes, but can be perused if the reader is interested. The ideas covered in this section will then be utilised in the formation of a new principle that could be used to generate models with contextual behavior, and perhaps the associated complexity that ALife is missing. We shall begin with a (shamefully) short summary of the functional formulation of Quantum Field Theory (QFT).

QFTs are constructed from a set of fields, ϕ, obeying a dynamics described by what is termed the *action*, $S[\phi]$. The action is generated

[2]For the sake of clarity this list of references has been kept to a minimum, consult the CLEA webpage (http://www.vub.ac.be/CLEA) for a far more extensive list of publications.

[3]To do so would require a monograph. There are many texts on the subject [302], but perhaps the most useful for the current discussion are those that cover some of the philosophical issues of these theories, and review their *form* [19; 10]. Interestingly, General Relativity, the other lauded theory of physics might also be considered a gauge field theory, which has led to the suggestion that these theories point to a new, more fundamental class of theories.

from taking some relevant Lagrangian[4], $L[\phi]$, and integrating it over time: $S[\phi] = \int dt L[\phi]$. Physically, a Lagrangian describes the difference between the kinetic and potential energies of the fields, $L[\phi] = K[\phi] - V[\phi]$. The Action is then used to define *correlation functions*[5] a process which is known as *quantization*, but which shall not be discussed here.

All realistic Lagrangians satisfy a conservation law of some type, a fact which is used to derive them. *Noether's theorems* link conservation laws with the *symmetries* apparent in both the configurations, and the dynamics of the fields. If a field satisfies Noether's theorem, then it also satisfies the Euler-Lagrange equations of motion.[6] Thus, one way in which to discover the dynamics of a system involves identifying its symmetries. Sometimes, however, not all of the symmetries of a field are immediately apparent. Often, a symmetry group is *broken* in a process called *spontaneous symmetry breaking* (SSB). Consider for example a thin metal rod, balanced on its end, as shown in figure 9.1. This configuration has an obvious rotational symmetry; at a given height, the rod looks the same regardless of which angle we examine it from. However, if we push down on the top of the rod so that it buckles, then the direction from which we look at the rod now changes the orientation of the buckle; the symmetry is lost. Thus, the rotational symmetry of our system has been destroyed by the force of the push on its top. This broken symmetry can be regained however, if we

[4]Instead of the Lagrangian, quantum field theorists often use what is termed the Lagrangian density $\mathcal{L}(\phi, \partial_\mu \phi)$, which describes the fields, and their behavior with respect to changes in space and time (represented by taking a derivative, $\partial_\mu \phi$). The action can be obtained by integrating the Lagrangian density over all spacetime: $S[\phi] = \int d^4\mathbf{x}\mathcal{L}(\phi, \partial_\mu \phi)$. A word of caution is in order for the uninitiated, both physicists and texts tend to use the terms Lagrangian and Lagrangian density, as well as the notation L and $\mathcal{L}$, interchangeably. Which term is in use must be deciphered from the context of the text, however, the two terms are related, $L = \int d^3x\mathcal{L}(\phi, \partial_\mu \phi)$, so this sloppy notation is not generally too hard to understand.

[5]Correlation functions describe the way in which a field at one point in space and time can influence other fields for the system of interest. For example the two point correlation function describes the influence of a field upon the (possibly distant) spacetime points $\mathbf{x}_1$ and $\mathbf{x}_2$:

$$\mathcal{G}(\mathbf{x}_1, \mathbf{x}_2) = \langle \psi_0, \hat{\phi}(\mathbf{x}_1)\hat{\phi}(\mathbf{x}_2)\psi_0 \rangle \tag{9.1}$$

$$= \frac{\int \mathcal{D}\phi \; \phi(x_1)\phi(x_2)e^{-S[\phi]}}{\int \mathcal{D}\phi \; e^{-S[\phi]}}. \tag{9.2}$$

It is possible to define far more complex correlation functions, with multiple fields, and many different spacetime points, although they may be impossible to solve.

[6]The Euler-Lagrange equations of motion, for the simple ϕ system under description are

$$\partial_\mu \left(\frac{\partial \mathcal{L}}{\partial(\partial_\mu \phi)} \right) - \frac{\partial \mathcal{L}}{\partial \phi} = 0. \tag{9.3}$$

Fig. 9.1 A thin metal rod (for example a needle) exemplifies the process of symmetry breaking, and the mathematical mechanism used to regain a broader conception of the symmetry. (a) The straight upright rod is rotationally invariant, as we move around the rod is appears the same, this system has a rotational symmetry. (b) If we push down on the top of the rod so that it buckles then it no longer has this rotational symmetry, however, (c) if we consider all possible bucklings then rotational symmetry is regained.

consider a larger system consisting of *all* possible directions in which the rod could bend. Thus, even though they are not immediately apparent, there may be relevant symmetries satisfied by a field of interest. Some forms of symmetry breaking, rather than being arbitrarily imposed upon the field of interest, as was the case in the simple example above when a force was applied to the rod, arise dynamically from within the system itself, in which case the process is called *dynamical symmetry breaking* (DSB) [55]. Two of the best known examples of symmetry breaking in physics take this dynamical form. These are the Bardeen–Cooper–Schrieffer (BCS) theory of superconductivity, and the 1961 theory of broken chiral symmetry created by Nambu and Jona-Lasinio [70]. This less *ad hoc* nature of DSB, where the fields that break the symmetry arise dynamically rather than being introduced, leads to the identification of DSB as an interesting alternative to SSB. Despite this promise, it has not been shown that DSB can reproduce physically important models such as the Standard Model, a fact which has led to its largely ignored status by the physics community. However, it is possible that this process will be much more important in 'higher level' phenomena such as biological systems where it could provide a dynamical explanation for the process of symmetry breaking that occurs during, for example, development.

A gauge symmetry is a *local internal symmetry*. When physicists fix a gauge they make an arbitrary choice which fixes the local internal symmetry to one of a continuous set of options.[7] Through the application of

[7]This has led to an interesting tension between the arbitrariness of the choice of gauge and the generally perceived importance of symmetry [258], which we shall not examine here.

gauge symmetry principles what generally starts as two sets of noninteracting fields, a matter field and an interaction field are coupled[8]; there is a deep sense in which symmetry is fundamentally linked with the concept of interaction. This idea will be fundamental to the Recursive Gauge Principle (RGP) that shall be proposed in the next section.

Like quantum mechanics, QFT is inherently contextual, more so as the localised particles of quantum mechanics become fields spread over all of space and time; everything interacts with everything else, leading to the irony that the difficulty in QFT is to *separate* the descriptions, rather than trying to incorporate missing contextuality. This places QFT in the interesting position of having a very well developed set of techniques that might perhaps be applied in new ways.

9.3 The Recursive Gauge Principle (RGP)

ALife models, and their theories, tend to reveal, and to describe what we might term (in the Baas sense) 1st level emergence, and yet getting any form of secondary level emergent behavior has proved to be rather elusive. Can we possibly generate higher levels emergence through the use of a more contextual formalism? In doing so, we may gain an understanding of the *form* our simulations should take. Gauge theories offer just such an opportunity.

Assume that we are interested in a system describable by a set of fields ϕ the dynamics of which are given by some Action S. A dynamical interaction between the fields which respects their basic dynamics could be instantiated through an appropriate application of local internal symmetry principles to this base system. This interaction, and its consequent behavior, form what should be considered as a set of second order phenomena. But these would then be describable by a set of fields themselves; the process could then be repeated. As long as there is behavior emergent from this process it would make sense to apply it.

Obviously, the choice of fields will be important throughout this process. There is a technique developed at Flinders University [63; 164], known as *Action Sequencing* where *Functional Integral Calculus* (FIC) techniques are used to derive nucleon dynamics (*i.e.* describe the dynamics of atomic nuclei) from the theory of quarks and gluons, quantum chromodynamics (QCD). FIC makes use of *dynamically determined* changes of variables. Thus the technique consists of what might be seen as two steps:

[8]Technically, matter fields are spin $\frac{1}{2}$, or fermion field, and interaction fields are spin-1, or boson fields. In QCD for example, quarks play the role of fermions, and gluons play a bosonic role.

(1) First, a new set of field variables ψ are identified which represent the new emergent variables. These variables are dynamically determined from the behavior of the original system.
(2) A new action S_ψ is then derived from the original action S_ϕ. This derivation usually consists of a number of intermediate steps, where the original fields ϕ are one by one replaced and the effect upon the action investigated.

In its original application, this process was used to jump four descriptive levels in QCD, thus it consistently describes the dynamics of what might be considered a number of different hierarchical levels. Action Sequencing can thus be used to describe the emergent nucleonic dynamics arising from the interactions of quarks.

Obviously, it will not be possible to use precisely the same formalism in any attempts to create artificial life. In particular it will be unlikely that one action will smoothly transform across as many levels as occurs in QCD. Physics is perhaps the discipline most amenable to mathematical analysis as the systems that it examines are among the most separable. The contextuality of physical systems is arguably the weakest; most of them are well solved by a classical analysis, it was only at the boundaries of physics that it became necessary to introduce the new quantum formalism. In contrast, biological systems are inherently contextual, and as such, far less separable. However there are enough parallels to make this avenue of investigation worthwhile. Firstly, the concept of a field transfers easily to areas other than physics. For example, the concept of a morphogenic field is again regaining ground in theoretical biology [144], and a number of different fields are widely used in economics [347; 20]. Commonly, an action is definable in these applications for a wide range of systems. Symmetry often plays an important role in the above procedure (as with all field theories), and again, this concept makes a lot of sense in fields other than physics. (Consider for example the use of symmetry principles, and symmetry breaking in developmental biology.)

We thus encounter a new procedure that might be used to discover, and even create, systems capable of displaying emergent behavior. First, we find a mathematical model which exhibits some sort of contextually dependent first order emergence. We can then assign new physically determined fields to the emergent objects of the theory, and determine the dynamics of the new fields using some sort of action sequencing. If we find any symmetries in the configuration, as well as the dynamics of the fields, then we might localise the symmetries, and use this to determine any extra interaction between the fields. Finally, we examine the behavior of the new system to determine if there is any new emergent behavior.

This is obviously no small task. However, this procedure offers a framework, which succeed or fail, will offer a number of new avenues by which our theoretical understanding of emergent systems might be enhanced. The most difficult aspect of such a procedure is the identification of the emergent objects within any model. This emergence will quite possibly have contextual characteristics which will add to the complication of of the analysis. The remainder of this paper will consist of a very brief discussion of two example systems, and then a return to our original theme of biological development.

9.3.1 *Cellular Automata and Solitons*

One obvious candidate for a mathematical model exhibiting first order emergence is the *soliton*. This is a solitary wave which results from a balance between nonlinearity and dispersion in a well understood set of wave equations. Specifically, solitons preserve their shape and speed in collision with one another, making them very stable emergent structures.[9] Solitons have been invoked to explain many emergent phenomena including: Jupiters long lived "giant red spot", energy storage and transfer in proteins (the Davydov soliton), and, the propagation of short laser pulses in optical fibres over long distances with negligible shape change. Even three dimensional spherical soliton-type solutions, light bullets, have now been identified.[10]

Because of their stability, solitons make ideal candidates for the first order emergent behavior that we are searching for, they can even be quantized, a process which induces a certain amount of contextuality into their interaction. Most importantly for the current discussion, there is a direct way in which they can be used in ALife; discrete soliton equations have already been implemented in a number of cellular automata [265]. There exist specific implementations CA which exhibit 'particles' attracting, repelling, and bouncing off of each other.[11] Often, the 'particles' come together in what might be seen as an interaction, forming complex structures before separating after a period of time. This model is particularly interesting with respect to the programme outlined above as it exhibits all of the dynamics required in order to implement the proposed method; stable structures emerge from a simple set of rules, which exhibit a number of interesting interactions, which themselves lead to higher order phenomena.

[9] A good source of general information about solitons is the collection [59].

[10] Strictly speaking, light bullets are not solitons, as they lose energy during collisions.

[11] See Pawel Siwak's page on iterons for an introduction to these models (http://www.cie.put.poznan.pl/Tutorials/Iterons/index-main.html).

9.3.2 *BCS Superconductivity and Nambu–Goldstone modes*

DSB provides another avenue of research in the current attempt to achieve a stronger form of emergent behavior in ALife. This is because of a physical mechanism that occurs in all SSB; symmetry breaking leads to the emergence of what are termed Nambu–Goldstone modes. The *Nambu–Goldstone theorem*[12] states that for every spontaneously broken continuous symmetry there must be a massless particle (or boson) associated with that broken symmetry. This is the process that drives all gauge theories, as bosons are the particles that are associated with interaction between fields; emergent bosons imply new coupling. With a careful analysis of DSB we start to see a general mechanism that might be applied, where a process of dynamically driven symmetry breaking leads to the generation of new emergent Nambu–Goldstone-type modes of behavior. This is a very promising avenue of research which could be used to develop a dynamical model of emergent behavior.

9.3.3 *Returning to Development*

Thus, this new methodology may provide a model of development, with all its complexity. This model is currently under construction, and I feel that such a model will be theoretically interesting, even if it turns out to be computationally intractable. In developing this model, a number of new techniques are being investigated, many of which could become useful to ALife, biology, and the analysis of highly complex systems in general.

[12]Any good textbook on QFT has a discussion of this theorem, sometimes called simply the Goldstone theorem, see for example [302].

Chapter 10

Localisation of Critical Transition Phenomena in Cellular Automata Rule-Space

A. Lafusa and T. Bossomaier

Centre for Research in Complex Systems
Charles Sturt University
Bathurst NSW 2795
Australia
alafusa@csu.edu.au

Several studies have shown the existence of critical transition phenomena in cellular automata rule-space (i.e. a sharp transition from order to chaos), but the problem of precisely localising such transitional regime in the rule-space has not been solved yet. Previous parametric approaches, such as classic lambda parameter, failed to individuate a precise parameter value at which the transition is found. In this work we localise the critical transition in a parameterised survey of multi-valued cellular automata rule-space, showing that there exists a hyperplane that precisely localises the critical transition and cuts the rule-space into two separate regions: the region of ordered rules and the region of chaotic rules. We also found that near this hyperplane, complex rules neither ordered nor chaotic, are likely to be found. Moreover, considering the normal vector of the hyperplane, we define a new critical parameter that is able to localise the critical transition.

10.1 Introduction

Complex systems are usually defined as collections of many interacting elements, from which a global coherent dynamics - the so called collective behaviour - may arise. This concept involves storing and processing of information at a macroscopic level, while no trace of the global behaviour is present at the microscopic level.

A number of studies has been carried out to understand the conditions that maximise the computational capabilities of complex systems, often related with self-organizative behaviour and with the quantitative level of "complexity" [232; 88; 87; 218; 105; 43; 143]. The hypothesis formulated by Langton in [232], well known as "computation at the edge of chaos", proposes that computational capabilities and self-organisation are achieved in the vicinity of a sharp order-chaos transition.

Cellular Automata (CA) are dynamical systems composed of many locally interacting, homogeneous elements, in which state, time and space are discrete. Many studies that investigated CAs led to very general ideas, like the edge-of-chaos hypothesis, originally formulated in this context.

Early evidence of the existence of abrupt order-chaos transitions in CA was given by Langton using the λ parameter, i.e. the density of non-quiescent state in the rule-table. He showed that varying λ from a very low value to a higher value leads to an abrupt transition from order to chaos, with the consequential discovery of highly complex behaviours. However, the critical value at which the transition occurs is not unique and depends on the particular path in the rule-space. Thus, the problem of precisely characterising the edge-of-chaos in the CA rule-space has remained unsolved.

In this work we develop a parameterised survey of CA rule-space, that individuates a clear frontier between the region of ordered rules and the region of chaotic rules. It turns out with our parameterisation that the separatrix between order and chaos is a very simple manifold, namely a hyperplane.

Taking a measure of order-chaos level and a complexity index, we will empirically show that near the critical transition, complex rules neither ordered nor chaotic, are likely to be found. Therefore, the characterisation of the edge-of-chaos will also localise complex rules.

Considering that the critical transition is precisely individuated as a hyperplane, we will define a new simple parameter (we called γ) by taking the normal vector of the hyperplane. We will verify with several experiments that $\gamma < 0$ individuates ordered rules and $\gamma > 0$ individuates chaotic rules. At the critical values $\gamma_c = 0$ there is an abrupt order-chaos transition, which also coincides with a sharp peak of the measured complexity.

10.2 Automatic classify rules with input-entropy

Let us now introduce the cellular automata rule-space and the statistical measures we used.

Cellular Automata are dynamical systems in which space and time are discrete. A CA consists of a regular lattice of cells in which each cell can be in one of k states, changing its state according to a local rule.

The rule is a function that maps from the states of the neighbours of a cell to the new state of the cell. In this work we consider two dimensional cellular automata with multiple states per cell. The neighborhood of a cell is identified by means of a local radius r, and consists of $I = (2r+1)^2$ cells (including the cell itself).

An exhaustive rule considers all $k^{(2r+1)^2}$ possible cases. As k and r increase, exhaustive rules become intractable and it is necessary to define a new simpler form.

In [45] and [46] we introduced k-totalistic rules for multi-valued CA, as a generalisation of the well known totalistic rules.

Let $h_s(t)$ be the number of cells of the neighborhood that are in the state s at time t. We denote by V the set of all possible configurations of the neighborhood, whose elements can be represented by a numerical string $(h_0 h_1 ... h_{k-1})$. h_i are positive integers and must satisfy the following constraint:

$$h_0 + h_1 + ... + h_{k-1} = (2r+1)^2 \tag{10.1}$$

Let Σ be the set of states, from 0 to $k-1$. By definition a k-totalistic rule T is an application that associates to any configuration $v \in V$ a Σ element:

$$T : V \longrightarrow \Sigma \tag{10.2}$$

A k-totalistic rule can be explicitly represented as a table (also known as look-up table) that associates to each configuration of the set V a number included between 0 and $k-1$.

In trying to classify the behaviour of different rules as order, complexity or chaos, we used Wuesche's *input-entropy,* S^t (defined in [431] for boolean exhaustive rules) that we generalised in [45] and [46] for k-totalistic rules, verifying that it is useful for classifying CA in ordered, complex and chaotic behaviour.

For a time-step of the simulation, the input-entropy S^t is defined as:

$$S^t = -\sum_{i=1}^{L} \left(\frac{Q_i^t}{n} \log \left(\frac{Q_i^t}{n} \right) \right) \tag{10.3}$$

where n is the number of cells in the CA lattice and Q_i is the look-up frequency of the ith neighborhood configuration.

As verified in [45] and [46], input-entropy on k-totalistic rules works in a similar way as found by Wuensche for exhaustive rules. In ordered systems, it goes immediately to a minimum constant value; in chaotic systems, it stays constantly at high values, with very little variations over time. On the contrary, in complex systems, input-entropy varies continuously over time without reaching an asympotic value.

Following Wuensche, we define an order-chaos measure taking the average input-entropy over a simulation ($\overline{S}$), and a complexity index taking the input-entropy variance ($\sigma(S)$).

In order to evaluate the input-entropy variance as an index of complexity, we report that in [45] we used an evolutionary process driven by this complexity index, with the consequential discovery of many complex CA. In particular, in many systems discovered by this method, self-replicating structures emerge from random initial conditions.

10.3 Parameterisation of cellular automata rule-space

We will now define a parameterised survey of the k-totalistic rule-space, similar to the mean field theory parameterisation developed in [169; 168], but in a simpler form, to deal with multi-valued CA.

Let $q \in S$ be an arbitrary quiescent state, and T a k-totalistic rule. Without loss of generality, we choose $q = 0$, i.e. where all cells are in the null state.

For the k-totalistic rule T, we define the parameter a_j as the number of configurations $(h_0 h_1 ... h_{k-1}) \in V$ which map to a non-quiescent state and contains j cells at a non-quiescent state (i.e. $h_q = h_0 = I - j$). In this definition, $0 \leq j \leq I$ and therefore there are $I + 1$ parameters.

10.4 Experimental determination of the edge-of-chaos

In the following experiments we consider k=4 and r=1 (I=9) rule-space, with the additional condition of quiescence for the state 0 (i.e all-0 neighbourhood maps to the state 0 in the rule table, or equivalently $a_0 = 0$).

For each experiment we allow two parameters to vary and keep the others to a fixed value (see Table 10.1). A sample of 100 rules was generated for each parameters setting, taking the average order-chaos degree and the average complexity index (i.e. $\sigma(S)$).

The results of the experiments are shown in Figures 10.1 - 10.12. In particular, Figures 10.1A - 10.12A correspond to the order-chaos degree and show that there is an abrupt transition between the ordered region (mean input-entropy near 0) and the chaotic region (mean input-entropy near to the maximum value).

Table 10.1 List of the experiments. Where not specified, a_i is fixed to 0.

Experiment	Parameters subspace
1	(a_4, a_5) fixing $a_3 = 10, a6 = 0$
2	(a_4, a_5) fixing $a_3 = 10, a6 = 28$
3	(a_5, a_6) fixing $a_3 = 10, a4 = 0$
4	(a_4, a_6) fixing $a_3 = 10, a5 = 0$
5	(a_4, a_6) fixing $a_3 = 10, a5 = 21$
6	(a_4, a_7) fixing $a_3 = 10$
7	(a_4, a_8) fixing $a_3 = 10$
8	(a_4, a_9) fixing $a_3 = 10$
9	(a_3, a_4) fixing $a_5 = 21$
10	(a_3, a_5) fixing $a_4 = 15$
11	(a_2, a_4) fixing $a_3 = 5, a7 = 5$
12	(a_1, a_3) fixing $a_4 = 15, a_5 = 21$

In Figures 10.1B - 10.12B the two-dimensional version of the same plot show that the boundary between these two regions describes a straight line, and corresponds to the edge of chaos.

Figures 10.1C - 10.12C correspond to the complexity index, and show that a sharp peak of complexity is found in correspondence of the transition to chaos. The two dimensional version of the same plot is shown in Figures 10.1D - 10.12D. It is easy to note that complex rules are identified by means of critical parameters values, forming a straight line.

The fact that for each experiment the critical transition from order to chaos is identified by a straight line, suggests that there is an hyperplane in the 9-dimensional parameter space, that cuts the rule-space into two regions: the region of ordered rules and the region of chaotic rules. Near these separatrix, in accordance with the edge-of-chaos hypothesis, complex rules are likely to be found.

To validate the hyperplane localisation of the edge-of-chaos, we performed a linear least-square fit of the data collected from all the previous experiments. In particular, for each experiment we focused on the complexity index by selecting points (i.e. parameters settings) with a complexity index greater than a given threshold, fixed empirically at 60% of the maximum measured value.

The obtained (affine) hyperplane has the following form:

$$\sum_{i=1}^{9} \beta_i a_i = 1 \tag{10.4}$$

The coefficients β_i are shown in table 10.2.

The standard deviation for this model is 0.025.

Table 10.2 Hyperplane coefficients of eqn. 10.4.

Coefficient	value
β_1	0.1911
β_2	0.1121
β_3	0.0655
β_4	0.0259
β_5	0.0118
β_6	0.0061
β_7	0.0036
β_8	0.0018
β_9	0.0003

In Figures 10.1B - 10.12B and 10.1D - 10.12D we plotted the line obtained by projecting the hyperplane on the two-dimensional sub-space relative to each experiment. The correspondence between the linear model and experimental data is easy to see on the figures, and confirms that the hyperplane individuates both the critical transition and the region occupied by complex rules.

10.5 Definition of a unique critical parameter

The hyperplane result suggests that if we consider the normal vector, it is possible to define a single parameter (a linear combination of a_i), that identifies the critical transition with a unique critical value.

We define the γ parameter as:

$$\gamma = \sum_{i=1}^{9} \beta_i a_i - 1 \tag{10.5}$$

Therefore, the eqn. 10.4 that defines the hyperplane is:

$$\gamma = 0 \tag{10.6}$$

and the critical value of gamma that corresponds to the edge-of-chaos is $\gamma_c = 0$. Moreover, ordered rules have $\gamma < 0$ and chaotic rules have $\gamma > 0$. The plots in Figure 10.13 and 10.14 show respectively the complexity index and the order-chaos measure for all the data of the experiments 1-12 plotted against γ. It is important to note that in Figure 10.13 the critical value $\gamma_c = 0$ corresponds to an abrupt transition from order to chaos. From the other hand, in Figure 10.14 it is easy to see that there is a sharp peak of complexity, reflecting the fact that complex rules are found near the critical value $\gamma_c = 0$, as expected from the edge-of-chaos hypothesis.

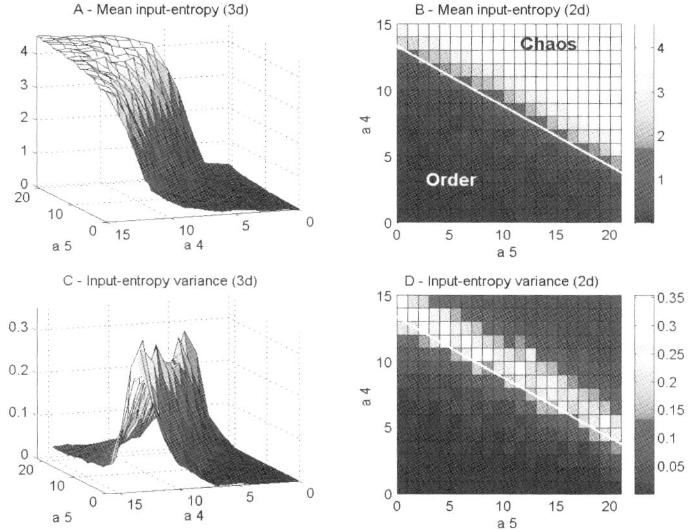

Fig. 10.1 Results of experiment 1. See text in section 10.4.

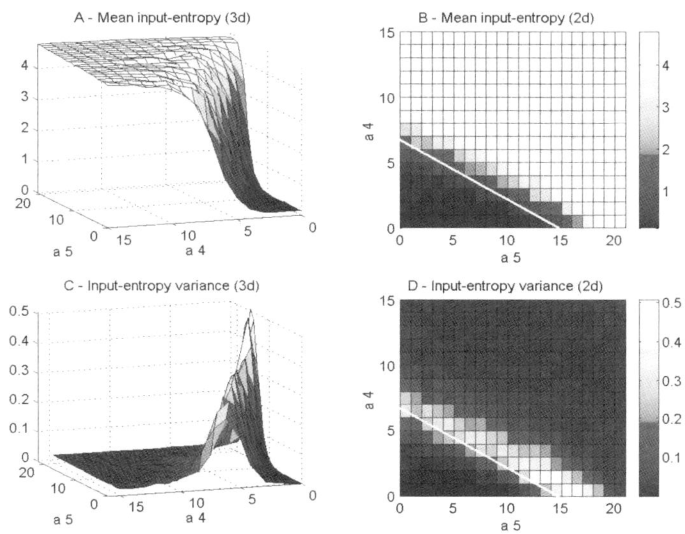

Fig. 10.2 Results of experiment 2. See text in section 10.4.

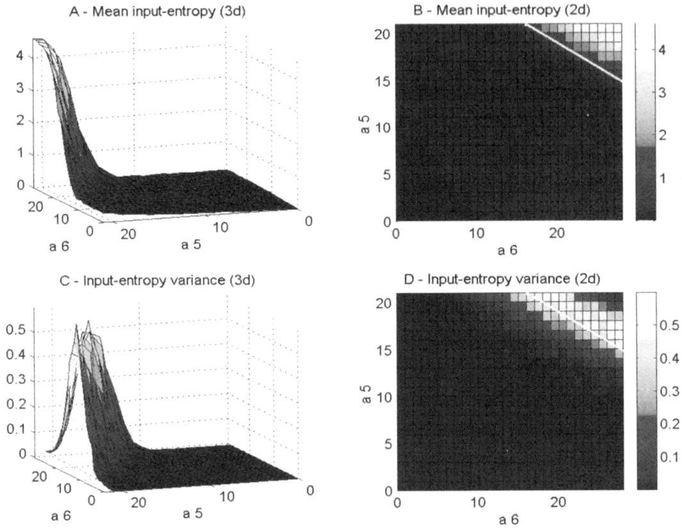

Fig. 10.3 Results of experiment 3. See text in section 10.4.

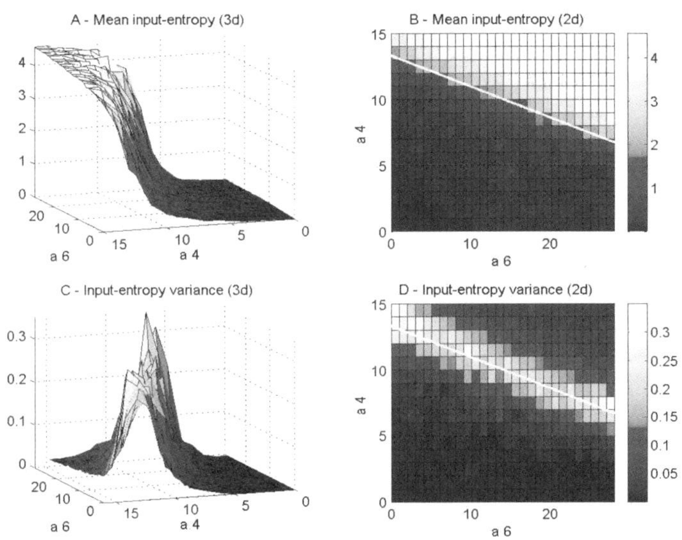

Fig. 10.4 Results of experiment 4. See text in section 10.4.

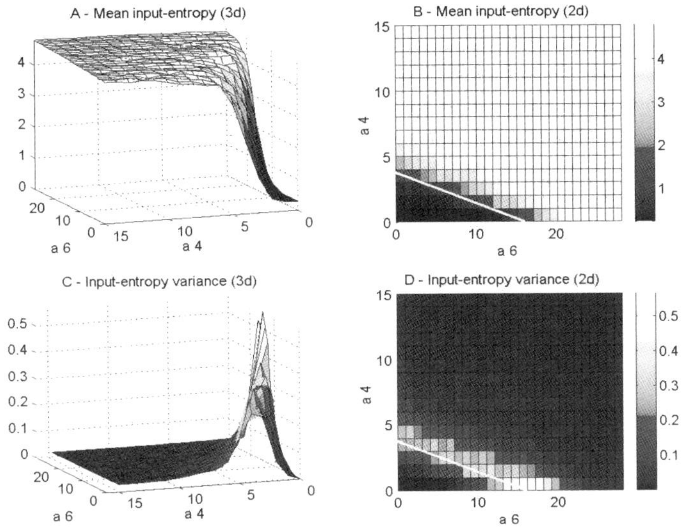

Fig. 10.5 Results of experiment 5. See text in section 10.4.

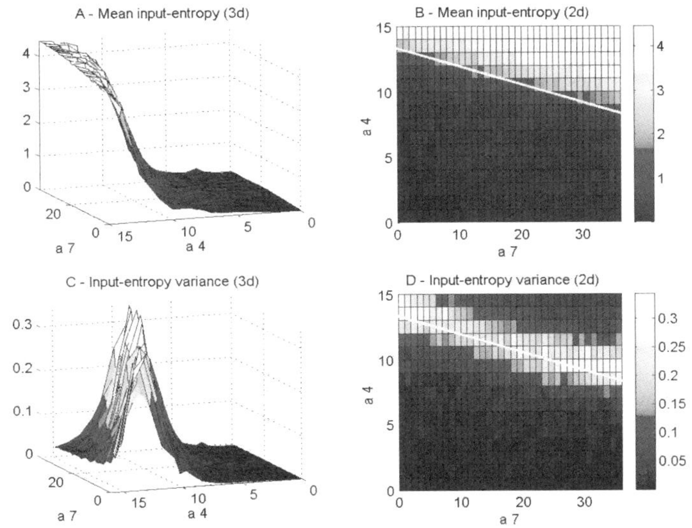

Fig. 10.6 Results of experiment 6. See text in section 10.4.

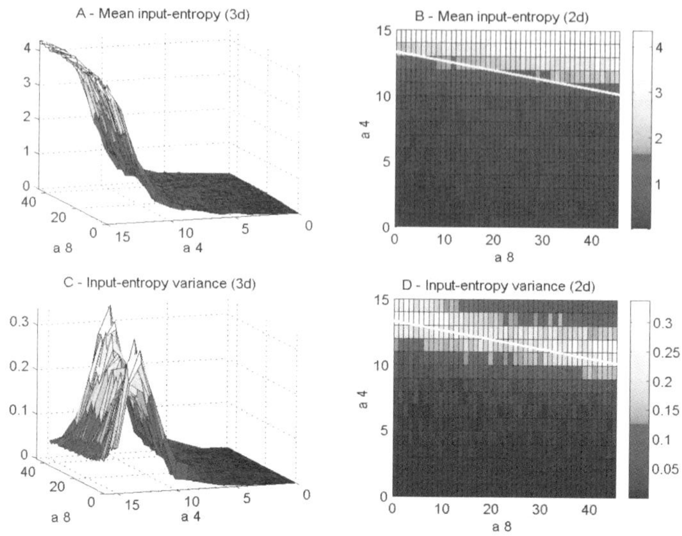

Fig. 10.7 Results of experiment 7. See text in section 10.4.

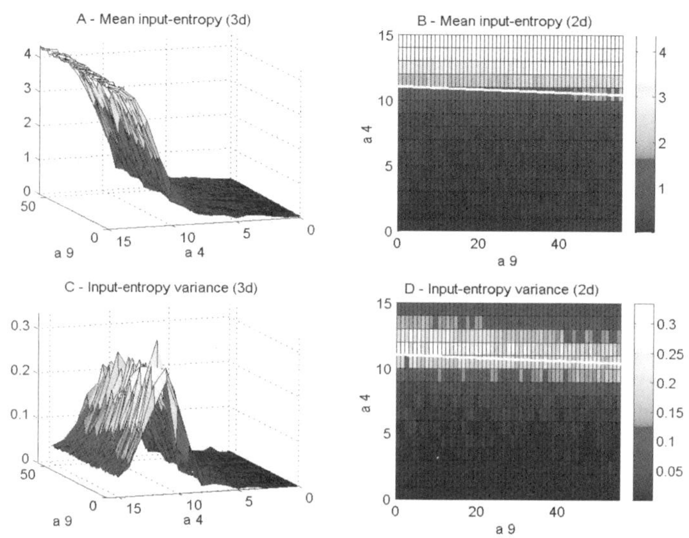

Fig. 10.8 Results of experiment 8. See text in section 10.4.

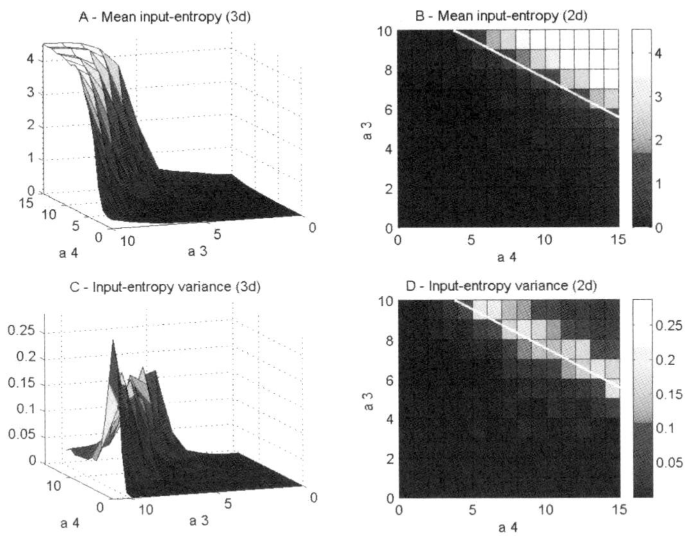

Fig. 10.9 Results of experiment 9. See text in section 10.4.

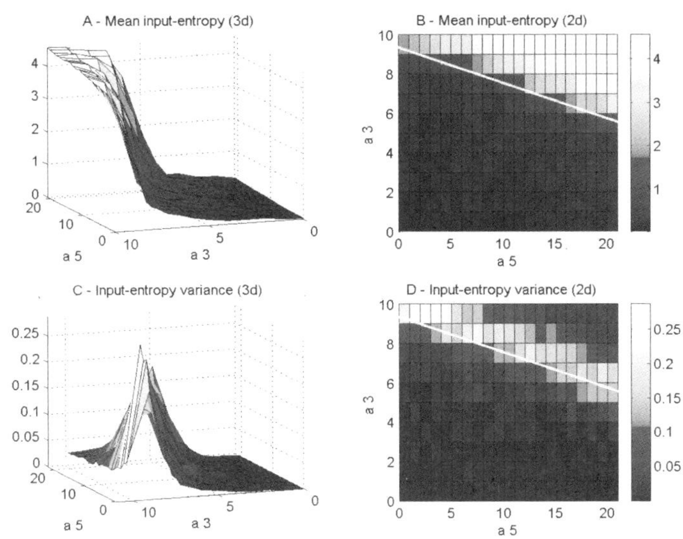

Fig. 10.10 Results of experiment 10. See text in section 10.4.

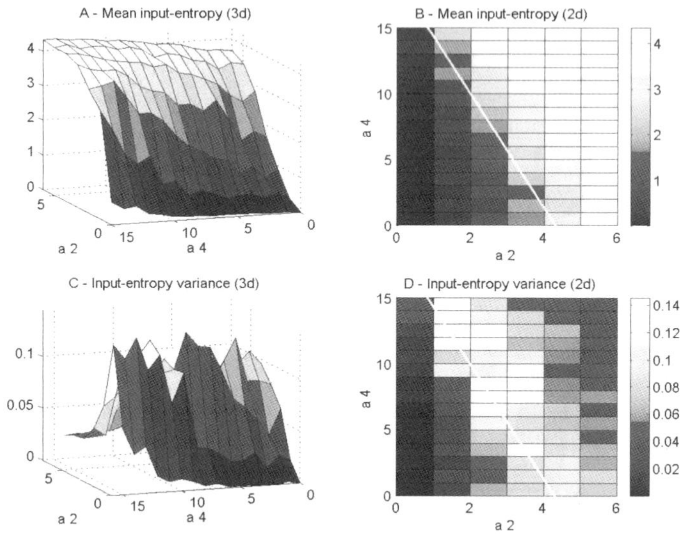

Fig. 10.11 Results of experiment 11. See text in section 10.4.

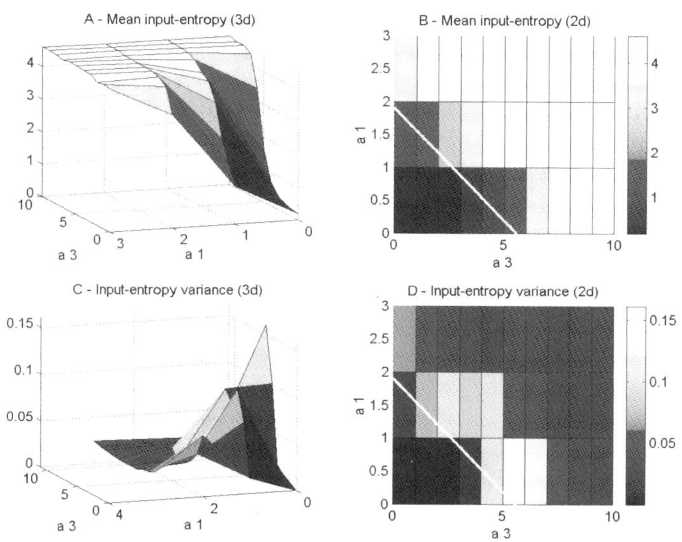

Fig. 10.12 Results of experiment 12. See text in section 10.4.

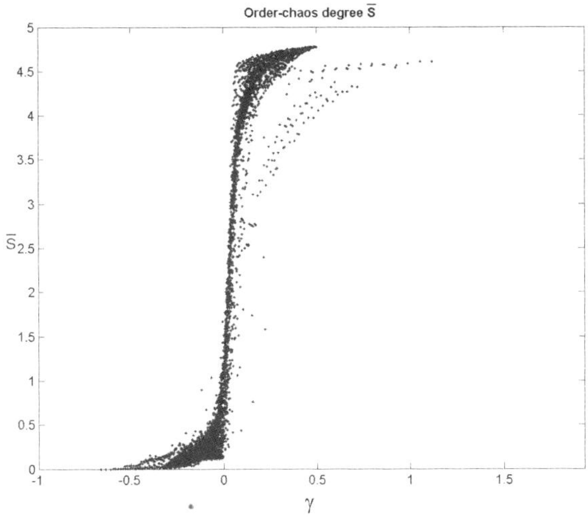

Fig. 10.13 Order-chaos level ($\overline{S}$) VS γ, including all the samples of experiments 1-12. It is important to note the abrupt transition from order to chaos in correspondence of γ=0.

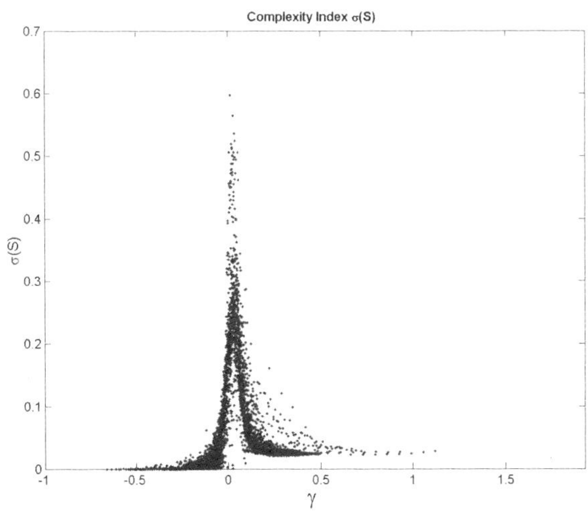

Fig. 10.14 Complexity index $\sigma(S)$ VS γ, including all the samples of experiments 1-12. It should be noted that there a sharp peak of complexity in correspondence of γ=0.

10.6 Conclusions

The existence of critical transition phenomena in cellular automata rule-space is well known, but the problem of precisely localising such transitional regime in the rule-space has remained illusive. Previous parametric approaches, like classic λ parameter, failed to individuate a precise parameter value at which the transition is found.

In this paper we localised a clear frontier between order and chaos in Cellular Automata k-totalistic rule-space. First, we defined a parameterised survey of the rule-space, localising the transition to chaos by means of a critical hyperplane in the parameter space, that separates the region of ordered rules from the region of chaotic rules. The statistical measure we used, led us to verify that complex rules are likely to be found near the "edge-of-chaos", characterised as the critical hyperplane we found.

The advantage of static parameters (i.e. directly calculated on the rule-table) is that they give useful information about the whole space of possible rules, and that it is possible to predict the dynamical regime of a given rule *without carrying out the simulations.*

Looking at the critical hyperplane and its normal direction, we defined a new static parameter, γ, that precisely localises an abrupt transition between order and chaos at its critical value $\gamma_c = 0$, that corresponds to a sharp peak of the measured complexity level.

Chapter 11

Issues in the Scalability of Gate-Level Morphogenetic Evolvable Hardware

J. Lee and J. Sitte

Smart Devices Laboratory
Faculty of Information Technology
Queensland University of Technology
GPO Box 2434, Brisbane, Qld 4001, Australia
E-mail: jm.lee@qut.edu.au, j.sitte@qut.edu.au

Traditional approaches to evolvable hardware (EHW), in which the field programmable gate array (FPGA) configuration is directly encoded, have not scaled well with increasing circuit and FPGA complexity. To overcome this there have been moves towards encoding a growth process, known as morphogenesis. Using a morphogenetic approach, has shown success in scaling gate-level EHW for a signal routing problem, however, when faced with a evolving a one-bit full adder, unforseen difficulties were encountered.

In this paper, we provide a measurement of EHW problem difficulty that takes into account the salient features of the problem, and when combined with a measure of feedback from the fitness function, we are able to estimate whether or not a given EHW problem is likely to be able to be solved successfully by our morphogenetic approach. Using these measurements we are also able to give an indication of the scalability of morphogenesis when applied to EHW.

11.1 Introduction

While evolvable hardware (EHW) has proven to be successful in the evolution of small novel circuits, generally on field programmable gate arrays (FPGAs), its applicability to complex problems has been limited, largely due to the use of direct encodings in which the chromosome directly represents the device's configuration. A practical approach to solving this problem for specific application domains has been function-level evolution, involving the use of higher-level primitives such as addition, subtraction, sine, etc. (see [190; 213] as examples). Although this scales the ability of EHW to solve more complex problems, it comes at the price of higher gate counts and designer bias [403], as well as the loss of potential novelty in solutions, thus countering some of the original motivations for EHW.

Another approach is through decomposing the problem into components or subtasks which are evolved first and then combined. Increased Complexity Evolution [392], and Bidirectional Incremental Evolution [212] are examples of this. However, these approaches are limited to applications with straightforward decompositions, without interdependencies.

A separation between genotype (the chromosome) and phenotype (the generated circuit), and a way of generating the phenotype from the genotype (a growth process), is the approach taken by nature to evolve complex organisms, and has increasingly been seen as a means of scaling EHW to more complex problems without losing its ability to generate novelty. By encoding a growth process, known as morphogenesis, rather than an explicit configuration, the complexity is moved from the genotype to the genotype-phenotype mapping.

Morphogenetic approaches have been successfully applied to generating neural networks [208; 118; 336] and tessallating patterns [40]. Recent work by the authors [236] showed that this approach could also be successfully applied to EHW at the gate level on a modern FPGA, specifically the Xilinx Virtex.

In this paper we briefly revisit the results of using morphogenesis to scale EHW on a signal routing problem, presented in Section 11.2. Then, in Section 11.3 we extend this work to evolving a one bit full adder. However, in the process of evolving adders, several obstacles were encountered, this forced us to come up with a measure of problem complexity, or more accurately *difficulty*, and fitness feedback as a means of identifying whether a given experiment is likely to succeed. This also allows us to quantify the scalability of our morphogenetic approach to EHW. This is presented in Section 11.4, after which we draw some conclusions.

11.2 Scaling with Morphogenesis

In this work biological cells correspond to functionally independent logic elements, comprised of a function generator and associated routing, within a configurable logic block (4 per CLB). Multiplexors (gates) are interpreted as proteins within a cell, with each protein representing that resource and its configuration. Inter-cellular signaling is done via shared lines, such that for each line that connects between logic elements, each cell has a signalling protein that corresponds to the configuration of the multiplexor (mux) in the other logic element. Each cell receives a copy of the chromosome, and implements a transcription level gene expression model, such that genes are regulated by the configuration state of the logic element, and in turn generate proteins, which then reconfigure this logic element. Through the interaction of genes with the FPGA configuration and inter-cell signalling, a morphogenesis process emerges. Details of the morphogenetic EHW system implementation can be found in [237].

To test the performance and scalability of the morphogenetic EHW approach this model was used to evolve signal routing circuits with severely constrained routing that disallowed simple connection rules (see [236] for details), and was compared with a standard EHW approach using a direct encoding on a fixed-length binary chromosome.

A first set of experiments required a signal to be routed horizontally across a 5x5 CLB matrix (100 cells) from an input in the center of the west edge to an output at the center of the east edge. Then to test the scalability of each approach, the size of the matrix was increased to an 8x8 CLBs (256 cells) and the number of inputs was increased to 4, placed in the center of the West edge of the CLB matrix, and outputs was also increased to 4, spread evenly across the East edge of the CLB matrix. This required evolution to learn not just how to connect horizontally across the matrix, but also how to spread vertically from the middle outwards. Fitness, in both cases, was based on how much progress was made in routing a signal, possibly inverted, from the inputs to the outputs. The relationship between the different inputs and outputs was disregarded, it was only required that all inputs are connected and one or more of these drive the outputs. Further details of the fitness function can be found in [236].

For each set of experiments twenty evolutionary runs were done with a population size of 100 and using a steady state genetic algorithm with tournament selection without replacement. The crossover rate was set at 80%, mutation at 2%, inversion at 5%, and for the variable length chromosomes used with the morphogenetic approach, a base insert/delete rate of 0.1% was used with 50-50 chance of insertion or deletion. Each run was continued until a solution with 100% fitness was found or until stagnation (defined as no improvement in the maximum fitness attained in 1000 or 1500 generations for the first and second problem, respectively). For the morphogenesis approach, growth is done for a minimum of 30 iterations, with fitness evaluated at each step. Growth is continued if the maximum phenotype fitness for this genotype increased in the last 15 iterations, or if phenotype fitness is increasing. The genotype's fitness is given by the

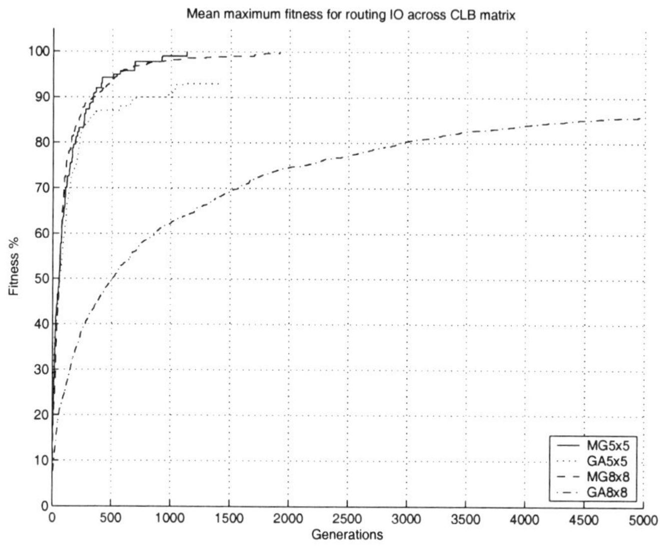

Fig. 11.1 Mean maximum fitness for routing IO across CLB matrix

maximum phenotype fitness achieved during growth.

Figure 11.1 show the mean maximum fitness over all runs for both approaches on the two experiments (up to generation 5000).

In the smaller problem morphogenesis approach was able to find a 100% solution every time, taking an average of 458.5 generations and 36.95 growth iterations. In comparison, the direct encoding approach was successful in only 13 of the 20 runs, with the average number of generations required for successful runs being 531.1 generations.

For the larger, more complex experiments, the morphogenetic approach was again successfully able to find a 100% solution in every run, taking an average of 1001.7 generations and 49.95 iterations. The direct encoding approach, however, was unable to solve this at all, with maximum fitness values reaching a mean of 86.6% (standard deviation was 3.1%), and taking on average 4647.1 generations. The best run, which reached 93.75% at generation 9954, was continued up to 35,000 generations, reaching a maximum of 96.875% at generation 16302.

11.3 Evolving One Bit Full Adders

While the previous section concentrated on evolving circuits with fitness primarily based on circuit structure, using a severely cut down set of gate-level resources, in this section the aim is to investigate the evolution of circuit structure *and* functionality on a more complete set of resources. This allows us to test the scalability of our morphogenetic approach to an increase in circuit functionality, and what amounts to an increase in platform complexity.

11.3.1 *Experimental Setup*

A one bit full adder has 3 inputs (x, y, cin) and 2 outputs $(sum, cout)$, with the relation between these being defined by Boolean Equations 11.1 and 11.2.

$$sum = x \oplus y \oplus cin \tag{11.1}$$

$$\begin{aligned} cout &= x \cdot y + x \cdot cin + y \cdot cin \\ &= x \cdot y + (x + y) \cdot cin \end{aligned} \tag{11.2}$$

The target for the adder circuit is a 2x2 CLB matrix in the center of the FPGA. This size was chosen as it contains sufficient logic and routing to provide evolution with freedom to explore a wide range of possible solutions to this problem. Although an adder could be easily defined using only two 4 input function generators (LUTs) in a single slice of a CLB, most functional circuits would require a combination of LUTs with routing lines connecting them. Thus, this approach allows the results from these experiments to be more readily generalised to other functional circuits.

Cells again correspond to logic elements (there are 4 per CLB), each containing a LUT-register pair, however the register is not used here due to the combinatorial nature of the circuit. Furthermore, each cell is allocated two out buses and six single lines per out bus line, such that there are connections available to each neighbouring CLB. This is shown in Fig. 11.2, with each logic element's LUT (F or G) and connection to the out bus (for e.g. S0_Y) shown, but the connections to and from the *Out To Single* multiplexor (which is responsible for driving the logic element outputs to neighbouring cells) are aggregated for the CLB. Each LUT has available the same set of input lines as the other three LUTs in the CLB, however, each LUT has different output lines. Only directly connecting single lines between neighbouring CLBs are used (dedicated horizontally connecting out bus lines are also used, however).[1]

[1]Note that 16 of the 48 single lines that we are using are able to be driven by the outputs of two neighbouring CLBs, a situation known as contention, which may damage the FPGA. To avoid this, prior to configuring an output multiplexor for driving a single line, its setting on the other end is checked, and if this configuration would cause contention, then it is not configured.

The evolvable region is then setup so that the x and y signals are provided to single lines that can be fed to the LUT inputs in the South West CLB, while the *cin* signals are fed towards the North West CLB's LUT inputs, and the *sum* and *cout* output signals are sampled from one logic element each on the two East CLBs. This layout is shown in Figure 11.2. Notice that several lines are available to the circuit inputs, while the outputs require signals to be routed to specific lines. Signals are routed directly from IO blocks on the perimeter of the FPGA to the CLBs adjacent to the evolvable region. This approach is an artefact of the original design aims, which were to allow the evolution of asynchronous circuits for robot control.

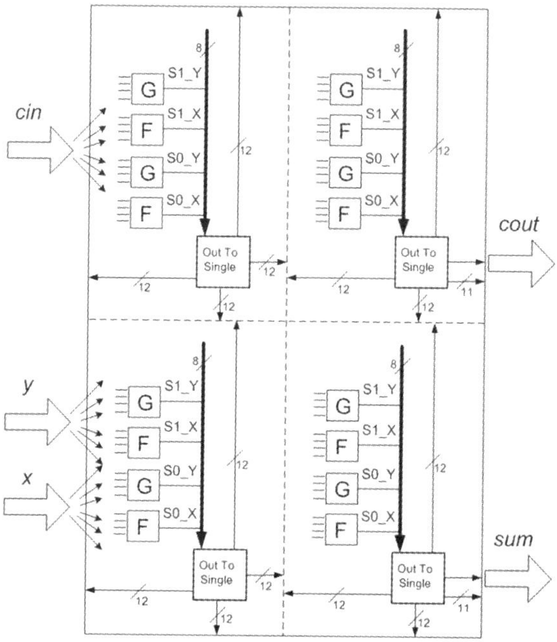

Fig. 11.2 Layout for Adder Experiments

11.3.2 *LUT Encoding*

The native encoding of the 4-input 1-output function generators (LUTs) on a Virtex is a 16-bit (2^4) inverted least significant bit first (LSB) truth table. While this encoding, which we refer to here as LUT bit functions, is sufficient for generating any required function there may be other more evolution-friendly encodings.

One such encoding is LUT active functions, which provides basic Boolean functions (AND, OR, XOR, NAND, NOR, XNOR, Majority with tie break to 0, and Majority with tie break to 1, all with optional inversion of selected lines) that are applied only to currently *active* input lines, that being inputs that have a line connected that is driven by an active CLB output. For the line to be considered active, it must be connected to a changing signal, originating either in the circuit inputs (and generally having undergone various function transformations), or in a feedback loop, which may produce an oscillating signal (for example a clock divider).

This is based on the idea that cell functionality should only use inputs that are active. In other words, unconnected inputs and undriven input lines, that are held high (i.e. at logic 1), should be eliminated from LUT functions, where they may possibly dominate the function applied to the other inputs. An example of this would be a simple OR function of all LUT inputs, which if any one of these input lines is disconnected, then the output from the LUT will always be a 1, no matter what signals are received on the other lines. In nature, it is unlikely that this would occur (for example in neurons or signalling pathways), as it would be undesirable for biological functionality to be crippled by a large state space of useless functions, especially when there are a large number of possible inputs (for example dendrites), which would often be inactive. In EHW, however, we are utilising LUTs which are designed for human or automated design software, such that they allow a great deal of flexibility in expressing Boolean functions within the constraints of an FPGA.

11.3.3 *Fitness Evaluation*

Fitness is based on how much progress is made in connecting from the inputs to the outputs, *and* on how closely the connected outputs match that of the desired function. Both connectivity (c) and functional adder fitness (a) components are given as a percentage (i.e. in the range 0 to 100 inclusive). The circuit fitness (f) is given by adding these together and then scaling the result back to a percentage (i.e. from 0 to 100), hence

$$f = (c + a)/2. \tag{11.3}$$

The circuit connectivity portion of fitness is produced in one of two ways, depending on which of the LUT encoding methods is used. If LUT active functions are used, then a recursive connectivity test is done based on the FPGA's current configuration. Fitness is calculated in the same manner as was done for the previous signal routing experiments. Briefly, connectivity fitness is calculated by testing how many layers are connected, and for each connected layer, how many elements in the layer are connected.

For active functions, there are 5 levels of connectivity measurable within a logic element. If LUT active functions aren't used (native encoding), then each logic element is only assigned a connectivity value based on whether or not there are signal changes (indicating connectivity) on the probe at the slice output, when signals are applied to the circuit input bus. Thus, for LUT bit functions, there are only 2 levels of connectivity measurable within a logic element.

The adder function fitness component is evaluated only when one or more circuit outputs are connected. To evaluate the circuit's functional fitness, each of the input signal combinations in the adder's truth table is iterated through and the connected circuit outputs are tested to see how closely they match the desired corresponding entry in the truth table.

Function equivalence is measured according to how close the circuit output matches the truth table's output signals, based on the Hamming distance between the outputs that *change* and the specified outputs, while unchanging outputs are assumed to be unconnected to the inputs, and for each of these 1 is added to the Hamming distance. Adder function fitness is then based on the proportion of matching signals, calculated as

$$a = 100(1 - \frac{h}{l}) \tag{11.4}$$

where h is the Hamming distance (with an integer value from 0 to l), and l is given by

$$l = On_c = 2n_c \tag{11.5}$$

with O being the width of the circuit output bus (i.e. the number of outputs from the adder function), and n_c is the number of input signal combinations required to specify each output's truth table, as given by

$$n_c = 2^I = 2^3 = 8 \tag{11.6}$$

where I is the width of the circuit input bus (i.e. the number of inputs to the adder function).

11.3.4 *Experiment Results*

Two evolutionary runs were done for both LUT encodings (time limitations prevented more than this), each using the same evolutionary and morphogenetic parameters as used in the previous signal routing experiments, with evolution being stopped at 5000 generations or after stagnation for 2000 generations. The results of all the runs is given in Fig. 11.3.

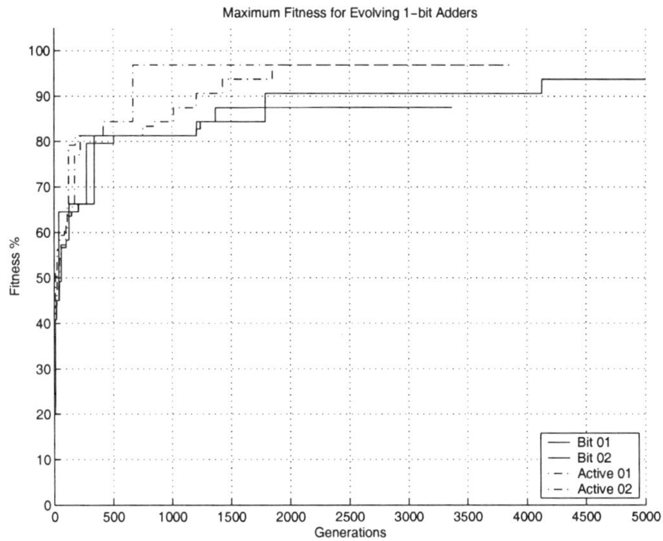

Fig. 11.3 Maximum Fitness for Evolving 1-bit Adders

Unfortunately, no approach was able to find a 100% solution, the best achieved was 96.875% (which means only 1 of the 16 output signals was incorrect), which was achieved by both active LUT runs (at generations 670 and 1849). The native (LUT bit function) encodings were only able to achieve fitnesses of 93.75% and 87.5% (corresponding to 2 and 4 wrong signals) at generations 4128 and 1369, respectively. Although these runs are not statistically significant, they can be seen as a rough indication of trends for the different approaches.

LUT active functions, not surprisingly, were slower off the mark at solving the problem, as each LUT is only able to encode a single simple Boolean function, requiring more LUTs to be configured and the routing between them to be developed, before they are able to offer any advantage. However, it is obvious that this is what occurred, and once this had been established, it rapidly overtook the other approach.

11.3.5 *Further Experiments*

To investigate where evolution fails in its task of creating a 1-bit full adder, we divided the problem into two separate components: circuit function, and circuit routing. To do this we ran two separate sets of experiments. In the first, we pre-connect the functional blocks together (i.e. fix the routing lines and the LUT inputs) and evolve the LUT functions, while in the second, we fix the LUT functions and evolve the routing between them.

Two evolutionary runs were done for both LUT encodings (again time limitations prevented more than this), each using the same evolutionary and morphogenetic parameters as before, with evolution being stopped at success or after 1000 generations of stagnation. The results of these are given in Table 11.1.

Table 11.1 Adder Fixed Run Results

Approach	Run 1		Run 2	
	Max Fitness	At Gen	Max Fitness	At Gen
BitFN fixed Mux	100.0	214	100.0	434
BitFN fixed LUT	91.25	1514	93.75	1417
ActiveFN fixed Mux	100.0	71	100.0	56
ActiveFN fixed LUT	94.8	728	90.625	993

This clearly shows that the problem lies in the evolution of the routing between LUTs, while LUT functions are able to be evolved quite rapidly.

11.4 Analysing Problem Difficulty

Here we analyse the difficulty of the various experiments, with the aim of developing a heuristic that will help us to determine whether or not any given experiment is likely to be solvable by evolution.

To do this, we need some measure of difficulty and of the feedback provided by the system to guide evolution towards a solution. The first measure is the state space of the problem, calculated as the total number of configuration states that the EHW system may configure the FPGA to. This is calculated for a logic element as the product of the number of settings (σ) for each of the m multiplexors:

$$M = \prod_{i=1}^{m} \sigma_i \tag{11.7}$$

then the number of configuration states is given as

$$S = M^n \tag{11.8}$$

where n is the number of logic elements being evolved (this could also be done per CLB).

The next measure is the number of solutions, or answers A, to the problem available within this configuration space. This is problem dependent, but in all cases, elements of the configuration space that don't affect the solution (i.e. redundancies), increase the number of solutions, as

$$A = \alpha M^{\rho} \tag{11.9}$$

where α is the initially calculated number of solutions for the essential (or actively participating) circuit components (logic elements), and ρ is the number of redundant logic elements.

With the state space (S) and number of solutions (A) calculated, then the probability (P) of a randomly generated configuration being a solution is given by

$$P = A/S \tag{11.10}$$

which can be seen as a measure of problem difficulty.

We also need to measure the size of the fitness feedback space, this being the amount of information provided by the system to the fitness function for guiding evolution. This can be calculated in a general manner using the amount of feedback from the circuit elements, given per logic element as μ, and overall function, given in terms of the number of circuit inputs (I) and outputs (O) required to completely describe the function. The total fitness feedback space is then given as

$$F = \mu^{n} 2^{(2^{I} O)} \tag{11.11}$$

where the 2's in the (latter) overall circuit function component of the equation are due to the binary nature of digital signals.

However, while this tells us the total amount of information available to the fitness function, it doesn't take into account how much of this information is actually used by the fitness function. For example, in our experiments, when we measure connectivity not all logic elements' feedback is utilised. In this case, in the circuit input and output layers (columns of CLBs) we only look at the connectivity of the circuit input and output logic elements, while all others in the layer have their connectivity measure dicarded in the fitness evaluation, reducing n to n'. This would reduce the above measure of F to an *effective* fitness feedback space (E) of

$$E = \mu^{n'} 2^{(2^{I} O)} \tag{11.12}$$

11.4.1 *Experiment Difficulty Comparisons*

With the measures introduced above, we are now able to analyse and compare the measures of difficulty for each set of experiments that were conducted. Due to space limitations, only the results of the calculations for the various measures are provided here.

It should be noted that as the number of solutions is increased by the amount of redundancy created by unused logic elements (that are able to have any configuration without affecting the circuit's function). This means that for the signal routing problems, only the shortest path needs to be found to give a first order approximation to the number of solutions (including the longer paths would have no affect on the calculated values here, however, due to their greatly reduced order of magnitude).

Note also, that for the unconstrained adder problem, where we evolve both the multiplexor settings and LUT functions together, an over approximation was used for the number of solutions. This was calculated starting with the number of solutions from multiplexor settings only evolution, where the LUT functions were fixed, and noting that for each funtion that is needed to generate an adder, there are several possible LUT functions that are able to achieve this, so the initial estimate is multiplied by the number of solutions from LUT only evolution. In other words the number of possible wiring connection settings for one particular valid set of adder function blocks is multiplied by the number of possible LUT function combinations for one of these particular wirings. Then, to take into account different routing and function combinations that could also work, this last estimate is then multiplied by the number of logic elements (n) that are being evolved. This gives

$$A = A_{mux} \cdot A_{lut} \cdot n \tag{11.13}$$

which is of course an over estimate, and not at all accurate, but is simply used to give some sort of estimate as to the upper limit of the order of magnitude.

For LUT active functions, we also need to take into account the fact that the other LUTs that weren't used with fixed routing (they are inactive) could take various configurations usually without upsetting the circuit's function, so n is squared for good measure, to give

$$A = A_{mux} \cdot A_{lut} \cdot n^2 \tag{11.14}$$

A comparison of the measures for each experiment set are given in Table 11.2, noting that $\sim$ indicates an approximation using an average for variables, and $<$ is estimated over approximation only. For the experiment names, *_lut* indicates evolving LUT functions (fixed mux), *_mux* indicates evolving mux settings (fixed LUT), *_act* indicates LUT active functions only used, *_bit* indicates LUT bit encoding only used, and *5x5:1-1* and *8x8:4-4* are used to denote the signal routing experiments (with generations taken

Table 11.2 Experiment Difficulty Measures

Experiment	S	A	P	E	E^*	$-P/E$	Gens
Adder_lut_act	38	17	−21	22	6 – 18	0.95	6
Adder_lut_bit	256	231	−25	22	7 – 22	1.14	8
Adder_mux_act	464	334	−130	28	37 – 114	4.64	-
Adder_mux_bit	464	385	−79	22	37 – 114	3.59	-
Adder_act	∼ 544	<375	< −169	30	48 – 149	5.63	-
Adder_bit	720	<623	< −97	22	28 – 85	4.41	-
5x5:1-1	999	963	−36	144	10 – 32	0.25	9
8x8:4-4	2558	2351	−207	464	59 – 182	0.45	10

for the morphogenetic approach only). Only experiments that were able to find a 100% solution are given entries for the *Gens* (average generations taken) column. Note that all entries are base 2 logarithms, rounded to the nearest integer.

From this it can be seen that the actual increase in problem difficulty between the smaller and larger routing problems was

$$P_{5x5} - P_{8x8} = 2^{-36+207} = 2^{171} \approx 3 \times 10^{51} \tag{11.15}$$

This shows that our morphogenetic approach is able to scale to a problem of more than 50 orders of magnitude greater difficulty, with only an increase of around 3 times more work (measured by the proportional increase in generations and growth iterations), while the standard direct encoding approach to EHW struggled (65% success rate) even with the simpler problem and totally failed to scale to the more difficult problem. Notice that, the morphogenetic approach has the desirable feature that, as P decreases (indicating an increase in problem difficulty), the number of generations required to solve the problem increases very slowly, in fact in a better than $log_2 N$ manner.

By examining this table, it seems that $-P/E$ correlates well with the solvability of the problem, by a *morphogenetic* approach to EHW. Tentatively, we can say that for the problem to be solvable this needs to be less than or equal to k, where from experimental evidence $1.136 \leq k < 3.593$ $(25/22 \leq k < 79/22)$. That is

$$-log_2(P)/log_2(E) \leq k \tag{11.16}$$

which can be expanded (noting that $P^{-1} = S/A$) and rearranged to give

$$log_2(E) \geq log_2((S/A)^{1/k}) \tag{11.17}$$

and as log_2 is a monotonic function, we can remove it from both sides while preserving the relation, thus

$$E \geq (S/A)^{1/k} \tag{11.18}$$

We can denote $(S/A)^{1/k}$ as E^*, which tells us the minimum amount of effective feedback required to successfully guide the morphogenetic system to a solution. Hence we require that $E \geq E^*$ for a problem to be solvable.

The range of values of E^*, using best (3.5) and worst (1.136) cases for k's value, for each experiment are given in Table 11.2. As can be seen, the routing experiments, while they are harder to solve, according to P, remain solvable as they have a more than sufficient amount of effective feedback ($E > E^*$) available to guide evolution. while, on the other hand, the unfixed adder problem and the adder problem where the multiplexor settings need to be evolved both lack sufficient feedback ($E < E^*$) to guide evolution to a complete solution. It can also be seen that evolving the adder LUTs was successful due to there being a larger proportional coverage of feedback to the problem, ensuring that there was sufficient feedback ($E \geq E^*$).

11.5 Conclusion

In this paper we have shown that morphogenesis scales extremely well to increases in circuit size and problem difficulty. This offers great promise to EHW, as it provides scalability without having to compromise the advantages of gate-level evolution.

We have also introduced a quantitative measure of problem difficulty, in terms of the probability of finding a solution, and a heuristic indicating whether the problem is solvable or not according to its difficulty and the amount of information provided by the fitness function for guiding evolution. This shows the importance of fitness feedback, and to a lesser degree problem difficulty, to a morphogenetic approach, while the state space size, reflecting FPGA platform complexity and circuit size, has little direct impact.

Ideally speaking, a problem should be analysed prior to applying evolution to solving it. If, according to the heuristics provided, it appears unsolvable, then either more feedback is required, or the probability of finding a solution should be increased, by decreasing the search space or by introducing more redundancies that effectively do the same.

Using morphogenesis coupled with these heuristics (refined through further experimentation), gate-level EHW remains a viable method and should continue to be able to provide novel solutions to hardware design problems.

Chapter 12

Phenotype Diversity Objectives for Graph Grammar Evolution

M.H. Luerssen

School of Informatics & Engineering
Flinders University of South Australia
GPO Box 2100 Adelaide 5001 Australia
Email: luer0001@infoeng.flinders.edu.au

Evolutionary algorithms are a practical means of optimising the topology of graphs. This paper explores the use of phenotype diversity measures as objectives in a graph grammar-based model of multi-objective graph evolution. Since the initial population in this model is exclusively constituted by empty productions, an active promotion of diversity is needed to establish the necessary building blocks from which optimal graphs can be constructed. Six diversity measures are evaluated on problems of symbolic regression, the 6-multiplexer, and neural control of double pole balancing. The highest success rates are obtained by defining diversity as the number of solutions that differ in at least one fitness case and do not Pareto-dominate each other.

12.1 Introduction

A directed graph is a system (V, E, s, t) where V is a finite set of vertices, E is a finite set of edges, and $s, t : E \rightarrow V$ assign a source $s(e)$ and target $t(e)$ to each $e \in E$. Natural and artificial structures that are representable as graphs are ubiquitous, and many problems of practical interest may be formulated as questions about graphs. Less commonly explored is the optimisation of graph topologies. Existing studies focus mainly on the optimisation of neural networks, but genetic programming (GP) [227], a widely applied method for optimising trees, is also of relevance in this context, since considerable research has gone into exploring the reuse of functions and modules, and other means of adding cycles back into the trees [429].

A graph grammar-based system for evolving general graphs has been presented previously [248]. Deriving graphs from a grammar mirrors the developmental process that gives complexity to biological organism and allows for an efficient representation of graphs. Genetic algorithms, including GP, commonly start with a population of random initial solutions from which all subsequent offspring is drawn. Randomly initialising a graph grammar, however, easily leads to disconnected and oversized graphs. A viable alternative is to establish complexity by evolving graphs towards greater diversity. This paper evaluates six phenotype diversity measures to be used as objectives in the multi-objective optimisation of graph grammars.

12.2 Background

12.2.1 *Evolution and Development*

Evolutionary algorithms (EAs) are a well-established class of methods for searching discontinuous spaces where little domain-specific knowledge is available [22]. EAs operate on a population of diverse solutions from which an offspring population is generated by applying mutations and/or recombinations; the fittest solutions are then selected to form a new population. The basic means of representing a graph for this purpose is an adjacency matrix folded into a binary string [278]. This representation is simple, but grows with the size of the graph rather than its complexity, inevitably leading to scalability issues.

Biological representations are more sophisticated. The genes of an organism are part of the genome, or genotype, which is encoded into several chromosomes of DNA. Natural selection does not apply to the genotype directly, but to the phenotype, which is an expression of the genotype within a given environment. Genes composed of DNA are transcribed into RNA, translated into polypeptides, and then processed into proteins which self-organize into phenotypic traits [133]. The ontogenetic process that defines the mapping of genotype to phenotype is called an embryogeny and involves complex feedback loops that control the expression of genes. These feedback loops can produce modular, iterative and recursive programs of development [354] and are characterized by polygeny (multiple genes define a single phenotypic variable) and pleiotropy (changes to a single gene affect multiple phenotypic variables). Consequently, when exploring the phenotype, large changes are possible through small variations to the genotype, while large neutral variations to the genotype will have little or no effect on the phenotype, although they may change the inductive bias on the search [393].

12.2.2 *Graph Ontogenesis*

There has been previous research into applying various aspects of biological ontogenesis, such as morphology [67] and chemistry [18], to the evolution of graphs. The drawback of realistic ontogenetic models is their necessary complexity, which increases computational cost and makes the systematic analysis of such systems difficult. A simpler, more transparent approach is to describe ontogenesis explicitly in terms of hierarchical modularity, iteration, and recursion. A popular instance of this is Cellular Encoding (CE) [162], which was developed for the optimisation of neural network topologies. CE explicitly represents each developmental step as a node in a tree of graph-transforming operators; the tree is evolved by GP. Another variant of GP, Cartesian Genetic Programming (CGP), has been used to directly construct graphs from nodes with labelled edges [279]. This approach is not developmental in nature, however, and requires extensions to incorporate concepts such as reuse and modularity [405].

An alternative model of ontogenesis is to derive a phenotype from a grammar of production rules. L-systems are commonly used to accomplish this [244]. L-systems rewrite a starting string into a new string by applying the set of production rules to all symbols of the string in parallel, reflecting the parallel division of cells in biology. The set of production rules can be optimised via evolutionary methods, and the string then translated into a graph. Early studies on this were also targeted at optimising neural network topologies [224; 49]; a more recent application is robot design [199].

12.2.3 *Evolving a Graph Grammar*

Instead of rewriting strings, it is possible to rewrite graphs directly by a process of hyperedge replacement [173]. Each hyperedge has multiple sources and targets $s, t : E \to V^*$; in contrast, binary egdes only have one of each. A graph with hyperedges is called a hypergraph, and a hypergraph with specially labelled begin and end nodes is called a multi-pointed hypergraph. A hyperedge can be replaced by a multi-pointed hypergraph by matching these nodes with hyperedge mappings s and t, respectively. Let $N \in C$ be the set of nonterminals over a label set C, $T \in C$ be a set of terminals and H be the set of all multi-pointed hypergraphs. Then a hypergraph production is an ordered pair $p = (lhs, rhs)$ with $lhs \in N$ and $rhs \in H$,

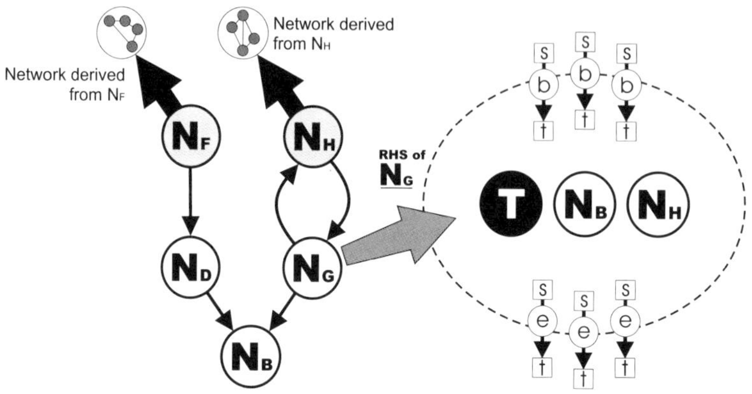

Fig. 12.1 Diagrammatic representation of an example grammar of cellular productions. Nonterminal nodes (N) identify productions of the grammar, which may also call each other (depicted as arrows). Nonterminals are always unique, with several nonterminals (depicted as grey N nodes) labelled as starting productions that produce known graphs. During graph derivation, nonterminal N_G would be replaced by the cellular graph on the right, where T is a terminal function, N_B and N_H are nonterminals (replaced subsequently), and b and e are terminal begin and end nodes, each with a source label s and a target label t.

and a hypergraph grammar is a system $hgg = (N, T, P, z)$ where P is a finite set of hypergraph productions over N and $z \in H$ is the axiom.

Hypergraph productions can be partitioned into several simpler *cellular productions* [248], which may be used as modular components for hyperedge replacement (see Figure 12.1). The right-hand side rhs of a cellular production does not correspond to a complete hypergraph but to a row in an adjacency list, which is extended by additional node labels to reduce the global side-effects of local changes to begin and end nodes [250]. Deriving a desirable graph from a cellular production grammar requires that the correct set of productions is determined, which can be accomplished by an EA.

A system has previously been developed for evolving a graph grammar with multiple starting nonterminals that match an intended population of graphs [249]. For every graph derived from its associated starting nonterminal, a single expressed production may be spontaneously replaced by a mutated copy. The mutation operators comprise the simple addition, deletion and replacement of all possible nonterminals, terminals, and their labels. After testing all the mutated graphs, the least fit solutions, both from the mutated and unmutated set, are eliminated, as are all productions not involved in any fitter solutions.

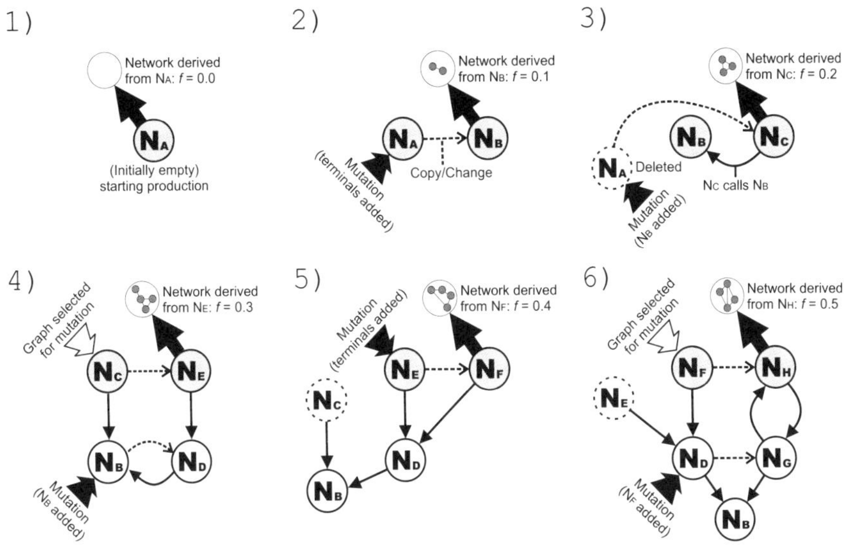

Fig. 12.2 Depiction of graph grammar evolution with a maximum population of two graphs. Starting with an empty production/graph N_A in generation (1), terminals are added to a copy N_B of this production in (2), then N_B is added to itself, producing N_C in (3), while the graph of N_A has least fitness f and is thus removed. N_B in the graph of N_C is then mutated in (4), producing N_D and a copy of N_C, N_E, with a reference to N_D. The graph of N_B is now uncompetitive, but remains as a production used by N_C. Further changes are applied in (5) and (6), leading to the graph grammar previously shown in Figure 12.1.

Conversely, if a mutation survived, the grammar is modified so that the mutated graph becomes one of the graphs derivable from the grammar. The mutated production is inserted into the grammar; then modified copies are made of all the productions that need to refer to the mutated production, not the original. This is recursively repeated for all the productions referring to the now modified productions, up to the starting production from which the new network can be derived. Evolution may thus be viewed as a repeated growing and pruning of the grammar, as shown in Figure 12.2.

12.2.4 *Diversity Objectives*

The GP algorithm requires an initial population of syntax trees, which can be generated by a variety of methods [251]. This population provides a reservoir of diverse building blocks from which further trees can be constructed. Building blocks are also essential for graph grammar evolution, as new graphs must be defined from productions that already exist. However, starting with random productions is not viable, as recursive relationships between these productions would make it difficult to control the size of the resulting graphs. Additionally, unlike with trees, vertices are not required to be adjacent to other vertices, so a random initialisation will likely produce disconnected graphs.

The alternative to obtaining diverse building blocks from an initialisation method is to generate them during evolution. Diversity refers to the differences between members of a population. Genotype diversity is the diversity among genomes in the population, whereas phenotype diversity is the diversity among fitness values in the population. Genetic lineages often reduce to one lineage early in the evolutionary process [273], so to maintain diversity a method of selecting for it must be devised.

Fitness sharing involves penalising the fitness of a solution if it is similar to other population members [147]. Rosca used fitness values to define an entropy and free energy measure for phenotype diversity [337]. High entropy reveals the presence of many unique fitness values, with the population evenly distributed over these. Bersano-Begey tracked the number of solutions that solved specific fitness cases, which was used to discover and promote more distinctive solutions [42]. Fitness sharing among different fitness cases has also been applied to GP, reducing the occurrence of premature convergence [272].

Another means of diversity facilitation is to add diversity as an objective to a multi-objective evolutionary algorithm (MOEA). MOEAs select for solutions that represent Pareto-optimal trade-offs between multiple objectives, with the fitness of a solution based on its Pareto-domination by others [98]. If multiple solutions have the same degree of domination, those residing in the most sparsely populated region of the search-space are preferred. Niching strategies of this kind can evenly spread solutions across the Pareto-boundary [101], but this is not guaranteed to lead to diverse building blocks and can indeed be detrimental to the scalability of the algorithm [346]. It is additionally possible to control solution size using MOEAs by applying a size objective [47]; however, without an active means of promoting diversity, selecting against size is known to lead to premature convergence on small solutions [94].

De Jong *et al.* achieved both smaller and more diverse trees by using tree distance as a genotype diversity objective in the multi-objective optimisation of n-parity problems [95]. Bue *et al.* explored several diversity objectives, including mean and minimum genotype distances; the latter was also implemented by Toffolo and Benini, and competitive results were achieved in all instances [388; 58]. The principal drawback of genotype distance measures is that their applicability to graph grammars is quite limited, as the extensive neutrality intrinsic to graph grammars would allow these to improve distance while remaining isomorphic. Since genotype and phenotype diversity are closely intertwined - a decrease in genotype diversity will often cause a decrease in phenotype diversity - a possible solution is to employ a phenotype diversity objective instead.

12.3 Experiments

12.3.1 *Measures of Phenotype Diversity*

The error returned by the objective function is the most available phenotypic trait of a solution and hence a solid basis for measuring phenotype diversity. To reduce any bias attributable to the nature of the specific objective function used, the solutions are ranked against each other on this function; distances are then computed as differences of ranks. Six different rank-based distance measures are suggested. The mean distance of solution i is the absolute difference between ranks,

$$D_{ij} = \frac{\sum_j |R_i - R_j|}{N} \tag{12.1}$$

where N is the number of other solutions. Since it is often easier to attain worst rank than best rank, using this measure encourages poor performance. A measure less biased towards poor performance is to compare whether two solutions i and j show identical performance,

$$S_{ij} = \begin{cases} 1 \text{ if } R_i = R_j \\ 0 \text{ otherwise} \end{cases}.$$

The diversity of solution i can be defined as the number of solutions that are not identical in performance,

$$D_{ij} = \frac{\sum S_{ij}}{N} \tag{12.3}$$

This 'difference measure' encourages solutions to be different but no worse than necessary to achieve this difference. For numeric optimisation, this would obviously lead to a population of very similar solutions; however, in the case of graph optimisation similar performance can be attained by very different graphs, so this is arguably less of a concern.

Solutions with equal mean performance can still be different, and the above approaches do not recognize this. Distinguishing these solutions without comparing their genotypes is only feasible if there are multiple fitness cases that can be compared separately. Then the mean rank distance can be averaged across each case c,

$$D_{ij} = \frac{\sum_c \sum_j |R_{ci} - R_{cj}|}{C \times N}, \tag{12.4}$$

where C is the number of fitness cases. Two solutions perform identically if

$$S_{ij} = \begin{cases} 1 \text{ if } \sum_c |R_{ci} - R_{cj}| = 0 \\ 0 \text{ otherwise} \end{cases},$$

so that diversity may again be defined as the number of non-identical solutions,

$$D_{ij} = \frac{\sum S_{ij}}{C \times N}. \tag{12.6}$$

Pareto-dominance across all fitness cases can also be established, so that dominated solutions can be excluded from the above measures. Thus, satisfying fewer fitness cases is only regarded as diversity if these fitness cases are different. The mean rank distance of a solution i is its distance to other solutions that do not dominate it,

$$S_{ij} = \begin{cases} |R_{ci} - R_{cj}| & \text{if } j \text{ does not } \in\text{-dominate } i \\ 0 & \text{otherwise} \end{cases},$$

and the proposed diversity measure is the mean of these,

$$D_{ij} = \frac{\sum S_{ij}}{C \times P_i}, \tag{12.8}$$

where P_i is the number of solutions that do not dominate i. Within this dominance framework, two solutions can also be defined to differ if

$$S_{ij} = \begin{cases} 1 \text{ if } \sum_c |R_{ci} - R_{cj}| = 0 \text{ and } j \text{ does not } \in\text{-dominate } i \\ 0 \text{ otherwise} \end{cases},$$

and diversity can be the proportion of non-identical solutions that do not dominate,

$$D_{ij} = \frac{\sum S_{ij}}{C \times P_i}. \tag{12.10}$$

For comparison, using each of the fitness cases as a separate objective in the MOEA will also be evaluated; solutions thereby remain non-dominated as long as they are superior to all other solutions in at least one fitness case. Ensuring diversity hence becomes the principal responsibility of the niching mechanism.

12.3.2 *Evaluation*

Diversity measures are evaluated on three tasks commonly used in GP and neuro-evolutionary research; this allows for easy comparison and provides a context in which to view the results. The first task is a symbolic regression of the sixth-order polynomial:

$$f(x) = x^6 - 2x^4 - x^2. \tag{12.11}$$

Fitness cases are 21 equidistant points generated by this function over the interval of $x = [-1, 1]$. The second task is the 6-bit Boolean multiplexer problem, which involves decoding a 2-bit binary address (00, 01, 10, 11) and returning the value of the corresponding data register (d0, d1, d2, d3). The final task is to evolve a neural network for double pole balancing. The pole balancing experiment is set up as described by [366] with position and velocity inputs. The Runge-Kutta fourth-order method is used to implement the dynamics of the system, with a step size of 0.01s. All state variables are scaled to $[-1, 1]$ before being fed to the network, which outputs a variable force to the cart. The initial position of the long pole is 1° and the short pole is upright; the track is 4.8m long, and poles are only regarded as balanced if between −36° and 36° from vertical. Fitness is the number of time steps that both poles remain balanced.

On all tasks, a $(\mu + \lambda)$ evolution strategy is used, with all parents producing a single offspring each ($\mu = \lambda$). For the symbolic regression, the population is 10, and the permitted terminals of the graph are the binary functions $\{+, -, \times, div\}$, where *div* returns 1 if the divisor is zero, otherwise it returns the normal result of the division. For the 6-multiplexer, the population is 30, and the terminals are *AND, OR, NOT, IF.* For the pole balancing, the population is 50, and the terminals are log-sigmoid neurons. Real-valued weight vectors are initialized randomly with a standard Gaussian distribution. New weight vectors are generated by adding the weighted difference vector between two weight vectors (of different neurons) to a third vector, adapted from Differential Evolution [311] with $F = 0.2$ and a crossover probability of 0.9.

On all tasks, recurrent relationships between terminals are disallowed; all graphs are feed-forward. Graphs are composed via a *soft matching* approach, and all cellular productions are modular [250]. One production is mutated for each graph at a time, with productions chosen randomly from those expressed by the graph. Graphs are evolved over 5000 generations. Selection occurs using a multi-objective NSGA-II [99] applied to three objectives: the function error, which is the mean squared error over all training samples; the size of the graph, which is a simple count of the expressed terminal and nonterminal nodes; and a diversity measure.

Table 12.1 Success rates for 100 runs on each of the three problem tasks for each of the applied diversity objectives.

Diversity Objective	Regression	6-Multiplexer	Pole Balancing
None	0.00	0.81	0.95
Distance	0.21	0.63	0.97
Difference	0.96	0.74	1.00
Case Distance	0.38	0.86	n/a
Case Difference	0.99	0.90	n/a
Non-Dominated Distance	0.64	0.85	n/a
Non-Dominated Difference	1.00	0.91	n/a
Case Objectives	0.85	0.54	n/a

For the symbolic regression and 6-multiplexer, 6(+1) diversity measures (as proposed in section 12.3.1) are evaluated, including the mean rank distance (equation 12.1), the mean rank difference (equation 12.3), the mean rank distance across fitness cases (equation 12.4), the mean rank difference across fitness cases (equation 12.6), the mean rank distance across fitness cases for non-dominated solutions only (equation 12.8), the mean rank difference across fitness cases for non-dominated solutions only (equation 12.10), and using each fitness case as a separate performance objective. For pole balancing, due to the absence of multiple fitness cases, only the first two measures are evaluated.

12.3.3 *Results and Discussion*

Results are averaged over 100 runs and shown in Table 12.1 and Figure 12.3. A few sample solutions are also shown in Figure 12.4 for illustration. Without applying a diversity objective, no convergence occurs on the symbolic regression. This contrasts with the 100% success rate that is attained with the non-dominated difference measure across fitness cases (12.10); a perfect solution is found after 6058 evaluations on average. Selecting for distance is overall significantly less effective ($p < .001$) than selecting for difference, but the distance measure improves significantly ($p < .001$) if applied to fitness cases, especially if only non-dominated solutions are considered. For comparison, the success rate of CGP is 61% for the same population size and 8000 generations [279], and up to 64% for GP with a population of 4000 and 50 generations [231].

The impact of diversity measures is less pronounced with the 6-multiplexer problem. 81% of runs produce an optimal solution in the absence of a diversity measure, which increases to a maximum of 91% when using the non-dominated case difference measure (12.10) ($p = .01$). Smaller improvements and in some instances reductions in performance are produced by the other diversity measures. This lack of strong results is sur-

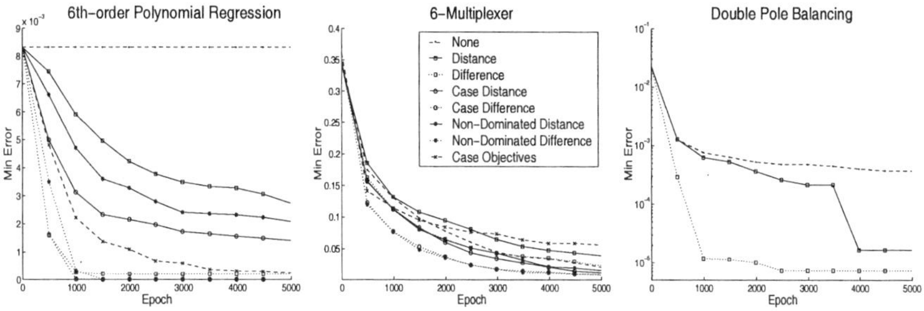

Fig. 12.3 Mean error over all runs of the minimum error solution at each generation across all problem tasks and diversity objectives.

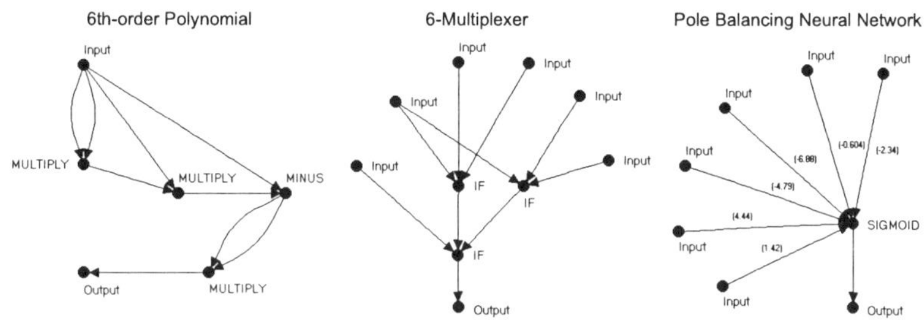

Fig. 12.4 Example graphs obtained through graph grammar evolution on the three problem tasks. (Bracketed values are input weights.)

prising given that performance benefits of diversity are quite large with GP on this problem [272]. However, the graph grammar system appears not to be very competitive on this problem in general. GP requires about 42000 evaluations to find a solution [231], as compared to 225100 evaluations with the graph grammar system.

Only a single fitness case is used for pole balancing, so none of the fitness case-based diversity measures can be applied. The evaluated diversity measures do not substantially improve the success rate on this problem, but the mean number of generations needed in a successful run is only 254 when using the difference measure (12.3), which is significantly better ($p < .001$) than 527 for the distance measure (12.1) and 467 for no diversity measure. The mean number of evaluations thus needed to find a perfect pole balancing solution is 12700, which is much less than the 80000 evaluations of Wieland [416], and also compares well to the 34000 evaluations of CE [163] and 12600 evaluations of Symbiotic Adaptive Neuro-Evolution [284], although it converges not nearly as fast as the 3600 evaluations required by Neuro-Evolution of Augmenting Topologies [366].

12.4 Conclusions

This paper presents several phenotype diversity objectives and applies these to the evolution of graph grammars. The starting population in this model is comprised solely of an empty grammar; since there are no initial building blocks, selecting for diversity helps establish the productions from which good graphs can be derived. Applying a diversity objective produces predominantly higher success rates on the evaluated problems. The sixth-order polynomial in particular cannot be regressed unless a diversity objective is provided, a phenomenon that is likely to be exhibited by any problem where the empty graph is a better solution than many non-trivial graphs.

The highest success rates are achieved by defining diversity as the number of solutions that differ in at least one fitness case and do not Pareto-dominate each other across fitness cases. Certain general trends are also observed: simply counting the number of different solutions is more effective than using a mean distance measure for diversity; and estimating diversity based on individual fitness cases is also more effective than otherwise, as is the use of a single diversity objective in place of multiple performance objectives.

The results show that merely by adding a simple phenotype diversity objective to a multi-objective optimisation framework the process of solution finding can be notably improved. This conclusion likely extends beyond graph grammars. For instance, the network required for double pole balancing has a trivial topology [163]; success is thus mostly dependent on the weight optimisation, and benefits from diversity are observed here as well.

Chapter 13

An ALife Investigation on the Origins of Dimorphic Parental Investments

S. Mascaro, K.B. Korb and A.E. Nicholson
School of Computer Science and Software Engineering
Monash University, VIC 3800, Australia,
Email: {stevenm,korb,annn}@csse.monash.edu.au

When Trivers [396] introduced the concept of parental investment to evolutionary theory, he clarified many of the issues surrounding sexual selection. In particular, he demonstrated how sex differences in parental investment can explain how sexually dimorphic structure and behaviour develops in a species. However, the origins of dimorphic parental investments also need explanation. Trivers and others have suggested several hypotheses, including ones based on prior investment, desertion, paternal uncertainty, association with the offspring and chance dimorphism. In this paper, we explore these hypotheses within the setting of an ALife simulation. We find support for all these alternatives, barring the prior investment hypothesis.

13.1 Introduction

The issue of sexual selection has been hotly debated ever since Darwin raised its possibility. Darwin noted that the females of most species tend to be choosier and less competitive than the males. Bateman later gave this observation further support with his experiments with Drosophila (fruit flies) [30]. He found that, under controlled settings, male Drosophila would mate as frequently as time allowed. Female Drosophila, on the other hand, would only mate once or twice, despite the opportunity for further mating. Bateman speculated that this behavioural difference was due to the difference in gamete sizes of the two sexes, with the female's being so much larger and, therefore, more costly. He also suggested this difference would lead to greater variability in reproductive success between the sexes — that is, females having roughly equal success, but with some males doing very well and other males doing poorly.

Trivers, taking inspiration from Bateman's work, generalised the idea of gamete cost to the idea of parental investment [396]. Parental investment covers any cost that a parent incurs in looking after an offspring, be it in gamete production, gestation or care after birth. As Trivers defines it, parental investment is "any investment by the parent in an individual offspring that increases the offspring's chance of surviving (and hence reproductive success) at the cost of the parent's ability to invest in other offspring" ([396], p.139). The definition specifically omits any effort put into attracting mates or competing with members of the same sex for mating opportunities. Other related concepts have been identified, such as parental care (which occurs strictly after birth); see [82] for a detailed review.

The concept of parental investment allows for the explanation of many cases involving sexual selection and the evolution of reproductive strategies. Trivers has used it to explain Darwin's observations of female choosiness and male competitiveness in species where females are the higher investors [396]. He has also used it to explain a parent's ability to vary offspring sex ratios in some species [398], and the period of conflict that will arise between a parent and its offspring during weaning [397]. Others have also found the concept helpful, using it to explain occurrences of infanticide and abortion [201; 253], the greater rate of child homicide amongst stepfathers and boyfriends [90], and the perpetration of rape principally by males [385; 384].

All such cases involve observing a sexual difference in parental investments, and explaining (or predicting) what evolves given such differences. However, this raises the question of how sexual differences in parental investments arise at all. Biologists have suggested several hypotheses. Trivers suggested that pre-existing differences in investment can cause further differences in investments to evolve [396]. Dawkins and Carlisle, in pointing out the faulty reasoning in Trivers' hypothesis, suggested a corrected hypothesis: that the sex that can quit investing first, will [93]. Trivers also suggested the idea that males who were less certain of their parentage would invest less [396]. Finally, Williams noted that the sex that remained with the offspring due to some pre-adaptation would be in a position to evolve parental care [418].

Interestingly, each of these hypotheses also depend on a pre-existing difference between the sexes. Trivers implicitly makes the case that pre-existing differences are likely, since such differences will be passed from species to species — that all that is necessary is a differentiation between the sexes in early evolutionary history [396]. Nevertheless, it is possible that some sexual dimorphism arises entirely anew, independent of existing sexual differences. If so, it would need to do so by chance, in much the same way that peripatric speciation occurs — that is, with small isolated populations.

Previously [264], we used an ALife simulation to investigate the effect of parental investment on various reproductive strategies – consensual mating, rape and abortion. In this paper, we use the same ALife simulation environment to explore each of the above hypotheses about the origins of parental investment – prior investment, desertion, paternal uncertainty, association with the offspring and chance dimorphism. We set up simulations according to the conditions of each hypothesis, and check how well the hypotheses predictions concur with our results. In some of the simulations, we directly evolve a numerical amount that stands in for parental investment; in other of the simulations, we evolve a period of parental care. As we will see, we find support for all of the hypotheses, barring (unsurprisingly) Trivers' original, fallacious, hypothesis.

In the next section, we cover the basic design of the simulation. We do so only briefly; for further detail, please see [262]. In subsequent sections, we cover each of the hypotheses in turn. Each such section describes the background, method and results pertinent to the hypothesis that we investigate.

13.2 ALife Simulation

Environment. The simulation is an agent-based ALife simulation. There are two entities: agents and food. These live on a board which is 25x25 cells in size, and bounded at the edges. The unit of time in the simulation is the cycle. A cycle consists of looping through all the entities currently on the board, and giving each a chance to do something. The simulations here run for 7000, 20,000 or 40,000 cycles. Another unit of time is the epoch, which defines a statistics collection period; an epoch is equal to 110 cycles.

Food is generated each cycle by the system and has a finite life time of 8 cycles on average. In most of the simulations, 50 pieces of food are generated each cycle. Each piece of food has roughly 70 units of health that is absorbed by any agent that eats it.

Agents. Agents have a numerical property called health. If the health of an agent falls below 0, it dies. An agent can gain health by eating food, and it will lose health when it moves about or mates. An agent will also gain health by resting, and lose health by continuing to exist (though both of these have a minor effect). Regardless of their health, agents have a maximum age limit of 130 cycles. Further, agents must have a minimum of 200 health units before they can reproduce.

Agents can perform one of the following five actions: eating, mating, resting, walking and turning. The exception is the simulation used for the association hypothesis, in which walking and turning are replaced with a generic 'Move' action (see Section 13.6). Agents choose an action on the basis of observations that they make of their environment. The variables that agents can observe consist of the following: self-health, self-age, self-sex, local food density, local population density and the presence of a mate request from another agent.

To choose an action, an agent passes its list of observations to its genetically inherited 'decision-maker' — essentially, a decision tree that is inherited via crossover from an agent's parents. The decision tree is structured so that each branch node splits on a single observation (an example branch node may be 'self-health > 0.1'), while each leaf node at the bottom of the tree consists of a probability distribution over the actions an agent can take. Thus, passing the observations down the tree will trigger a single leaf node. Once a leaf node is triggered, the distribution it holds is sampled to determine which action the agent will perform.

Parental investment. For this investigation on the origins of dimorphic parental investment, the key agent properties are, obviously, the amount of parental investment, pi, and the parental investment term, it. In the simulations reported in this paper, agents can either evolve their parental investment or their investment term, but not both. Each agent stores genetic information about what (or for how long) it invests in offspring, and genes for both male and female investment are stored. A child inherits both these genes from a randomly chosen parent, but only expresses the gene corresponding to its own sex (of course).

These genes are mutated by a mutation variable — itself, stored with each agent. This is so that the system can meta-mutate these mutation variables, allowing for adaptive mutation levels to evolve.

Statistics. The main statistics in the following experiments involve averages of pi and it. Another important statistic is the action rate, which is defined as follows:

$$\frac{Count_e(a_n)}{\sum_k Count_e(a_k)} \tag{13.1}$$

where e is a given epoch, a_k is one of the acts available to agents, a_n is the act of interest, and $Count_e(a)$ is a count of the number of times the act a was performed in the epoch e. As noted earlier, an epoch is simply a period of cycles in which statistics are gathered. One last important statistic is reproductive success, which is the number of offspring that an agent has.

Usually, average statistics will be collected from a *run set* — that is, a set of runs with identical parameters, that differ only in the initial random number seed. The run sets here consist of 15, 30 or 50 runs, as indicated. Some of the graphs for run sets are displayed with confidence intervals — these use the between run variance of a parameter, *not* the within run variance of the agent population.

13.3 Prior investment hypothesis

The first hypothesis we investigate is the one implied by Trivers in his seminal essay on parental investment [396]. Namely, that the sex that commits the most investment has the more to lose — and thus is the sex more likely to evolve further investment. Therefore, if correct, an arbitrary initial difference in parental investments may lead to greater differences of the same kind. In this paper, we call this hypothesis the *prior investment hypothesis.*

While at first this may seem a plausible hypothesis, it was criticised by Dawkins and Carlisle [93], who noted that it involved fallacious reasoning — of the sort used to justify continued spending on a project based on how much has been invested, rather than what future investment will likely return. They used the then topical example of government spending on a supersonic airliner based on past spending, and the fallacy is now often referred to as the 'Concorde fallacy'.

Method. To test the prior investment hypothesis, we set up the simulation as follows. An agent can invest in just one way — that is, by transferring some of its health to its offspring at birth. We call this investment *total parental investment* (or *tpi*). As noted earlier, there is a *tpi* for each sex — tpi_f and tpi_m — the genes for which each agent inherits from a randomly chosen parent. The test of the hypothesis is then quite simple: we initially set $tpi_f > tpi_m$ for all agents at $t = 0$ (i.e. $tpi_{f,0} > tpi_{m,0}$), and then allow them to evolve. If the prior investment hypothesis holds, then $tpi_f - tpi_m$ measured late in the simulation should be greater than the same difference at the beginning.

Results. Using an initial male investment of zero ($tpi_{m,0} = 0$), we ran two experiments with different initial female investment, $tpi_{f,0} = 20$ and $tpi_{f,0} = 100$. The evolution of the male and female health investments (*tpi*) for these experiments is shown in Figure 13.1(a) and (b) respectively. Clearly, regardless of the initial settings for *tpi*, $tpi_f - tpi_m$ does not evolve to be greater than it was at first. Indeed, quite the opposite happens —

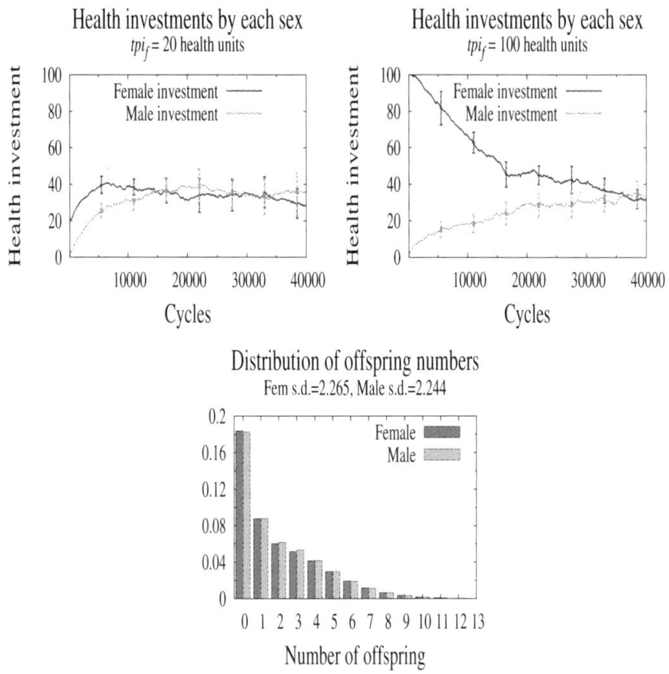

Fig. 13.1 (top left) Evolved *tpi* made by males and females when $tpi_{m,0} = 0$ and $tpi_{f,0} = 20$ and (top right) $tpi_{m,0} = 0$ and $tpi_{f,0} = 100$. Also, (bottom) distribution of reproductive success by sex for the run set shown in **b**. (Average of 15 runs.)

that is, sexually dimorphic investment disappears entirely. We also ran experiments with different initial values for tpi_f and tpi_m (ranging from 0 to 100) with the same result.

As noted earlier, Bateman identified a key idea in parental investment theory: that the sex that invests more will evolve to have less variance in its reproductive success [30]. In contrast, we would expect there to be *no* difference in reproductive variability if both sexes invest equally. We check this prediction in Figure 13.1c, which is taken from the last 7000 cycles of the $tpi_{f,0} = 100$ run set. The graph is a frequency distribution of the number of offspring agents have, split by sex. As we can see, the distributions are near identical. While the distributions are significantly different on a chi-square test, ($\chi^2 = 161$, $p < 0.001$), the Kullback-Leibler (KL) distance between the distributions is negligible (7.5×10^{-5}).

In addition to parental investment, we can also see whether any sexually dimorphic *behaviour* is evolving by looking at action rates, as shown in Table 13.1. The top row shows the female minus male difference in action

Table 13.1 Female minus male action rates for the prior investment experiments. Female action rates in parentheses.

	Eat	Mate
$tpi_{f,0} = 20$	1.1% 0.5 (81.4%)	-1.2% 0.5 (15.3%)
$tpi_{f,0} = 100$	-0.54% 0.5 (81.7%)	0.54% 0.5 (15.6%)

rates that evolves for the $tpi_{f,0} = 20$ run set, and the bottom row shows the same for the $tpi_{f,0} = 100$ run set (the numbers in parentheses are the female action rates alone). As we can see, females evolve to eat 82% of the time, while they evolve to mate 16% of the time (the remaining 2% is due to resting). Further, there is little to no dimorphism in both run sets. In fact, there is an initial rapid move toward dimorphism in both eating and mating (not shown), with males mating more, and females instead eating more. This is almost certainly due to the sexual difference in investments at the beginning of the simulations. However, this dimorphism disappears, resulting in no stable dimorphism by the end.[1]

While no stable dimorphism develops for the prior investment hypothesis, we will see an example of stable dimorphism at the end of the next section on the desertion hypothesis.

13.4 Desertion hypothesis

The desertion hypothesis was born from Dawkins and Carlisle's criticism of Trivers' prior investment hypothesis [93]. Dawkins and Carlisle noted that dimorphic investments may evolve when exactly one parent is required to raise a viable offspring. In particular, if one sex has a chance to desert the offspring first, then it will. Dawkins and Carlisle cited parental investment amongst fish as an example of this: in many species of fish, it is the male who looks after the offspring. They suggested that this is because females spawn their eggs first and males fertilize them after — by which time, of course, the female has the opportunity to leave. In contrast, male mammals fertilize female eggs internally, producing zygotes that are stored within the female. Thus, the male clearly has the first opportunity to desert, potentially explaining why *maternal* care (which occurs after birth, of course) is predominant amongst mammals.

[1] We also ran an experiment in which we set a minimum — non-evolvable — amount that females must make. This simulates investment methods such as gestation, which, once evolved, are difficult to evolve away. We then left each sex free to evolve additional investment. On doing this, we found that females did *not* evolve to make greater additional investments. In some cases, both sexes evolved the same additional investments, while in others, males evolved the greater additional investments.

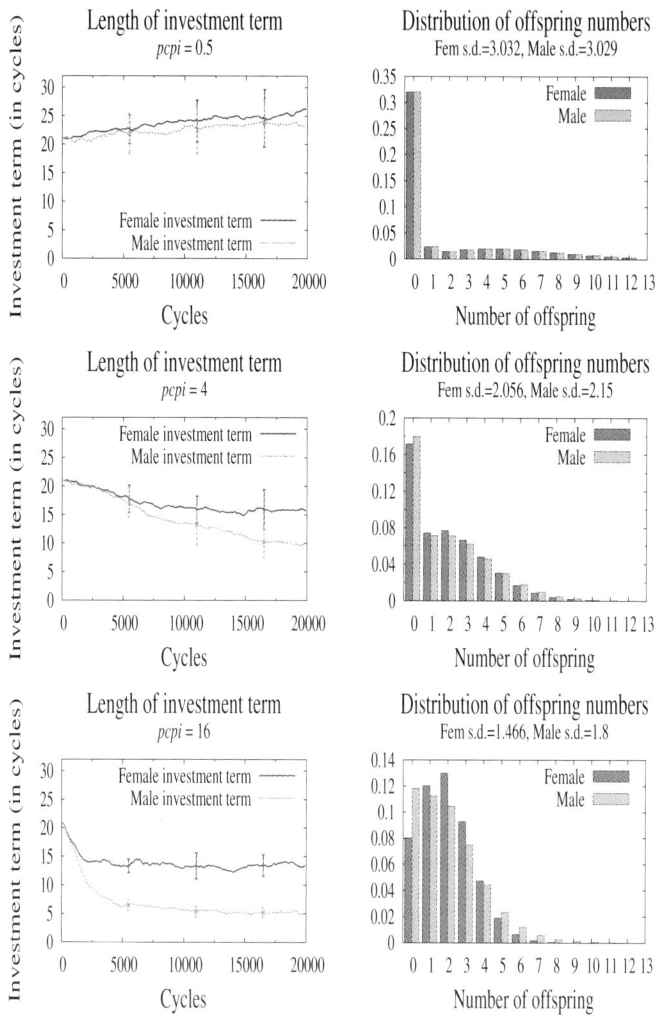

Fig. 13.2 The evolved *eit* for males and females when (top) $pcpi = 0.5$, (middle) $pcpi = 4$ and (bottom) $pcpi = 16$. Distributions of reproductive success by sex for the simulations in (top right) **top left**, (middle right) **middle left** and (bottom right) **bottom left**. (Average of 30 runs.)

Method. To test this hypothesis, we allow parents to invest for an evolvable period after birth (the evolvable investment term, or *eit*). For females, we set the minimum eit_f to 5 cycles; in contrast, males have no minimum

Table 13.2 Female minus male action rates for the desertion experiments. Female action rates in parentheses.

	Eat	Mate
$pcpi = 0.5$	0.38% 0.5 (59.1%)	-0.52% 0.5 (29.6%)
$pcpi = 16$	7.5% 0.5 (71.6%)	-7.2% 0.5 (20.1%)

period (other than 0, of course). The child needs a minimum investment of 32 cycles — so if both parents invest for the same terms, they would each invest at least 16 cycles. If one parent quits investing before 16 cycles, the other parent is forced to make up the other parent's investments. We force the remaining investment for simplicity, rather than try to produce environments in which full investment by at least one parent is needed. Finally, we fix the *per cycle parental investment* (or *pcpi*) as a parameter of each run set. In the simulations shown here, the *pcpi* takes on one of 3 values: 0.5, 4 and 16 health units per cycle.

Results. Figure 13.2 shows the results of our tests of the desertion hypothesis. When the *pcpi* is lowest, no dimorphic investments result (Figure 13.2a). In this case, relatively high periods of investment are needed from both parents: each tries to invest for ~25 cycles, which results in ~50 cycles of combined investment — well above the minimum 32 cycles of investment needed by the child. Thus, the female's minimum *eit* of 5 cycles becomes irrelevant. In the reproductive success distributions for this run set, shown in Figure 13.2b, we can see that no substantial sexual difference exists (KL distance of 4.8×10^{-5}). Furthermore, there is no sexually dimorphic behaviour evident either (first row, Table 13.2).

For the run set in which *pcpi* sits at the higher level of 4 health units per cycle, the result is very different. Here, eit_f reaches an average of 15 cycles, while eit_m reaches an average of ~10 cycles. Since $eit_m < 16$, females must make up the remaining investment, so that females invest for the greater of $eit_f = 15$ and $16 + (16 - eit_m) = 22$ — which, of course, is the latter. It is interesting that the minimum eit_f of 5 cycles can have an effect here. In fact, the average standard deviations of eit_m and eit_f (not shown in the graphs) fall between 5 and 7 cycles, allowing the minimum eit_f to influence the evolution of investments.[2] Note that Figure 13.2d shows that dimorphism in reproductive success begins to develop in this run set.

Finally, in the $pcpi = 16$ run set shown in Figure 13.2e, eit_m reaches 5 cycles and eit_f reaches 15. That is, females come to invest for ~27 cycles. This establishes strong conditions for dimorphism, which indeed evolves — as can be seen quite obviously in Figure 13.2f and the bottom row of

[2]Keep in mind that the confidence intervals in the graphs only use the variance in *runs*, not the variance in the underlying populations.

Table 13.2. This dimorphism is exactly the kind that parental investment theory predicts — that is, that the sex that invests less will evolve to try mating more often. Of course, trying does not equate with succeeding — males (and females) must average 2 offspring in a stable population. Instead, the eagerness of males leads some to greater success, and this in turn causes others to have lesser success; which is exactly what we see in Figure 13.2f.

13.5 Paternal uncertainty hypothesis

The paternal uncertainty hypothesis is again due to Trivers [396]. He suggested that males are often in a situation of being uncertain about their parentage, particularly in species where females go through a gestation period. In contrast, uncertain female parentage is very unlikely. In that case, it may pay males to spend less effort on an offspring, and instead spend more effort trying to mate. There is some evidence in humans that paternal uncertainty has an effect on how parents and their families interact. Daly and Wilson report that the mother's family will make comments about how similar the child looks to the father more frequently than reciprocal comments are made by the father's family [89]. Further, Fox and Bruce report that paternal certitude affects how fathers take to their roles as fathers [129].

Method. We test the paternal uncertainty hypothesis by fixing the probability of paternity, pp, as a parameter of the simulation. In particular, females always invest in their own offspring; in contrast, females choose males from the neighbourhood to invest in their offspring according to pp. At one extreme, if $pp = 1$ for a simulation, the chosen male is *certainly* the father; at the other extreme, if $pp = 0$ for a simulation, the chosen male is *never* the father. We set pp to 101 equally spaced values between 0 and 1 inclusive. As for the prior investment experiments, the parental investments that both sexes make, tpi_f and tpi_m, are free to evolve.

Results. Figure 13.3a shows the main result of this experiment. Each point in the scatter plot represents the average tpi_m of the last 1000 cycles (of 7000 total) in a single run. The horizontal axis shows the setting of the pp parameter for each run, and the vertical axis indicates the investment amount. The result here is clear: the lower the probability of being the actual father, the less males invest in the offspring. Indeed, pp and tpi_m have a correlation coefficient of 0.848 ($t(100) = 15.81$, $p < 0.001$). Thus, this result provides strong support for the paternal uncertainty hypothesis.

We can also see how females evolve tpi_f for different pp from Fig-

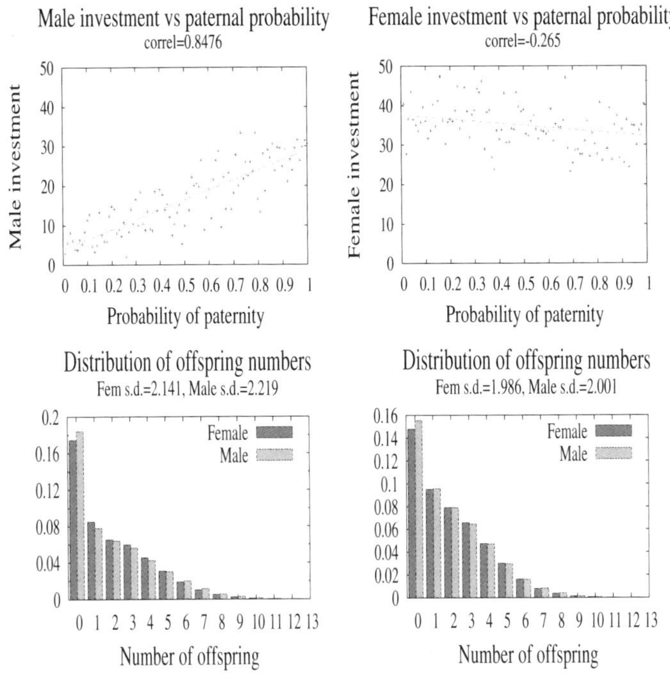

Fig. 13.3 (top left) Evolved investments made by males as a function of pp. (top right) as per **top left**, but for females. Distributions of reproductive success by sex for (bottom left) $pp < 0.1$ and (bottom right) $pp > 0.9$. (1 run per graph point.)

Table 13.3 Female minus male action rates for the paternal uncertainty experiments. Female action rates in parentheses.

	Eat	Mate
$pp < 0.1$	3.0% 0.5 (67.2%)	-3.1% 0.5 (19.2%)
$pp > 0.9$	0.72% 0.5 (66.9%)	-0.86% 0.5 (19.4%)

ure 13.3b. As pp increases, and therefore as males invest an increasing amount, tpi_f falls away slightly. The negative correlation is not large ($r = -0.265$), but is significant ($t(100) = -2.72$, $p < 0.004$). Thus, the more males invest, the more females take advantage by investing less.

To assess the level of dimorphism (in behaviour and reproductive success) that evolves in these runs, we take the runs in which $pp < 0.1$ as one group and $pp > 0.9$ as another. The former should exhibit more dimorphic behaviour, while the latter should exhibit less. Figure 13.3c and

Figure 13.3d shows the reproductive success distributions for the last ~1200 cycles of runs with $pp < 0.1$ and $pp > 0.9$, respectively. We can see that there is a slight dimorphism evident in the $pp < 0.1$ runs (KL distance of 0.0011) that is not evident in the $pp > 0.9$ runs (KL distance of 0.00019). More tellingly, we can see a reasonably strong behavioural dimorphism in Table 13.3 for the $pp < 0.1$ runs, that is much reduced in the $pp > 0.9$ runs.

13.6 Association hypothesis

The association hypothesis or, more generally, the pre-adaptation hypothesis was suggested by Williams [418]. He noted the perhaps obvious point that if only one sex remains in the vicinity of the offspring after birth — due to some pre-adaptation of that sex — then that sex has the opportunity to evolve parental care, while the other sex does not.

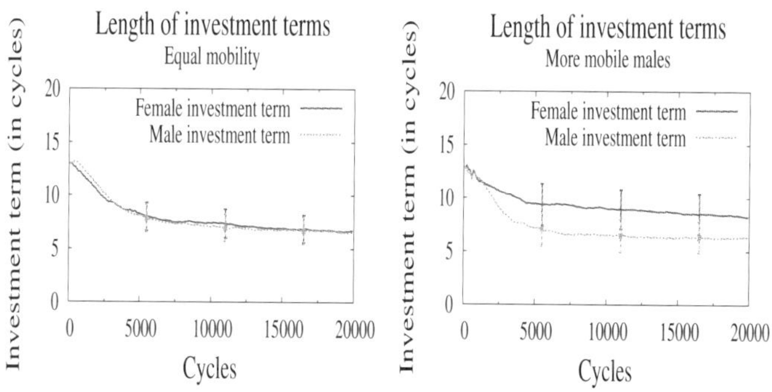

Fig. 13.4 The evolved after birth investment terms for both males and females for (on left) no sex differences and (on right) males as the more mobile sex. (Average of 50 runs.)

Method. As it stands, the association hypothesis is almost tautological. However, this need not be so: the sex that does *not* remain with the offspring — which we will take to be the male in these experiments — could evolve to return every so often to make parental investments. There is no a priori reason why males cannot continue investing. Nevertheless, males will find it *harder* to invest in offspring. 'Harder' here simply means that males have to do more to invest at the same rate as females. In this case, it is not immediately obvious that females will invest more than males, though we would expect them to do so since they find investment easier.

We choose to test this form of the hypothesis by having a non-evolvable 'Move' action that causes males to move about more actively. Specifically,

Table 13.4 Female minus male action rates for the association experiments. Female action rates in parentheses.

	Eat	Mate
Equal mobility	0.20% 0.5 (68.0%)	-0.18% 0.5 (23.8%)
More mobile male	1.1% 0.5 (66.2%)	-1.1% 0.5 (25.2%)

males move about randomly in a 9x9 neighbourhood with 0.6 probability each cycle, while females move about randomly in a 3x3 neighbourhood with 0.2 probability each cycle. In addition, we established that parental investments have a certain 'efficiency', dependent on the distance from the child. That is, the closer one is to a child, the more of one's investment reaches the child. The function of efficiency, e, over distance, d, that we used is a simple linear inverse function of distance from the parent: $e = 1 - \frac{d}{20}$ if $d < 20$ and $e = 0$ otherwise. The distance is the minimum number of cells in either the horizontal or vertical direction.

Similar to experiments in previous sections, the investments are in the form of per-cycle investments after birth. Here, *pcpi* = 8 and agents are free to evolve the term for which they invest (the *eit*).

Results. Figures 13.4a and b show the results of 2 run sets, the first in which the 'Move' action is the same for both sexes, and the second in which the 'Move' action makes males more mobile. The graphs show the *eit* for both sexes. In the first graph, no dimorphism evolves — as we would expect. In contrast, the second graph shows that females — the sex that can invest more efficiently — evolve to invest for longer periods.

Surprisingly, the degree of behavioural dimorphism that evolves is very slight. The bottom row of Table 13.4 shows that a difference of only 1% in action rates evolves — in contrast to experiments in previous sections that showed differences of between 3% and 7%. Further, dimorphism in reproductive success (not shown) is not evident (KL distance of 6.6×10^{-5} for the more mobile male run set).

13.7 Chance dimorphism hypothesis

All of the above hypotheses on the evolution of sexual dimorphism assume that there is a pre-existing sexual difference. But there may be cases in which there is no pre-existing difference or, perhaps more likely, that an existing difference is not sufficient to cause the evolution of further dimorphism. Trivers suggested that the sexes differentiated very early on due to positive selective pressure acting on gametes whose sizes fell in the tails of the normal curve [396]. That is, smaller, mobile, gametes would be selected

Table 13.5 Averages and standard deviations of degree of dimorphism (*dd*) for runs of differing population sizes. (Average of 50 runs.)

Avg stable pop'n size	Mean dimorphism (*dd*)	S.D. of *dd*
237	$\mu = 44.32$	$\sigma = 83.96$
409	$\mu = 29.19$	$\sigma = 54.98$
648	$\mu = 9.25$	$\sigma = 11.91$
901	$\mu = 11.19$	$\sigma = 17.12$

for since they can fertilize other cells more easily, while larger, immobile, gametes would be selected for since they increase the probability of a viable offspring. In contrast, those with intermediate sizes would not fare so well. Trivers does not seem to regard this process as occurring anew in new species, but rather occurring amongst progenitor species, from which dimorphism is inherited.

However, perhaps it is possible, as Gould might hold, that sexual differences in parental investment can arise by chance. If a chance difference in investments persisted for long enough, the sexes may adapt to the difference. This could then 'lock them in' — that is, chance reductions in dimorphic investment could cause agents to become less fit, and thus be selected against. We would expect such events to be most likely amongst small populations, since the genetic variance *within* such populations will be small, while the genetic variance *between* such populations will be large.[3]

Method. To see if dimorphism may arise at all, we run several runs in which the sexes are initially identical, and then see whether substantial dimorphic investments ($|eit_f - eit_m|$) and behaviour can develop. Further, to discover if the size of the population has an effect on the frequency with which dimorphism develops, we run the simulations with different sized populations — which we achieve by regulating the food supply. To assess the degree of dimorphism, *dd*, for a single run, we take the mean $|eit_f - eit_m|$ in that run, and divide by the pooled standard deviation of eit_f and eit_m within that run; this is so as to counter the run to run differences. In essence, *dd* is the number of standard deviations of difference between eit_f and eit_m.

Results. Figure 13.5 summarises the results of run sets, each with different average population sizes. The table shows the mean *dd* for a run set with a given population size (along with the standard deviation). The first thing to note is that dimorphism evolves quite regularly. If we focus on those cases in which there are 3 standard deviations or more of difference (i.e. $dd \geq 3$), we note that dimorphism results in half or more of all cases

[3]This is similar to the argument supporting peripatric speciation.

(not shown). Further, there seems to be an inverse correlation between the size of the population and the average *dd* that evolves.[4] There also seems to be an inverse correlation between the variance of the *dd* and population size.

13.8 Conclusion

We have explored various hypotheses on the origins of sexually dimorphic investments through simulation, and found support for those that we would expect to be correct. Our simulation results concur with the view that the prior investment hypothesis is wrong, given initial sex differences in investments (and also minimum sex differences in investments). We found strong support for the desertion hypothesis and for the paternal uncertainty hypothesis. While our results also agreed with the association hypothesis, the level of dimorphic behaviour and reproductive success in these experiments was lowest. Finally, we had little difficulty in finding simulations that produced dimorphism by chance, and confirmed that smaller populations do indeed lead to greater levels of dimorphism.

[4]The last run set here defies this trend, but runs that we further tested, using larger populations, continue the correlation.

Chapter 14

Local Structure and Stability of Model and Real World Ecosystems

[1]D. Newth, and [2]D. Cornforth

[1]*CSIRO Centre for Complex Systems Science*
GPO Box 284 Canberra, ACT 2601, Australia
Email: david.newth@csiro.au

[2]*School of Environmental and Information Sciences*
Charles Sturt University
PO Box 789, Albury, NSW 2601, Australia
Email: dcornforth@csu.edu.au

For over a century, the analysis of community food webs has been central to ecology. Community food webs describe the feeding relationships between species within an ecosystem. Over the past five years, many complex systems —including community food webs— have been shown to exhibit similar global statistical properties (such as higher than expected degree of clustering) in the arrangement of their underlying components. Recent studies have attempted to go beyond these global features, and understand the local structural regularities specific to a given system. This is done by detecting nontrivial, recurring patterns of interconnections known as motifs. Theoretical studies on the complexity and stability of ecosystems generally concluded that model ecosystems tend to be unstable. However this is contradicted by empirical studies. Here we attempt to resolve this paradox by examining the stability of common motifs, and show that the most stable motifs are most frequently encountered in real ecosystems. The presences of these motifs could explain why complex ecosystems are stable and able to persist.

14.1 Introduction

Community food webs describe who eats whom within an ecosystem. For almost a century, they have been central to community ecology as they provide a complex yet tractable description of biodiversity, ecosystem structure, function [115] and fragility [358]. For over fifty years there has been an ongoing debate over the relationship between the stability and complexity of community assemblages [270]. MacArthur [254] was one of the first to suggest that the more complex an ecosystem was, the more likely it was to be stable, as population fluctuations have a greater chance of being corrected. However, theoretical studies on the matter usually conclude that systems with more species, and stronger interactions between species, are more likely to be unstable than those with fewer species and weaker interactions [139; 266; 267; 335; 304; 305; 395].

Recently there has been a renewed effort towards the understanding of food web topology. This interest was sparked by the finding that many complex networks share common topological features [12; 408; 407]. In previous work [288; 290], we have argued that complex dynamical systems like community food webs evolve through the process of invasion (where new species are added) and collapse (where species become extinct). The collapse of a food web results in a "stable core" around which a more complex system can evolve. In this paper we examine the dynamical stability properties of small sub-networks, or "motifs", that represent small ensembles of species that are candidates for these stable cores. We will then examine real world food webs for the occurrence of stable motifs.

The remainder of this paper is structured as follows. The following section outlines the theory of stability applicable to the study of ecosystems and of motifs. Section 14.3 describes the experiments used to determine the stability properties of motifs, and their frequency within real ecosystems. Section 14.4 provides the results of these experiments. A discussion of the implications and significance of this study is given in section 14.5. Finally section 14.6 provides some closing comments and possible future directions.

14.2 Ecological stability and patterns of interaction

The behavior of an ecosystem is subject to external and internal perturbations. External perturbations arise from macroscopic uncertainties in the environment. Internal perturbations arise from spontaneous noise processes from microscopic fluctuations in the environment. Much of ecological theory is based on the underlying assumption of equilibrium population dynamics. An ecosystem's stability properties are determined by the way in which it responds to internal and external perturbations. A system is said to be stable if the system returns to an equilibrium point after being perturbed away from its steady state. In the following section we will outline linear stability analysis, a common approach for determining the stability of a community assemblage.

14.2.1 *Community Stability*

Landmark studies by May [266] introduced notions of linear stability analsysis into theoretical ecology. Typically, models of ecological interactions are inspired by equations similar to that proposed in [247]:

$$\frac{dN_i}{dt} = F_i(N_1(t), N_2(t), \ldots, N_n(t)), \tag{14.1}$$

where F_i is an empirically inspired, nonlinear function of the effect on the ith population on the remaining n populations within the ecosystem. Most commonly the function F_i takes on the form of the Lotka-Volterra equations:

$$F_i = N_i \left(b_i - \sum_{j=1}^{n} N_j \alpha_{ij} \right), \tag{14.2}$$

where N_i is the biomass of the ith species; b_i is the rate of change of the biomass of species N_i in the absence of prey and predators; and α_{ij} is the per unit effect of species j's biomass on the growth rate of species i's biomass. Of particular interest in ecology is steady state of the system, in which all growth rates are zero, giving the fixed point or steady state populations N_i^*. This occurs when:

$$0 = F_i((N_1(t), N_2(t), \ldots, N_n(t)). \tag{14.3}$$

The local population dynamics and stability in the neighborhood of the fixed point can be determined by expanding equation 14.1, in a Taylor series about the steady state populations,

$$\frac{dN_i(t)}{dt} = F_i|_* + \sum_{j=1}^{n} \left[\frac{\partial F_i}{\partial N_j} \right|_* x_j(t) + \frac{1}{2} \sum_{k=1}^{n} \left[\frac{\partial^2 F_i}{\partial N_j \partial N_k} \right|_* x_j(t) x_k(t) + \ldots \tag{14.4}$$

where $x_i(t) = N_i(t) - N_i^*$ and $*$ denotes the steady state. Since $F_i|_* = 0$, and close to the steady state x_i are small, all terms that are second order and higher, need not be considered in determining the stability of the system. This gives a linearized approximation that can be expressed in matrix form as:

$$\dot{x}(t) = Ax(t) \tag{14.5}$$

where $x(t)$ is an $n \times 1$ column vector of the population deviations from the steady state, and the community matrix A has the elements a_{ij}

$$a_{ij} = \left.\frac{\partial F_i}{\partial N_j}\right|_* \tag{14.6}$$

which represents the effects of population j on the rate of change of population i near the steady state. As May [267] demonstrates, solving the algebraic eigenvalue problem for A reveals the systems dynamical response to perturbations . Of particular interest here is the stability of the system near the stead states. A system will return to a steady state if (and thus said to be stable) if all the real parts of the eigenvalues associated with A are negative.

14.2.2 *Local patterns of interaction*

Over the past 10 years, it has been shown that many natural systems including community food webs share a number of common statistical properties. For example the distribution of links within a network are often found to follow a power-law distribution, in which some nodes are more highly connected than other nodes. Other common patterns found in many complex networks include the small-world properties, of high clustering and short path-lengths [408]. While many systems may share global characteristics, they may vary greatly in their local patterns of interaction. Recently it has been found that many networks display local patterns of interaction or "network motifs" at a much higher frequency than expected in random networks [280; 351]. A motif is a small sub-graph that defines the local interactions between a set of nodes [280]. Figure 14.1 illustrates a series of three node motifs. Previous studies have shown that different types of complex networks are constructed from different combinations of motifs. For example social systems tend to have a high number of complete three node subgraphs, while linguistic networks tend to have highly expressed tree structures [280]. Over expressed motifs can be interpreted as structures that arise because of the special constraints under which the network has evolved. In previous work it has been shown that local food web assemblages change as a function of latitude, environment, and habitat [289].

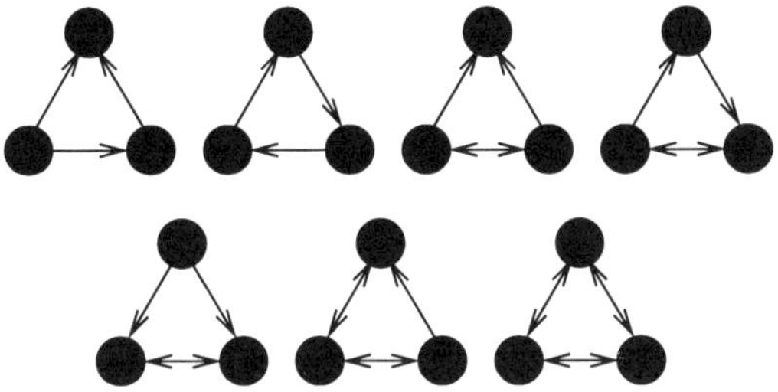

Fig. 14.1 The seven three node motifs, where each node interacts with each other node.

Here we are concerned with the frequency of occurrence of structural motifs and their stability properties. In general a motif defines a dynamical system, where the nodes are state variables and the links define relationships between the state variables and link weights define the rate of change of a state variable. We will attempt to reconcile the stability properties of a given motif (via the use of the linear stability criteria outlined above), with its frequency of occurrence in real world food webs.

14.3 Experiments

In this section we describe the experiments used to determine the stability properties of motifs, and to determine the frequency of these motifs in real ecosystems.

14.3.1 *Stability properties of motifs*

To determine the most stable motifs, we have devised a simple systematic numerical experiment that tests the stability of each motif. The stability of a motif is influenced by three factors: (1) the minimum on-diagonal term (which determines the lower bound for all eigenvalues); (2) the interaction strengths between elements (which determines the rate of change of a state variable); and (3) the loop structure (which influences the sign of the real part of the eigenvalues). Each of the motifs defines a particular loop structure, so it is necessary to test systematically the effects of varying interaction magnitudes and self-regulatory terms. For each motif, the on-diagonal terms were systematically varied between 0 and -1 in steps of 0.05. All non-diagonal terms were randomly drawn from uniform distribution centered around zero, and systematically varied in increments of 0.05. The range of the non-diagonal term was $[-1, 1]$. This results in 441 parameter combinations for each of the motifs. To gain a probability of stability,

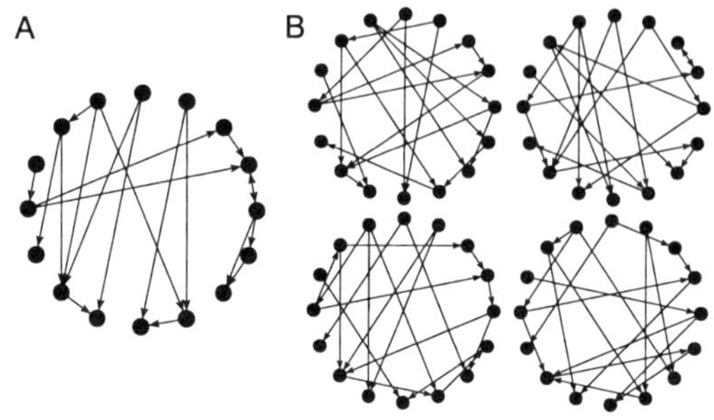

Fig. 14.2 Network randomization procedure. (A) The original network. (B) An ensemble of four networks that have the same degree distribution as (A), however the local patterns of interactions are destroyed.

each parameter configuration was executed 50,000 times. To determine the stability of a given motif, we have calculated the eigenvalues for each of the 50,000 randomly generated weights for a specific motif configuration. The probability of stability was calculated as the number of times the motif was found to be stable across all parameter configurations.

14.3.2 *Motif frequency*

To measure the frequency of the motifs making up the networks, we have implemented an algorithm for detecting recurring patterns of interconnections, or motifs. A detailed overview of the algorithm used here and its application to a gene regulation network was presented in [351]. Each community food web (see §14.3.3) was scanned for all possible n-node sub graphs (in the present study, $n = 3$ and 4), and the number of occurrences of each sub graph was recorded. To focus on those motifs that are likely to be important, we compared the frequency of occurrence of these motifs with their occurrence in randomized networks. The randomization procedure employed here preserves the degree distribution of the original network; that is, each node in the randomized ensemble has the same in- and out-degree as the corresponding node in the original network. This allows for a stringent comparison between the randomized and the observed networks, as the randomized ensembles account for patterns that appear because of the degree distribution. Fig 14.2 depicts the result of the randomization procedure for a simple 16 species food web. The ratio between the motif frequencies in the real network and the randomized ensemble provides a measure of how over or under expressed each motif is.

14.3.3 *Community food web data*

To reconcile the stability properties of motifs to their frequency in real world ecosystems, we analyzed a set of 184 community food web. A majority of the food webs were taken from the data set compiled by Briand and Cohen [57]. This database contains 113 food webs, from a wide range of habitats including: salt marshes, deserts, swamps, costal areas, estuaries, lakes and pack ice zones. Other communities include the Grasslands in Great Britain [260]; Silwood Park, England [275]; two variations of the Ythan Estuary, in Scotland [175; 202]; Little Rock Lake, in Wisconsin [259]; Chesapeake Bay [24]; St Marks Seagrass, an estuarine seagrass community in Florida [78]; St Martin Island in the Caribbean [148]; Skipwidth pond, England [406]; Coachella valley, in southern Californian desert[306]; 10 grassland communities surrounded by pastures in southern New Zealand [394]; and 50 New York Adirondack lake food webs [183].

14.4 Results

In this section we describe the results of the two experiments: stability properties of motif, and frequency of motifs in real ecosystems.

14.4.1 *Stability properties of motifs*

Figure 14.3 shows the probability of a given motif being stable for a given parameter configuration for a fully connected three cycle (14.3A), and a feed forward loop (14.3B). As can be seen, the fully connected three cycle is only stable when the self regulatory term is sufficing high, and the interaction terms are low. By contrast, the feed forward loop is completely stable regardless of the self regulatory and interaction terms. The feed forward loop could be considered to be sign stable. That is regardless of the nature and magnitude of the interactions within the motif, the system is always stable.

These results demonstrate that certain motifs are only stable under certain conditions, while other motifs exhibit stability over a wide range of the interaction terms. This results may allow us to speculate on the nature of the interactions taking place when these motifs are observed in nature.

14.4.2 *Stability and occurrence of three node motifs*

In this section we will compare the stability properties of the community food webs with their occurrence in community assemblages and their rate of expression. Figure 14.4 compares these results. In Figure 14.4 the top graph

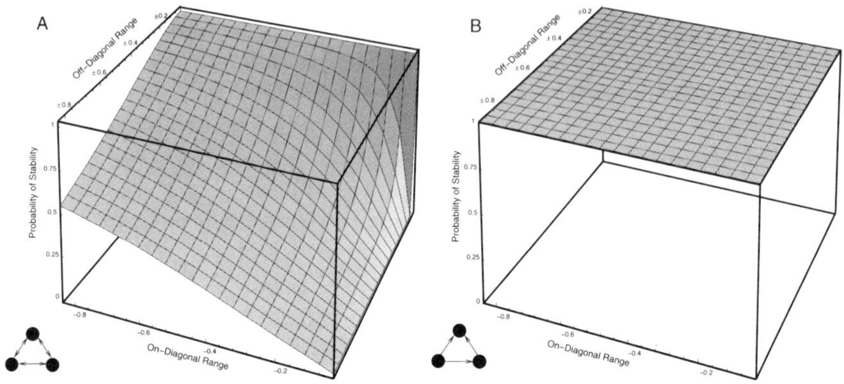

Fig. 14.3 The relationship between stability and magnitude of the matrix elements a_{ij} for two motifs, a fully connected three cycle (A), and a feed forward loop (B)

shows the probability of stability of the various 3 species motifs, with more unidirectional vertices on the left, and more bidirectional vertices on the right. It can be seen from this figure that the presence of cycles in the graph is associated with instability. The second, fourth, sixth and seventh motifs contain cycles, and have lower probability of being stable. The middle graph in figure 14.4 shows the percentage of real food webs examined in this study that contained the respective motif. Although the most stable motifs (see Figure 14.4 Top) tend to be expressed more frequently than more unstable ones, these frequencies are not normalized for the effects of the link (degree) distribution of the food webs. Figure 14.4 bottom shows the rate of expression of each motif in actual ecosystems compared to those in randomized networks, so these figures have been normalized for the effects of the degree distribution. It can be seen that a number of motifs that are less stable tend to be under expressed, as is the case with the second, fourth and sixth motif. However the seventh motif, the fully connected three cycle, (while only present in a small fraction of real world food webs) is highly over expressed by a factor of almost two. This motif is highly unstable. So why is it so highly over expressed in real-world food webs? An inspection of the data reveals that this motif is the result of the life cycle of a trophic species. Given two species A and B, species B eats the young of species A. However when species A matures, it eats species B. In this case the aggregation of species life stages into a single trophic species may appear to be theoretically unstable, but within natural systems this is a stable configuration. This implies that ecosystems models should take account the life stages of species.

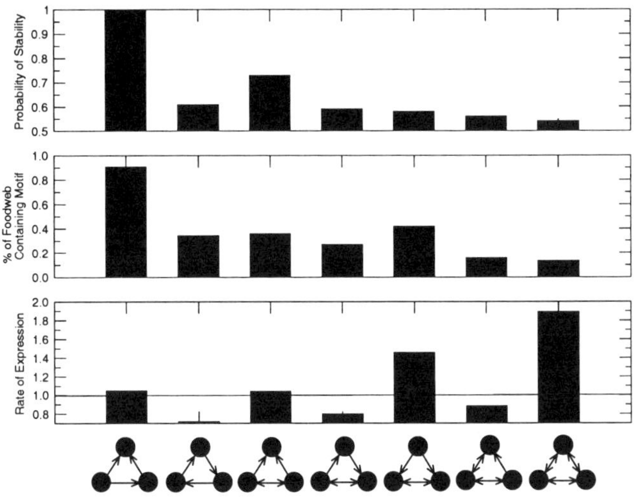

Fig. 14.4 Properties of three species motifs. Top: stability of motif, middle: relative frequency of motif in actual food webs, bottom: normalized occurrence, compared to randomized controls.

14.4.3 *Stability and occurrence of four node motifs*

We now turn our attention to 4-node motif stability and frequency. Figure 14.5 shows the results of this analysis. Again we see that simple motifs are more stable than those with elaborate loop structures. Like the feed forward loop of the 3-node motifs, the bi-parallel feed forward, and the bi-parallel fan (first and second motifs) are sign stable. It is interesting to note that these motifs are found in a wide cross section of community food webs (see Figure 14.5 middle). In contrast, other more complex structures (with the exception of the fully connected 4-node motif) are only found in a small fraction of communities. This may indicate that these motifs are specific to a particular type of habitat or they may provide a specific ecosystem function. Again we see that the most complex motif (the fully connected 4-node motif) is the least stable, but is found more in many community assemblages. This may be the result of the aggregation of life stages into a single trophic species. Again this warrants further investigation. Figure 14.5 bottom illustrates the rate of expression of each motif. It is interesting to note that almost all appear to be expressed with approximately the frequency found in the randomized controls, although in most cases slightly higher. This suggests that many of the local patterns can be attributed to the degree distribution of the community assemblage. The fully connected 4-node motif occurs more frequently than expected.

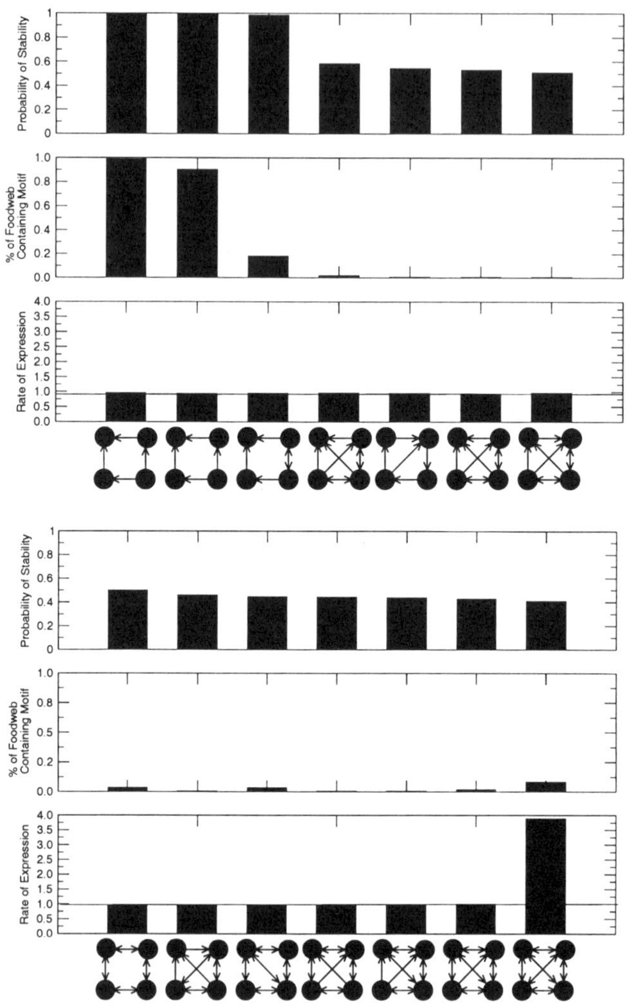

Fig. 14.5 Properties of four species motifs. Top: stability of motif, middle: relative frequency of motif in actual food webs, bottom: normalized occurrence, compared to randomized controls.

14.5 Discussion

We are remarkably ignorant of the dynamics that govern the natural systems that surround us. The southern oceans around Antarctica; for example, produce about 3% of the world's phytoplankton. An insufficient amount, it would seem, to support complex ecosystems; yet they do. Even more surprisingly, these ecosystems are notoriously top-heavy. Some 15 million Crab eater seals—possibly the world's second most abundant large animal after humans—, two million Weddel seals, half a million Emperor penguins and four million Adelie penguins all live on the pack ice around Antarctica. While seeming to be awkwardly top-heavy these ecosystems work, and more importantly they persist.

Earlier we noted that theoretical studies of community food webs based on randomly assembled systems with Lotka-Volterra type dynamics generally find these systems to be mostly unstable [139; 266; 267; 335; 304; 305; 395], but studies of the structural properties of real world ecosystems suggest that they are not randomly assembled [115; 24; 289; 358; 419]. More recent theoretical studies based on the analysis of small species models show trends where certain topological regularities are preferred —in some sense selected for— over others [290; 419]. The results presented in this paper also support this notion. It appears that the more stable motifs are being selected for, and as a result occur more frequently, and across a wide range of communities. It is also important to note that the environment in which an ecosystems is embedded places constraints on resources —such as food, water, space and temperature. These constraints drive the evolution of the community towards a robust topology. Knowledge of the context of an environment coupled with an understanding of the dynamics and the structure of the interactions may be able to shed light on the natural laws that govern complex ecological communities, and explain why the seemingly infeasible systems such as the Antarctic pack ice food web described above could persist.

From the analysis presented here we can speculate on some heuristics that govern the assemblage of community food webs. First, those motifs with no cycle components are the most stable (we will call these structures tree like). Of the tree like structures feed forward loops, (along with other linear structures) are the most stable and appear most frequently in community assemblages. This result is also supported by the analysis of the spectra of graphs, which show these structures are the most linearly stable. Second, long cycles seem to be important. Those motifs that have long cycles tend to be more stable. Again, this is supported by more detailed eigenvalue analysis that suggests that long weakly coupled loops have a stabilizing influence on dynamical systems. While loops within loops seem to have a stabilizing effect, this is only up until a certain extent. Fully connected motifs seem to be less stable over certain parameter ranges. This notion also suggests that the inclusion of a tree-like component that breaks several loops, or joins loops in a certain way, plays a key role in promoting stability. These results suggest that there is a trade-off between stability and complexity, but if we are clever about the arrangement of the interactions within a system, large complex community assemblages are feasible.

14.6 Closing comments

In this paper we have attempted to reconcile the stability properties of small groups of interacting species, with observed local patterns of interaction (motifs) in community food webs. It appears that those motifs which are most stable are most frequently encountered in community food webs. This suggests that these stable motifs may be the basic building blocks of complex ecosystems.

The most dramatic structural change that many ecosystems face is from human-driven biodiversity loss. Following our natural feeding patterns and needs for material resources, we have historically tended to impact on higher trophic levels through over fishing and hunting. Often the exhaustion of a natural species population results in a string of extinctions that cascade through the community. In this paper we have investigated the stability properties of fixed motifs. One future avenue for further research is to investigate how the stability properties of a motif change as nodes/arcs are removed, and the nature of the interactions change. Such studies could provide insights into the possible effects of human impact (and natural extinction) on ecosystem services.

Finally, the findings presented here open at least three lines of further inquiry. First, how can these ecological generalizations be explained in terms of the large scale behaviour and population dynamics of individual and collective species ensembles? Second, do these topological features of ecological organization explain other significant feature of food webs, such as species turnover? Third, what structural and dynamical characteristics of the individual communities account for their deviations from these overall trends? Understanding these questions in the context of the structural properties of community food webs may provide insights into the dynamical behavior they can support.

Acknowledgements

The authors wish to thank Ross Thompson, Karl Havens and David Raffaelli for providing the data upon which this study was based.

Chapter 15

Quantification of Emergent Behaviors Induced by Feedback Resonance of Chaos

A. Pitti, M. Lungarella, and Y. Kuniyoshi

Lab. for Intelligent Systems and Informatics
Dept. of Mechano-Informatics
The University of Tokyo, Tokyo, Japan
Email: {alex,maxl,kuniyosh}@isi.imi.i.u-tokyo.ac.jp

We address the issue of how an embodied system can autonomously explore and discover the action possibilities inherent to its body. Our basic assumption is that the intrinsic dynamics of a system can be explored by perturbing the system through small but well-timed feedback actions and by exploiting a mechanism of feedback resonance. We hypothesize that such perturbations, if appropriately chosen, can favor the transitions from one stable attractor to another, and the discovery of stable postural configurations. To test our ideas, we realize an experimental system consisting of a ring-shaped mass-spring structure driven by a network of coupled chaotic pattern generators (called coupled chaotic fields). We study the role played by the chaoticity of the neural system as the control parameter governing phase transitions in movement space. Through a frequency-domain analysis of the emergent behavioral patterns, we show that the system discovers regions of its state space exhibiting notable properties.

15.1 Introduction

How do infants explore their bodies and acquire motor skills? How do humans and other animals adapt to unexpected contingencies and opportunities in a dynamic and ever changing environment such as the real world? Or more in general, what are the mechanisms that allow a complex embodied system consisting of a multitude of coupled and potentially heterogenous elements to autonomously explore, discover, and select possibilities for action and movement? These are difficult issues whose answers despite intensive efforts still elude us. The main goal of this paper is to shed some light on how body dynamics might be explored. Exploration and emergence represent important first steps towards gaining further insights into how higher level cognitive skills are bootstrapped.

Nikolaus Bernstein was probably the first to address in a systematic way the question of how humans purposefully exploit the interaction between neural and body-environment dynamics to solve the complex equations of motion involved in the coordination of the large number of mechanical degrees of freedom of our bodies [41]. In the last decade or so, Bernstein's degrees of freedom problem has been tackled many times through the framework of dynamical systems (e.g. [214; 381]).

Such research has three important implications which are relevant for this paper. First, movements are dynamically soft-assembled by the continuous and mutual interaction between the neural and the body-environment dynamics. Second, embodiment and task dynamics impose consistent and invariant (i.e. learnable) structure on sensorimotor patterns. Third, when the neural dynamics of a system is coupled with its natural intrinsic dynamics, even a complicated body can exhibit very robust and flexible behavior, mainly as a result of mutual entrainment (e.g. neural oscillator based biped walking [378] and pendulation [252]).

In this paper, we pursue further the idea of a network of chaotic units used as a model for exploration of body dynamics [229]. One of the core features of our model is that it allows to switch between different attractors while maintaining adaptivity. We make two main contributions: The first one is the introduction of a mechanism of feedback resonance of chaos in our model. The second contribution is a set of tools for analyzing the resulting spatio-temporal dynamics. In the following section, we will present the three pillars on which our augmented model rests: (a) dynamical systems approach, in particular the notions of global dynamics and interaction dynamics; (b) mechanism of feedback resonance thanks to which the neural system tunes into the natural frequencies of the intrinsic dynamics of the mechanical system; and (c) concept of coupled chaotic fields which is responsible for exploration of the neural dynamics. We then introduce a set of methods used to analyze the spatio-temporal dynamics of the neural and mechanical system. Subsequently, we describe our experimental setup and report the outcome of our experiments. In the last section, we discuss our results and conclude by pointing to some future work.

15.2 Model System

In this section, we introduce the three key elements of our model of exploration. We briefly describe dynamical systems theory, and the concepts of feedback resonance and coupled chaotic field.

15.2.1 *Dynamical Systems Approach*

Dynamic systems theory is a well-established framework primarily geared towards modeling and describing nonlinear phenomena that involve change over time [372; 383]. From a dynamical point of view, systems are typically described in terms of their evolution over time, their robustness against internal and external perturbations, number and type of attractors and repellors, as well as bifurcations, that is, qualitative changes of the dynamics of the system occurring at critical states.

A dynamical system – initialized in a particular stable attractor state and affected by noise – fluctuates irregularly inside it despite internal and external perturbations. The system cannot evolve to a new state until the pertubations reach a certain level triggering a transition to a new (possibly more stable) attractor. Here, we conceive of perturbations (internal and external ones) (a) as a means to explore and discover stable as well as unstable action possibilities of a mechanical system, and (b) as a mechanism of adaptation against environmental changes. Our approach is reminiscent of the process of chaotic itinerancy which can be defined as the chaotic transition dynamics resulting from a weak instability of the attractor landscape of the dynamical system [214].

By contrast, control theoretical approaches (e.g. [360]), including adaptive methodologies, are typically framed as abstract mathematical problems and avoid exploiting any nonlinear physical aspect in the control process such as interactions, perturbations, body dynamics, and entrainment.

15.2.2 *Feedback Resonance*

The second key element of our approach is feedback resonance [130; 298]. This mechanism indicates that small but timely feedback actions can dramatically affect the dynamics of a nonlinear system, e.g. turn chaotic motions into periodic ones. The rationale is that by having the feedback actions occur at a specific instant in time, it is possible either to entrain a system to the action or to destabilize it and induce a transition to a new behavioral pattern. The phenomenon can be conceptualized as the energy-efficient excitation through resonance of the many degrees of freedom of a system. Resonance is pivotal because when the frequency of oscillation of the system matches its natural vibration frequency, the system absorbs injected oscillatory energy more effectively.

Action on the system using feedback resonance is described as:

$$F_i(t+1) = F_i(t) + \gamma x_i(t), \tag{15.1}$$

where $F_i(t) >> \gamma x_i(t)$ expresses the dynamics of the system (here, the force acting on motor i at time t), and $x_i(t)$ is the controlled variable (here, the neural excitation) scaled by a parameter γ. By exploïting resonance, the system can amplify the small perturbations and dramatically affect the global dynamics of the system inducing bifurcations and new postural configurations (in the case of a mechanical system). The resonant forces are used to transfer energy to new behaviors. The general idea is akin to the concepts of global dynamics and intervention introduced by [434].

The work done in physics has mainly focused on idealized pendular systems and chaotic models [130]. By contrast, our framework explicitly includes information about the morphology of the embodied system (i.e. distribution and type of actuators and sensors), the properties of the environment, as well as the coupling between body and neural system. The dynamics of the body embedded in the environment is exploited and natural resonant states are discovered.

15.2.3 *Coupled Chaotic Field*

We used a chaotic pattern generator as an internal source of perturbation (external perturbations are gravity, impactive forces acting on the body, and so on). As a model of neural activity we chose the globally coupled map (GCM) which is a network of chaotic elements instantiating a minimal model that has local chaotic dynamics but also global interaction between the elements [214]. Although simple, such maps exhibit a rich and complex dynamics. A coupled chaotic field (CCF) is essentially a globally coupled map in which every chaotic element is connected to the embodied system through sensors (supplying the input) and motors (to which the output is relayed). In other words, the global coupling between the chaotic units is provided by the environmental interaction (see Fig. 15.1).

Formally, the neural system can be described as follows:

$$x_i(t+1) = (1-\epsilon) f_\alpha(x_i(t)) + \frac{\epsilon}{N}\sum_{j=1}^{N} f_\alpha(x_j(t)) \tag{15.2}$$

$$f_\alpha(x_i(t)) = 1 - \alpha x_i(t)^2 + \eta F_i(t) \tag{15.3}$$

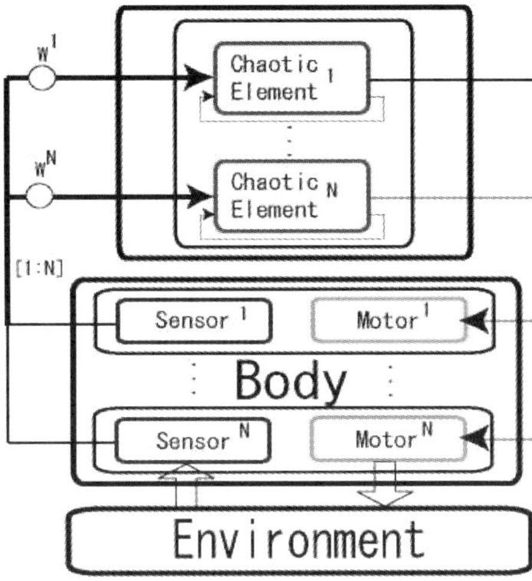

Fig. 15.1 Outline of our model.

$$1 - \alpha x_i(t)^2 >> \eta F_i(t) \text{ with } \alpha \in]0.0, 2.0] \text{ and } \epsilon \in]0.0, 0.4]$$

The parameters α and ϵ are the two control parameters of the CCF, and $f_\alpha(x)$ realizes the chaotic dynamics of the neural unit at time t. The variable α determines the chaoticity of each neural unit and ϵ controls the level of synchronization among the units. In our experiments, the only control parameter was the chaoticity. The level of synchronization was fixed to a small number ($\epsilon < 0.05$) so as not to bias the emergence of synchronized and coherent states. In fact, rather than initializing the synchronization parameter to a specific value among the units and have a static coupling among the chaotic system, the coupling was dynamically altered through internal and external perturbations (here, the CCF was perturbed by the output of force sensors located in the joints). The coupling constant η (eq. 15.2.3) was selected empirically. Its value was small enough so as not too affect too much the intrinsic dynamics of the chaotic units. The output of the CCF was then fed – after appropriate scaling – to the motors (eq.15.1).

By letting neural and body-environment dynamics interact many instances of mutual entrainment among the two dynamical systems could be observed. This kind of mutual interaction has been called global entrainment, and has been hypothesized to generate stable and flexible movements in a self-organized manner despite unpredictable changes in environmental conditions [378].

15.3 Methods

In this section, we introduce the tools used to analyze the data generated in our experiments. The main purpose of our methods is to quantitatively assess behavioral patterns. We describe three methods: (1) the "Spectral Bifurcation Diagram, (2) the "Wavelet Transform", and (3) a novel visualization method inspired by methods 1 and 2, which we shall refer to as the "Wavelet Bifurcation Diagram."

The Spectral Bifurcation Diagram is a recently introduced method for investigating the qualitative changes of the dynamics of high-dimensional systems [296]. Essentially, it displays the power density spectrum of a system variable as a function of a control parameter of the system (the power spectra of the individual variables are superposed). The control parameter is a variable to which the behavior of the system is sensitive and that moves the system through different (attractor) states. The representation illustrates how the neural system affects the coordination among the different parts of the body in frequency space. This method allows to identify resonant states which are characterized by sharp frequency components, chaotic states having rather broad power spectra, as well as bifurcations, that is, abrupt transitions from one attractor state to another.

The second method used was the Wavelet transform [257]. In the Wavelet space, one variable of the system is projected onto a space spanning time and frequency in which it is possible to identify changes of behavior at different time scales, short-range as well as long-range temporal correlations. The Wavelet transform thus allows to analyze temporal bifurcations, and consequently the evolution of the dynamics of each unit of the embedded neural system.

Because we are also interested in understanding the spatial correlations between neural units while performing a certain movement, as well as the type of interactions at different spatio-temporal scales, we introduced a novel tool for the analysis of high-dimensional systems spanning frequency, time, and index of the neural unit. In this space, we are not only able to identify spatio-temporal correlations but also changes over time of each unit or groups of units as well as bifurcations in the dynamics of systems with a large number of degrees of freedom. We refer to this method as Wavelet Bifurcation Diagram.

15.4 Experiments

For our experiments we used a ring-shaped mass-spring robotic system composed of 10 prismatic elements connected by 10 force-controlled sliding

γ	ϵ	η	density	width	damping	stiffness
0.001	0.001	0.01	2.0	0.1	0.3	0.5

joints (Fig. 15.2). Each element was connected to its two neighbours by compression springs. The spring-motor complex had 3 degrees of freedom (DOF), and the complete mechanical system had a total of 30 DOF. The neural and body dynamics were mutually coupled as explained previously (Eqs. 15.1 and 15.3). The robot was placed in a plane devoid of obstacles and of other external perturbations except gravity, ground reaction forces, and friction. We realized our ring-shaped robot using a physics-based simulator (see [355]). Table 15.1 shows the parameters we used for all our simulations. It is important to stress that despite its simplicity, our robot model can display a sufficiently large set of behaviors, and is therefore appropriate for investigating the mechanisms underlying the emergence of stable as well as as unstable behavioral patterns.

Our analysis was mainly focused at understanding how the body dynamics evolved in time as a function of the chaoticity α of the neural units. The initial value of α was zero. For this value the output of the neural system was a constant, and the system did not move. By increasing by a small amount the chaoticity of the neural units ($\alpha \in]0.0, 0.1]$), the "ring"

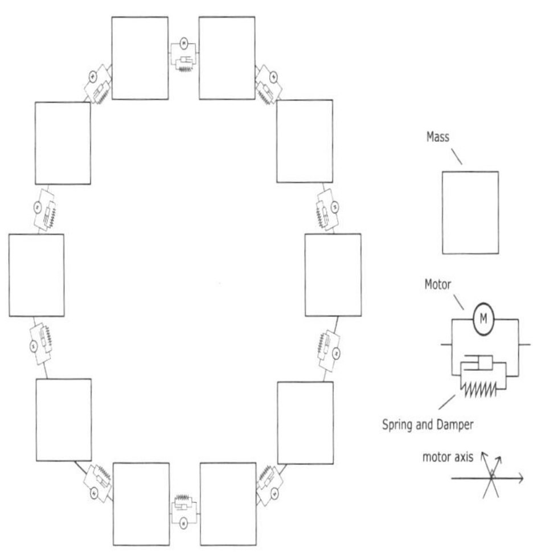

Fig. 15.2 Schematic representation of our robot model. The masses are prisms and are connected to each other by force-controlled compliant linear actuators. The system has 30 degrees of freedom.

started to vibrate almost unperceivably at a very small spatial scale. After a certain amount of time, the seemingly random movements converged to a slow rhythmical rocking movement, that is, the system's dynamics had found a stable attractor. Figure 15.3 shows the phase space trajectories of

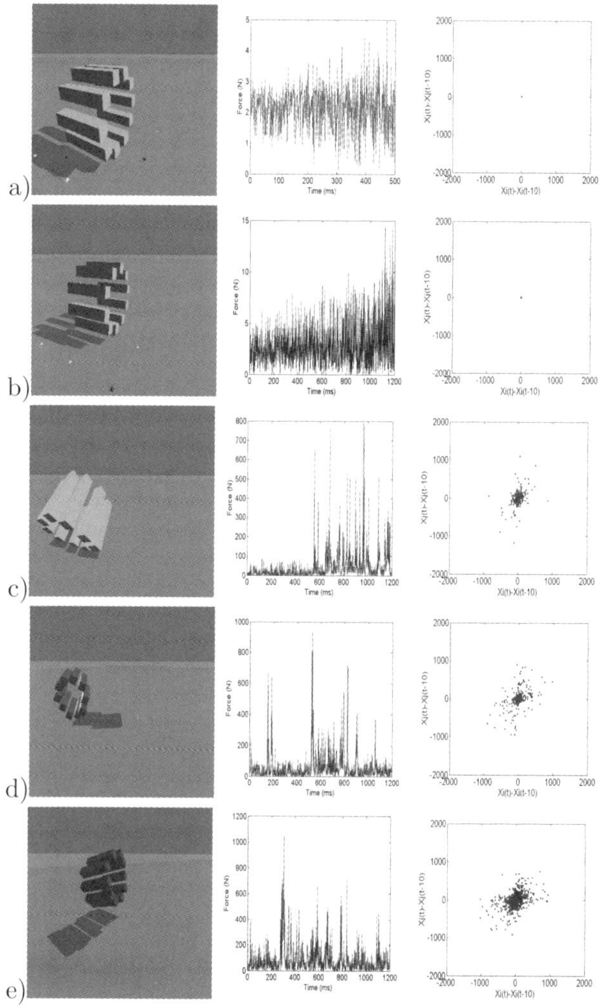

Fig. 15.3 Left: Snapshots of different postures for different α; (a) balance ($\alpha = 0.05$), (b) rolling ($\alpha = 0.10$), (c and d) quick and unstable movement ($\alpha = 1.0$ and $\alpha = 1.5$), and (e) uncoordinated movement pattern ($\alpha = 1.9$). Center: Time series of one joint force pressure (unit: N). Right: Time-delayed phase potraits of two arbitrarily selected neighboring joints as a function of α.

two arbitrarily selected neighbouring joints for different values of α. It plots $X_i(t) - X_i(t - t')$ against $X_j(t) - X_j(t - t')$ giving some information concerning the spatial correlations (X_i vs. X_j) and the temporal correlations between joint variables ($t' = 10$).

Up to a very specific level of chaoticity ($\alpha = 0.097477$), the system stroke a perfectly poised balanced posture, and did not have sufficient energy to start rolling. By slightly increasing the control parameter ($\alpha = 0.1$), a phase transition occurred, and the system started to roll. Interestingly, for values of $\alpha < 0.15$, the system oscillated quite unpredictably between rolling and balancing. We hypothesize that the emergence of particular movement patterns depends on the presence or absence of entrainment between neural and body-environment dynamics.

For levels of neural chaoticity $\alpha \in [0.15; 0.4]$, novel behaviors and patterns of locomotion emerged. The neural system seemed to exploit the natural dynamics of the ring-shaped body to balance, rock, roll, accelerate and decelerate, and in rare occasions, even to jump. We suggest that the unstabilities in the movements patterns were mainly caused by micro-scale perturbations acting on the neural dynamics with a consequent disruption of the entrainment between neural and body-environment dynamics and the emergence of new locomotion patterns.

In the range $\alpha \in [0.4; 1.2]$, the perturbations were larger and thus had a more pronounced effect on the system's dynamics. While still displaying coherence (that is, rather strong correlations between neighbouring segments of the ring), the movements were generally more complex and characterized by abrupt changes. Most notable was the high sensitivity of the system to internal (neural) and external perturbations (due to sensory feedback), and the emergence of a number of different behavioral patterns. The ring rolled quickly, accelerated, decelerated, and displayed many unstable postural configurations (such as balancing on an edge).

Finally, for α in the interval $[1.2; 2]$, the system did not display any coherent or organized movement patterns. The activation levels of the chaotic system were too large to be influenced (perturbed) by sensory feedback causing a disruption of entrainment between neural and body-environment dynamics.

15.5 Analysis

One immediate implication of our experiments is that in an embodied system, the exploration of the space of possible coherent and stable postural modes is induced by the mutual adaptation between neural and body-environment dynamics. In other words, the neural system, the body, and the environment are all responsible for the emergence of particular movement patterns. Note that this global view contrasts with the one that sees only neural parameters responsible for exploration of the movement space.

15.5.1 *Analysis 1: Body movements*

The Spectral Bifurcation Diagram for varying levels of chaoticity is reproduced in Figure 15.4. Low levels of chaoticity ($\alpha < 0.05$) are characterized by sharp peaks in the power spectral density of the force sensors located in the joints, and given a particular value of α the resonance response is close to the one of a damped oscillator. The low frequency component around $10\,Hz$ dominates the interaction dynamics between the neural system and the ring. This frequency corresponds to the fundamental mode of the coupled system, that is, its eigen-frequency. For this frequency, the joints are highly synchronized and the system displays a high degree of coordination. A minimal amount of energy is required to move the system and to transfer energy to the different parts of the body.

When the chaoticity increases higher harmonics appear introducing discontinuities in the resonance response. The main resonance persists for all values of α but we observe abrupt changes and bifurcations in the magnitude of other peaks. The new harmonic peaks are located at integer and fractional multiples of the first eigen-frequency [383]. The latter peaks are caused by small damped actions of the chaotic system and affect the joint properties, in the sense that a change of chaoticity of the neural system can induce a change in the stiffness of the springs in the joints. As a result the system is able to generate a large variety of patterns (stable, weakly stable, and unstable).

When the amplitude of the harmonics is too large, it negatively affects the groups formed in different regions of the body generating decoherence and destroying stable activity patterns. Note that the harmonic states are intrinsic to the coupling between neural, body, and environmental dynamics, and even if the spectral patterns seem complex, they should not be considered to be the outcome of yet another kind of neural noise.

We have previously suggested that behavioral changes are a complex function of the coupling between neural and body-environment dynamics. By using the Spectral Bifurcation Diagram we can now shed light on the patterns of neural activity leading to such changes. For example for a level of chaoticity $\alpha = 0.097477$ (that is, when the ring starts to roll) the power spectrum has a second harmonic which disappears in the interval $[0.1, 0.13]$ (that is, when the system present difficulties to roll again) – see 15.4. The more harmonics there are, the more complex the behavior, despite preservation of coherence of behavior.

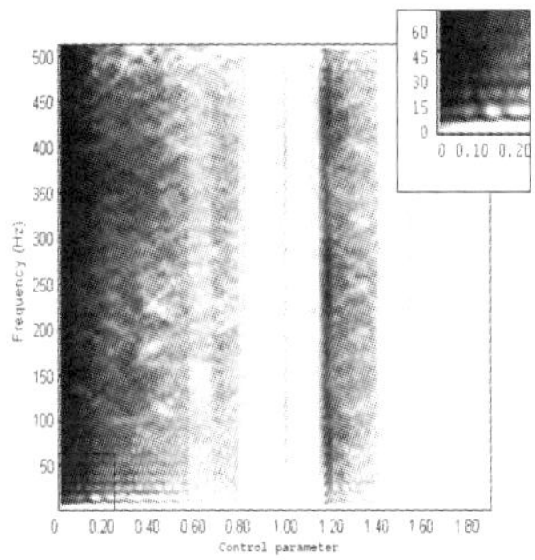

Fig. 15.4 Spectral Bifurcation Diagram. Inset shows spectral peaks for low values of neural chaoticity. The control parameter is α.

15.5.2 *Analysis 2: Neural coupling*

We are interested in the spatio-temporal interaction patterns that emerge in the neural system. To get a better grasp on these patterns, we applied the Wavelet Transform to the activity of an arbitrarily selected chaotic unit for different values of the control parameter (see Figure 15.5). In the Wavelet space, the activities of the units disclose temporal correlations at different scales. The larger is the value of the control parameter, the higher is the complexity of the temporal patterns. Further analysis reveals a scale-free fractal-like structure in the neural activity and temporal coherence bridging multiple time scales. This result can be easily explained by considering the harmonics produced through feedback resonance. Long-range movements resulting from highly chaotic neural activity are composed of short-range movements triggered by lower values of chaoticity (Figure 15.5 c).

One surprising effect is that the higher harmonics can dynamically alter the stiffness of the springs and thus their temporal responses. "Positive" resonance hardens the springs, and "negative" resonance softens them (see also [383]). We note also that the temporal scale of correlations seem to be quantified with respect to the control parameter (see horizontal lines in Figures 15.5 a-d). In other words, changes of "locomotion pattern" occur for specific values of neural chaoticity.

The Wavelet Bifurcation Diagrams of the neural activity for different values of the control parameter are shown in Figures 15.6 and 15.7. Through the mutual interactions between the neural sytem and the body, groups of neural units synchronize at multiple spatial scales (vertical axes of the figures). As in the case of the Wavelet Transform, the scale at which synchronization takes place and the type of emerging patterns depends on the amount of chaoticity in the neural system.

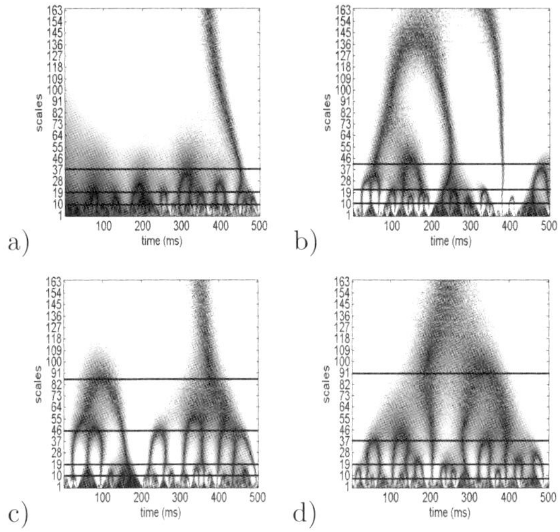

Fig. 15.5 Wavelet Transform. The spectro-temporal correlations between chaotic units are displayed. The control parameter α varies linearly in the time interval 0-500 msec from (a) [0.0;0.5], (b) [0.5;1.0], (c) [1.0;1.5] and (d) [1.5;2.0].

Interactions between neural system and body for low values of chaoticity form long-range correlations at a low spatial scale (rolling and balancing behavior). The "low-scale" spatio-temporal patterns correspond to disconnected short-range movements and the "high-scale" ones correspond to long-range movements. For higher chaoticity values, the same fractal-like spatio-temporal organizations are formed. Note that the control parameter is correlated to the complexity degree of the emergent behaviors. For low values of α, stable activity groups of neurons are generated in the chaotic system. For higher values of α, we can observe chaotic itinerancy in the system with an higher spatio-temporal complexity structure in the units. The groups are unstable and bifurcate to new transient configurations.

15.6 Discussion and Conclusion

In this paper, we introduced and discussed a novel framework for exploring action possibilities of complex mechanical systems. In particular, we studied (a) how chaotic neural activity can drive the exploration of movement patterns, and (b) how feedback resonance can be used to "tune into" particularly efficient movements. We also provided a set of tools to quantitatively measure the spatio-temporal organization of the neural system,

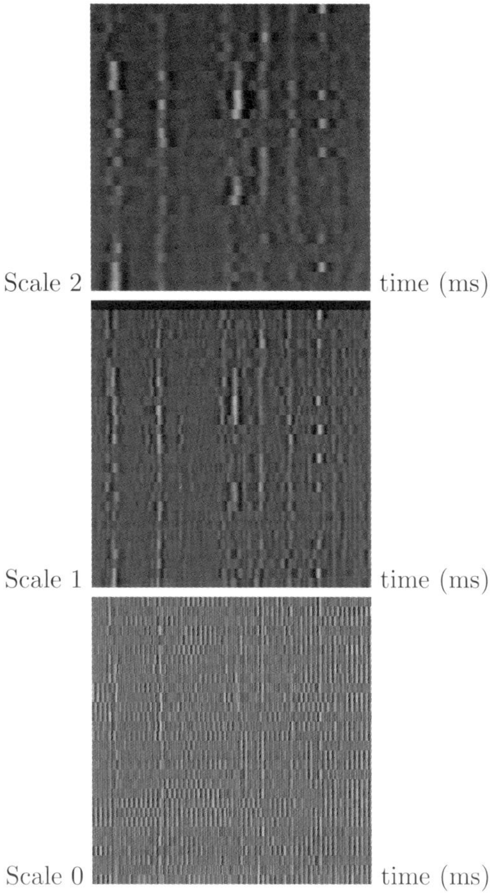

Fig. 15.6 Wavelet bifurcation diagram at different scales. In all three scale-plots, the horizontal axis denotes time (0-500 msec), and the vertical axis is the index of the chaotic unit.

and the stability of the emerging behavioral patterns.

We suggest that resonance plays a pivotal role for learning to control our bodies. Resonant states act as some kind of amplifier guiding the exploration and discovery of intrinsic modes of the body dynamics. One important side-result is the reduction of the number of degrees of freedom despite an increase in the overall complexity of the system. Another result is that resonance pushes the compliant actuators composing the body to dynamically alter their properties (e.g. stiffness) and to cooperate. In a sense, resonance also satisfies the principle of cheap design [303]. This principles states that when designing a system it is better to exploit physics and a)

b)

c)

d)

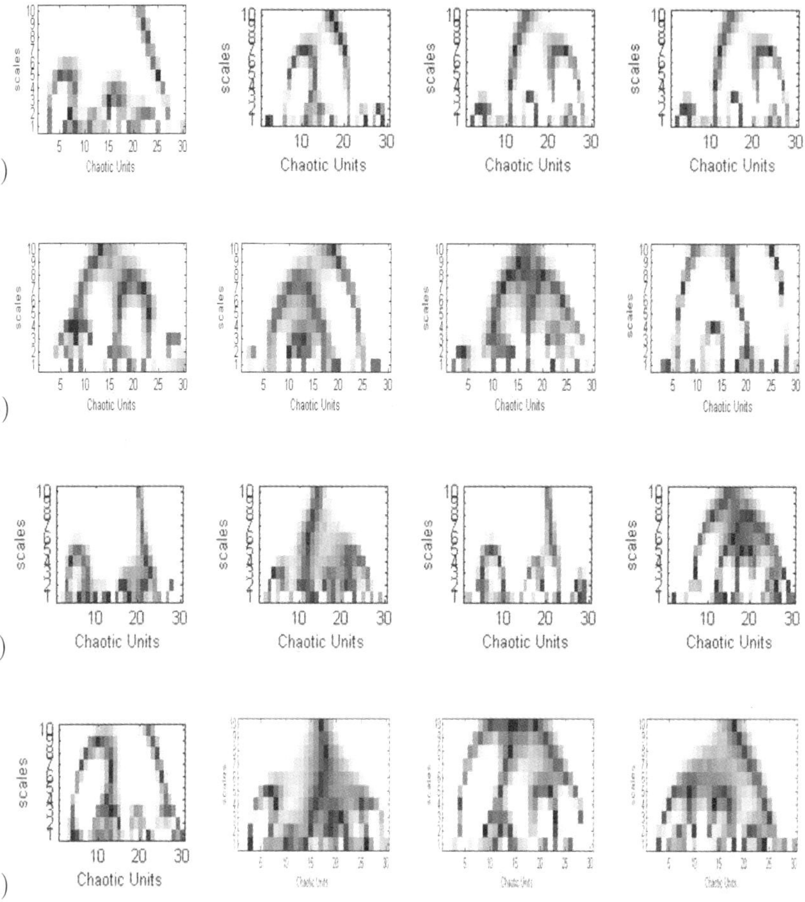

Fig. 15.7 Wavelet bifurcation diagram at different time instants. Snapshots of different neural configurations taken at different time instants for varying levels of chaoticity: a) $\alpha = 0.5$, b) $\alpha = 1.0$, c) $\alpha = 1.5$, d) $\alpha = 1.9$. The horizontal axis denotes the index of the chaotic unit, the vertical one is the scale (adimensional).

the dynamics of the system-environment interaction. Mapped onto our case study it means that resonance guarantees the emergence of energy-efficient movement patterns. As for learning or planning, this property can also be useful to understand when to increase or decrease the coupling between parts of the body, and to understand which parts have to be linked rather than testing all possible combinations. In addition, critical states (e.g. corresponding to unstable activity patterns or states where bifurcations occur) can also be identified and analyzed. We hypothesize that a mechanism of feedback resonance is responsible for combining unstable short-range patterns into stable long-range ones.

In future works we intend to implement our exploration model in a real robot situated in a dynamic environment. The robot will hopefully autonomously explore its body, and over time acquire a repertoire of complex, adaptive, and highly dynamic movements.

15.7 Acknowledgements

The authors would like to thank the Japan Society for the Promotion of Science for funding, as well as K. Shiozumi and S. Suzuki for valuable discussions.

Chapter 16

A Dynamic Optimisation Approach for Ant Colony Optimisation Using the Multidimensional Knapsack Problem

M. Randall

School of Information Technology
Bond University, QLD 4229, Australia
E-mail: mrandall@bond.edu.au

Meta-heuristic search techniques have been extensively applied to static optimisation problems. These are problems in which the definition and/or the data remain fixed throughout the process of solving the problem. Many real-world problems, particularly in transportation, telecommunications and manufacturing, change over time as new events occur, thus altering the solution space. This paper explores methods for solving these problems with ant colony optimisation. A method of adapting the general algorithm to a range of problems is presented. This paper shows the development of a small prototype system to solve dynamic multidimensional knapsack problems. This system is found to be able to rapidly adapt to problem changes.

16.1 Introduction

Many industrial optimisation problems are solved in non-static environments. These problems are referred to as *dynamic optimisation problems* and are characterised by an initial problem definition and a series of "events" that occur over time. An event defines some change either to the data of the problem or its structural definition. A dynamic operating environment can be modelled in two ways. Either a series of events may be determined *a priori* in which case the optimum solution can be pre-determined, or a probabilistic discrete event simulator is used to create events. While the latter mimics real world problems more closely, the former is better for testing and development purposes.

In comparison to static optimisation problems, dynamic optimisation problems often lack well defined objective functions, test data sets, criteria for comparing solutions and standard formulations [32; 114; 341]. At the present time, the main strategies used to solve these problems have been specialised heuristics, operations research and manual solving [33; 341]. Commercial implementations of operations research software can also be used by providing additional constraints at each problem change in order to lock some existing solution components in place. An example of manual solving is given by aircraft controllers at London's Heathrow airport determining appropriate aircraft landing schedules [33].

Evolutionary algorithms such as genetic algorithms have been modified to accommodate dynamic optimisation problems. A survey of these approaches is given by Branke [56]. However, for a group of meta-heuristic strategies known collectively as *Ant colony Optimisation* (ACO) [110], relatively little work has been done in this area. This is despite the fact that ACO offers possibilities in this direction. Natural ants are adaptive agents that must source food in a continually changing environment and supply this back to a nest. Food sources constantly change, obstacles appear and disappear and new routes become available, while old routes become cut off or impractical to traverse.

One of the major differences between ACO meta-heuristics and other evolutionary algorithms is that they are constructive, i.e., they build solutions a component[1] at a time. This paper therefore develops new generalisable strategies that are applicable to ACO. In terms of artificial ant systems, some successful work (in the main) has already been done on producing specialised ACO implementations for the travelling salesman problem [16; 123] and telecommunication and routing problems (see Dorigo, Di Caro and Gambardella [111] for an overview). This paper, however, looks at the issue related to adapting the standard algorithm to suit a wide range of combinatorial optimisation problems. It is an initial investigation demonstrating the effective implementation of a system to solve dynamic multidimensional knapsack problems .

Another attempt to generalise ACO for dynamic problems has been made. Population ACO (P-ACO) [165; 166] is an ACO strategy that is capable of processing dynamic optimisation problems. It achieves this by using a different pheromone updating strategy. Only a set of 'elite' solutions are used as part of the pheromone updating rules. At each iteration, one solution leaves the population and a new one (from the current

[1] A solution component is the building block of the solution. Two examples are a city for the travelling salesman problem and a facility for the quadratic assignment problem.

iteration) enters. The candidate solution to be removed can be selected on age, quality, a probability function or a combination of these factors. The authors argue that this arrangement lends itself to dynamic optimisation as extensive adjustments (due the problem change) need not be made to the pheromone matrix. Instead, a solution modified to suit the new problem is used to compute the new pheromone information. This modification process is a heuristic called *KeepElite* [167] that works for full solutions only and is tailored for particular problems.

The approach proposed herein allows for the automatic adaptation of a solution to suit a changed version of the problem. The main difference between this and P-ACO is that it incorporates a general process that adapts partial solutions to new problem definitions. This has the advantage that the solver can process a change to the problem at any time without having to wait for the ants' constructive process to finish.

The remainder of the paper is organised as follows. Section 16.2 describes how standard ACO may be adapted to solve dynamic optimisation problems. Preliminary results of this approach (using the dynamic multi-dimensional knapsack problem) are given in Section 16.3 while Section 16.4 gives the conclusions of this work. Note that a thorough description of the ACO meta-heuristic (and its variants) is given by Dorigo and Di Caro [110].

16.2 Adapting ACO to Dynamic Problems

16.2.1 *Overview*

Beasley, Krishnamoorthy, Shariah and Abramson [32] describe a generic displacement model that allows a static integer linear program (ILP) to be transformed into another static ILP, given that some change takes place to the original ILP. The model encourages (by the use of penalty functions) new solutions to be close to the previous solution. The rationale is that high quality solutions to the new problem should be relatively close to those of the old problem as usually only a small structural/data change to the problem has been made. The ideas presented as part of the displacement problem can be adapted to form part of a generic strategy that is suitable for ACO. The following stages define a framework to solve dynamic optimisation problems in a real-time/production environment for constructive meta-heuristics (in particular ACO). The following is a set of generic rules, the details of which (such as when to use improved solutions in the production environment) may vary according to industry and management strategy.

(1) While no changes occur to alter the problem definition/data, allow

the meta-heuristic to solve the problem. Improved solutions are sent to the production environment. For instance, at an airport the production system would correspond to the control tower while for dynamic vehicle routing it would correspond to the dispatch section of the depot. As it may be costly or impractical to "reset" the production environment too frequently, the above may be done in a number of ways. For instance, the system may receive the revised solution at certain time intervals or when a significant improvement has been found (bounded by either an absolute or relative amount).

(2) Suspend Stage 1 *if* either a number of events has occurred or an event of sufficient magnitude has occurred. This is very environment specific and in an industrial setting would need to be decided with management input. For instance, given the problem of scheduling aircraft to land on a runway, how many aircraft would need to enter the airspace of an airport before landing schedules need to be changed?

(3) Use a *solution deconstruction* process to determine which components of an ant's solution need to be discarded so that the feasibility of the "new" part of the problem is not violated. The ACO process is then restarted from this partial solution. It is best only to perform this deconstruction process on one solution as executing it on a population of solutions may yield different length partial solutions. This is potentially computationally costly and means that ants could not synchronously complete their solutions within an iteration. The most reasonable solution on which to perform this process would be the best solution from the current colony. This new partial solution is then copied to all the ants in the colony for the next iteration. This process is described in detail in Section 16.2.2.

(4) Go back to Stage 1.

It is critical that Stage 3 is very efficient. This is because the production environment must stop or be put in some 'holding pattern' while the solution deconstruction process and new set of solutions are being computed and evaluated respectively. However, the advantage is that the solver system does not have to be stopped and restarted from the beginning. The existing "best" solution to the problem should serve as a good starting point to the new problem (as has been found by Beasley et al. [32]).

16.2.2 *The Solution Deconstruction Process*

When an event occurs, it is probable that current solutions will become infeasible. However, as an event will typically change only part of the problem data/structure, it is likely that current solutions will only require small modifications to make them feasible [165; 166]. Rather than restarting the solution construction process, it is possible to modify a partial solution instead. In this way, it is similar to the Beasley et al.'s [32] process as it ensures that the solution to the modified problem will be relatively close to the previous best solution. The solution deconstruction technique progressively removes components from a solution until it becomes a feasible solution to the changed problem. In the generalised form presented herein, deconstruction is applicable to a wide range of dynamic combinatorial optimisation problems (some examples being dynamic travelling salesman and vehicle routing problems).

The first part of the deconstruction process determines which constraints need to be satisfied at each step of the construction process and which constraints need only be satisfied on the termination of the process [321]. More specifically, the two types of constraints can be described as:

(1) *Constraints that must be satisfied at each step.* An example of such a constraint is a knapsack capacity constraint of the form of $\sum_{i=1}^{N} C_i x_i < b$. Regardless of the number of steps that have elapsed within an iteration, it is always necessary to satisfy such a constraint (given positive C coefficients and a positive value of b).
(2) *Constraints that cannot be satisfied at each step of the algorithm.* Consider the case where a constraint of the form of $\sum_{i=1}^{N} C_i x_i > b$ is present in the problem model. It is apparent that in the beginning stages of the construction, the length of the solution vector is small and it is unlikely that this constraint will be satisfied. Hence, these types of constraints must be treated as special. Only once these types of constraints have been satisfied can the algorithm consider the termination of the solution augmentation process.

The analysis to determine which constraints fall into which categories can be done before the dynamic optimisation problem is solved (i.e., before Stage 1 from Section 16.2). If an event occurs that adds a constraint (within Stage 2), this analysis will need to be partially redone. Algorithm 16.1 shows the generic deconstruction algorithm. In this algorithm, only constraints of type 1 need to be processed. This algorithm simply takes the most recently added solution component and removes it, calculating the amount of feasibility violation. If this increases the infeasibility, the component is added back and the next newest component is tried. This process is repeated until a feasible solution to the new problem is produced (i.e., the value of the constraints' violation is 0). The algorithm has a complexity of $O(aM)$, where a is the length of the partial solution and M is the number

Algorithm 16.1 The solution deconstruction process. Note: 'constraints' refer to constraints of type 1 and X is the solution.

```
c = Calculate the amount of constraint violation of X
if c > 0 then
    a = the length of X
    X' = X
    while c > 0 do
        component = X'(a)
        X' = Remove component from X'
        c_prev = c
        c = Calculate the amount of constraint violation of X'
        if c > c_prev then
            X' = Add component to X'
        end if
        a = a - 1
    end while
    X = X'
end if
```

of constraints.

Constraint violation is calculated according to the relational operators that are present in the constraints [321]. For instance, if the sign of a constraint is $\leq$ and the left-hand side is larger than the right-hand side, the net difference is the amount of constraint violation. This is shown in Equation 16.1. The constraint violations (16.2-16.6) for the other signs are calculated in a similar manner. The sum of the constraints' violations is given in Equation 16.7.

$$(\leq) \qquad c_i = \mathrm{MAX}(0, lhs_i - rhs_i) \tag{16.1}$$

$$(<) \qquad c_i = \mathrm{MAX}(0, lhs_i - rhs_i + 1) \tag{16.2}$$

$$(\geq) \qquad c_i = \mathrm{MAX}(0, rhs_i - lhs_i) \tag{16.3}$$

$$(>) \qquad c_i = \mathrm{MAX}(0, rhs_i - lhs_i + 1) \tag{16.4}$$

$$(=) \qquad c_i = |lhs_i - rhs_i| \tag{16.5}$$

$$(\neq) \qquad c_i = \begin{cases} \text{if } (rhs_i = lhs_i) \; 1 & \\ 0 & \text{otherwise.} \end{cases} \tag{16.6}$$

$$v = \sum_{i=1}^{M} c_i \tag{16.7}$$

Where:

M is the number of constraints,
$|a|$ is the absolute value of a,
$\text{MAX}(a, b)$ returns the larger value of a and b,
c_i is the constraint violation of constraint $i, 1 \leq i \leq M$,
lhs_i is the evaluation of the left hand side of constraint $i, 1 \leq i \leq M$,
rhs_i is the evaluation of the right hand side of constraint $i, 1 \leq i \leq M$ and
v is the total amount of constraint violation.

16.2.2.1 *Event Descriptors*

Algorithm 16.1 is a naïve heuristic that does not take into account the type of change made to the problem. As such, it is likely that that the algorithm will remove more components than is necessary to provide a new feasible partial solution. In order to ensure that deconstruction is efficient, the nature of possible events must be defined. These *event descriptors* can be used in conjunction with the algorithm in Algorithm 16.1 to efficiently adapt a solution to suit the altered problem (i.e., determine the most appropriate component to remove). An event descriptor may be any of:

- `ADD_COMPONENT`
- `REMOVE_COMPONENT`
- `ADD_CONSTRAINT`
- `REMOVE_CONSTRAINT`
- `MODIFY_CONSTRAINT`
- `MODIFY_OBJECTIVE`

The last descriptor is superfluous (but is present for completeness) as such an event will not change the feasibility of the solution, only its cost. An example of an event descriptor for the previously described knapsack event

Algorithm 16.2 The modified solution deconstruction process to include event descriptors.

$c =$ Calculate the amount of constraint violation of X
if $c > 0$ **then**
  $a =$ the length of X
  $X' = X$
  while $c > 0$ **do**
    component $=$ evaluate event descriptor($event, X$)
    $X' =$ Remove *component* from X'
    $c_{prev} = c$
    $c =$ Calculate the amount of constraint violation of X'
    if $c > c_{prev}$ **then**
      $X' =$ Add *component* to X'
    end if
    $a = a - 1$
  end while
  $X = X'$
end if

would be: "`On REMOVE_COMPONENT` a,remove $+a$ `from` X". This has the advantage of being an $O(1)$ operation.

The modified deconstruction algorithm is given in Algorithm 16.2. In a general solver system, the description could be implemented as a set of high level rules, without the need to reprogram the solver.

16.3 Computational Experience

In order to test the concepts outlined in the previous section, a relatively straightforward problem has been chosen rather than an industrial application (which is in the future development of this project). The multidimensional knapsack problem (MKP) [80] is an extension of the classical knapsack problem. In it, a mix of items must be chosen that satisfy a series of weighting constraints whilst maximising the collective utility of the items. Equations 16.8 - 16.10 show the 0-1 ILP model.

$$\text{Maximise} \sum_{i=1}^{N} P_i x_i \tag{16.8}$$

s.t.

$$\sum_{j=1}^{N} w_{ij}x_j \leq b_i \quad \forall i \quad 1 \leq i \leq M \tag{16.9}$$

$$x_i \in \{0,1\} \quad \forall i \quad 1 \leq i \leq N \tag{16.10}$$

Where:

x_i is 1 if item i is included in the knapsack, 0 otherwise,
P_i is the profit of including item i in the knapsack,
N is the total number of items,
w_{ij} is the weight of item j in constraint i,
M is the number of constraints, and
b_i is the total allowable weight according to constraint i.

In order to make this problem dynamic, MKP problem instances from OR-Library [31] have been adapted (hereafter referred to as *variants*). This has been done by taking the instance of the problem and modifying it in the following six ways:

(a) *No Change* - The original problem.
(b) *Modify the capacity of each constraint* - This is achieved by setting each capacity to 80% of its original level.
(c) *Remove a constraint* - Removes a particular constraint.
(d) *Add a constraint* - Adds a particular constraint.
(e) *Remove an item* - Removes a particular item from the item mix.
(f) *Add an item* - Adds a particular item to the item mix.

Any and all of these changes occur within a particular run of the ACO solver. They take place at the step level of the ACO algorithm and the likelihood of a problem change is given by a Poisson distribution. To calculate the number of events that will occur within an interval (defined by a number of ACO steps) ζ, the average number of events that will occur within this period, λ, is required. From this, the probability of an event occurring at each ACO step can be calculated. Within the solver system, this probability is generated each ζ steps.

The simulated environment notifies the solver (i.e., the ACO engine) if a change to the problem has been made. As the problem variants are fixed, the results can be compared to a standard (static) ACO solver. The problem instance descriptions[2] are as follows:

[2]These are available online at http://www.it.bond.edu.au/randall/dynmkp.tar

Problem Name	N	M	Optimal Value
mknap1	6	10	3800
mknap2	20	10	6120
mknap3	28	10	12400
mknap4	39	5	10618
mknap5	50	5	6339
mknap6	60	5	6954

The only event descriptor that will be required is: "`On REMOVE_COMPONENT, remove`+$N+1$ `from` X". It is necessary because the identifier number of each knapsack item will be decremented for all the items above the item dropped. Therefore, if item $N+1$ (where N is the number of items in the new problem) is present in the solution, it is no longer valid and must be removed to ensure a feasible solution.

The computing platform used to perform the experiments is a Sun Ultra 5 (rated at 400 MHz). Each problem instance is run across 10 random seeds. The Ant Colony System (ACS) approach [112] will be used as the solver. Its parameter settings are as follows: $\tau_0 = 0.01, \beta = 2, \gamma = 0.1, \rho = 0.1, q_0 = 0.9, \zeta = 200, \lambda = 5, ants = 10, iterations = 3000$. Note that the ACO parameters τ_0, γ, ρ, Q and q_0 are defined in Dorigo and Gambardella [112].

In order to describe the range of objective costs gained by these experiments, the minimum (denoted "Min"), median (denoted "Med"), maximum (denoted "Max") and Inter Quartile Range (denoted "IQR") are given. Non-parametric descriptive statistics are used as the data are highly non-normally distributed. The first half of Table 16.1 shows the results of solving each of the five variants (and the original) of the test problems. This was done using the static version of the ACS solver. The second half shows the results of the dynamic system. In it, the best objective cost achieved for each variation of each problem is recorded. To determine the degree of adaptation from one problem variant to the next, the number of ACS iterations between each change and attaining the best objective cost (within that period) is also recorded.

The results from Table 16.1 indicate that the dynamic ACS engine can very quickly adapt to a new problem, usually within a small number of iterations. In some cases, after the solution deconstruction procedure has been performed, ACS constructs a solution whose cost is the best known for that particular problem within that iteration.

Unlike the other problems, improved objective costs were received for mknap5 and mknap6 using the dynamic solver over the static solver. A possible reason for this is that the change in pheromone values due to changing problem definitions may allow ants to become free of attractive

Table 16.1 The results of solving each of the test problems (and their variants) using the static and dynamic solvers.

Problem	Variants	Static Solver Objective Costs				Dynamic Solver Best	Trials			
		Min	Med	Max	IQR	Cost	Min	Med	Max	IQR
mknap1	a	3800	3800	3800	0	3800	0	1	10	2
	b	3300	3300	3300	0	3300	0	1	11	2
	c	3300	3300	3300	0	3300	0	1	12	2
	d	2400	2400	2400	0	2400	0	2	20	3
	e	3700	3700	3700	0	3700	0	1	11	1
	f	3900	3900	3900	0	3900	0	1	18	1
mknap2	a	6120	6120	6120	0	6120	0	0	4	1
	b	5200	5200	5200	0	5200	0	1	7	1
	c	4475	4475	4475	0	4475	0	0	2	0
	d	6450	6450	6450	0	6450	0	0	3	1
	e	6120	6120	6120	0	6120	0	0	4	1
	f	6120	6120	6120	0	6120	0	0	3	1
mknap3	a	12380	12380	12380	0	12380	0	0	1	0
	b	10750	10750	10750	0	10750	0	0	4	0
	c	10310	10310	10310	0	10310	0	0	1	0
	d	9380	9440	9440	0	9380	0	0	5	1
	e	11950	11970	11970	0	11970	0	0	1	0
	f	13070	13070	13070	0	13070	0	0	2	0
mknap4	a	9909	9981.5	10130	97	10120	0	0	1	0
	b	4260	4260	4260	0	4260	1	3	41	4
	c	9936	10080	10202	178.25	10084	0	0	1	0
	d	9909	10061	10157	90	10141	0	0	0	0
	e	9723	9730	9771	5.25	9777	0	0	1	0
	f	11533	11546	11600	47	11534	0	0	2	0
mknap5	a	6285	6285	6285	0	6339	0	0	1	0
	b	5739	5739	5739	0	5810	0	0	1	0
	c	5389	5389	5389	0	5389	0	0	3	1
	d	6285	6285	6285	0	6339	0	0	1	0
	e	6110	6110	6110	0	6110	0	0	1	0
	f	6711	6711	6711	0	6765	0	0	1	0
mknap6	a	6923	6923	6923	0	6954	0	0	1	0
	b	6161	6176	6193	17	6250	0	0	1	0
	c	6923	6923	6923	0	6954	0	0	1	0
	d	6665	6665	6665	0	6783	0	0	1	0
	e	6765	6765	6765	0	6765	0	1	1	0
	f	7101	7101	7101	0	7101	0	0	1	0

local optima. A similar effect has been observed in Kawamura, Yamamoto and Ohuchi [220].

16.4 Conclusions

This paper is a preliminary investigation that has presented methods for generalising ACO so that it can solve dynamic optimisation problems. It was shown that a modified ACS engine, that incorporates solution deconstruction and event descriptors, can solve dynamic multidimensional knapsack problems, within the framework given in Section 16.2. In many cases, the ACS engine can adapt to a change in problem definition within an ant colony iteration. The dynamic solver also managed to receive better solution objective costs than the static solver on the largest problems. The solution deconstruction and subsequent adaptation is very efficient for this class of problems.

Due to the complimentary nature of P-ACO and the new approach, it would be interesting to combine P-ACO's pheromone updating mechanism with solution deconstruction and event descriptors. Implicit in this would be to conduct performance comparisons between the new approach, P-ACO and the hybrid of both. Beyond this, the ultimate aim is to apply these techniques to industrial dynamic problems using a probabilistic event simulator. A suitable application would be the aircraft landing problem [32; 322]. From here, a generic system capable of processing a range of problem definitions and utilising other constructive meta-heuristics (such as GRASP [125]) will be developed.

Chapter 17

Maintaining Explicit Diversity Within Individual Ant Colonies

M. Randall

School of Information Technology
Bond University, QLD 4229
Australia
E-mail: mrandall@bond.edu.au

Natural ants have the property that they will follow one another along a trail between the nest and the food source (and vice versa). While this is a desirable biological property, it can lead to stagnation behaviour within artificial systems that solve combinatorial optimisation problems. Although the evaporation of pheromone within local update rules, mutating pheromone values or the bounding of pheromone values may alleviate this, they are only *implicit* forms of diversification within a colony. Hence, there is no guarantee that stagnation will not occur. In this paper, a new *explicit* diversification measure is devised that balances between the restriction and freedom of incorporating various solution components. In terms of the target applications, the travelling salesman problem and quadratic assignment problem, this form of diversification allows for the comparison of sequences of common solution components. If an ant is considered too close to another member of the colony, it is explicitly forced to select another component. This restriction may also be lifted if necessary as part of the aspiration criteria. The results reveal improved performance over a control ant colony system.

17.1 Introduction

Ant Colony Optimisation (ACO) represents a group of powerful techniques for solving combinatorial optimisation problems (COPs) [110]. A property of these systems is that ants will be attracted to select solution components[1] that are associated with a combination of large pheromone value and high quality. While this helps to exploit promising areas of the solution space, it may also lead to a lack of diversity within explored solutions.

Ensuring that solutions within a colony are sufficiently different in order to adequately sample the search space of a COP is essential to the operation of ant colony techniques. If this is not achieved, ants will tend to produce the same small set of solutions, thus leading to stagnation behaviour. In essence, this becomes an inefficient utilisation of computational resources. ACO meta-heuristics have in some way addressed this problem by using implicit means based on pheromone. For example, ant colony system (ACS) has a local update rule that evaporates the pheromone of frequently incorporated solution components while $\mathcal{MAX} - \mathcal{MIN}$ Ant System limits the allowed range of possible trail strengths between problem specific constants τ_{min} and τ_{max} [374]. Additionally, $\mathcal{MAX} - \mathcal{MIN}$ Ant System uses a technique known as pheromone trail smoothing. This goes further by redistributing pheromone when the trail strength limit has been exceeded [375]. Another ant colony meta-heuristic, Best-Worse Ant System [84; 85], mutates the pheromone matrix with a low probability. Over time the strength of the mutation increases to encourage diversity. Additionally, it will periodically reset values of the pheromone matrix to the initial value if the makeup of the best and worse solutions becomes too close.

In this paper, a new form of explicit diversification within colonies is proposed. It uses the tabu search notion of balancing the restriction and freedom of incorporating components into solutions to achieve this. For example, a solution component (for an ant) becomes tabu if it gives the same partial solution as another ant. The restriction can be lifted (through an aspiration function) if the partial solution is better than that obtained previously or if no other choices are available. The experimental evidence herein shows that more diversification is achieved within individual colonies and subsequently improved solution costs are obtained (over a control strategy).

The remainder of the paper is organised as follows. Section 17.2 gives a technical description of one of the ACO meta-heuristics, ACS, while Section 17.3 describes some of the existing diversification approaches that have been used for ACO. Section 17.4 outlines the mechanics of a new colony diversification measure while Section 17.5 describes an implementation of the scheme that is suitable for the travelling salesman problem (TSP) and quadratic assignment problem (QAP). Additionally, a comparison is made against a control technique. Finally, Section 17.6 provides the conclusions

[1]A solution component is the building block of the solution. Two examples are a city for the travelling salesman problem and a facility for the quadratic assignment problem.

and suggests some extensions to this new technique.

17.2 Ant Colony System

ACS is one of the common varieties of ACO that has been shown to be a robust optimisation technique [112; 188]. The following is a description of the operation of ACS in terms of one of the test problems, the TSP.

Consider a TSP with N cities. Each pair of cities r and s is separated by distance $d(r,s)$. Place m ants randomly on these cities. In discrete time steps, all ants select their next city s and then simultaneously move to their next city. Ants deposit a substance known as *pheromone* to communicate the utility (goodness) of the edges to the rest of the colony. The quality of a solution component is given by the visibility heuristic $\eta(r,s)$ (for this problem $\eta(r,s) = d(r,s)$). Denote the accumulated strength of pheromone on edge (r,s) by $\tau(r,s)$.

At the commencement of each time step, Equations 17.1 and 17.2 are used to select the next city s for ant k currently at city r. Equation 17.1 is a greedy selection technique that will choose the city that has the best combination of short distance and large pheromone levels. Using the first branch of Equation 17.1 exclusively will lead to sub-optimal solutions due to its greediness. To compensate, there is a probability that Equation 17.2 will be used to select the next city instead. This equation generates a probability for each candidate city and then roulette wheel selection is used to select s.

$$s = \begin{cases} \arg\max_{u \in J_k(r)} \left\{ \tau(r,u)[\eta(r,u)]^{\beta} \right\} & \text{if } q \le q_0 \\ \text{Equation 17.2} & \text{otherwise} \end{cases} \tag{17.1}$$

$$s = R\left(\begin{cases} \frac{\tau(r,s)[\eta(r,s)]^{\beta}}{\sum_{u \in J_k(r)} \tau(r,u)[\eta(r,u)]^{\beta}} & \text{if } s \in J_k(r) \\ 0 & \text{otherwise} \end{cases} \right) \tag{17.2}$$

Note that $q \in [0,1]$ is a uniform random number, q_0 is a parameter and R represents the roulette wheel selection function. To maintain the restriction of unique visitation, ant k is prohibited from selecting a city that it has already used. The cities that have not yet been visited by ant k are indexed by $J_k(r)$. As TSP is a minimisation problem, β is negative so that shorter edges are favoured. The use of $\tau(r,s)$ ensures preference is given to edges that are well traversed (i.e., have a high pheromone level). The pheromone level on the selected edge is updated according to the local updating rule in Equation 17.3. This has the effect of decreasing the pheromone level slightly on the selected edge to ensure some implicit measure of diversification amongst the members of the colony is achieved.

$$\tau(r,s) \leftarrow (1-\rho)\cdot\tau(r,s) + \rho\cdot\tau_0 \tag{17.3}$$

Where:

(r, s) is an edge within the TSP graph,
ρ is the local pheromone decay parameter, $0 < \rho < 1$ and
τ_0 is the initial amount of pheromone deposited on each of the edges.

Upon conclusion of an iteration (i.e., once all ants have constructed a solution), global updating of the pheromone takes place. Edges that compose the best solution (so far) are rewarded with an increase in their pheromone level while the pheromone on the other edges is evaporated (decreased). This is expressed in Equation 17.4.

$$\tau(r,s) \leftarrow (1-\gamma)\cdot\tau(r,s) + \gamma\cdot\Delta\tau(r,s) \tag{17.4}$$

$$\Delta\tau(r,s) = \begin{cases} \frac{Q}{L} \text{ if } (r,s)\in S \\ 0 \ \text{ otherwise.} \end{cases} \tag{17.5}$$

Where:

$\Delta\tau(r, s)$ is used to reinforce the pheromone on the edges of the iteration best solution (see Equation 17.5),
L is the length of the best (shortest) tour to date while Q is a constant,
γ is the global pheromone decay parameter, $0 < \gamma < 1$ and
S is the set of edges that comprise the best solution found to date.

It is typical that a local search phase is performed on each ant's solution, before global pheromone updating takes place.

17.3 Explicit Diversification Strategies for ACO

There have been a few attempts to explicitly incorporate diversification strategies into ACO techniques. Gambardella, Taillard and Dorigo [137] propose a hybrid ant system in which the usual constructive component is replaced by local search instead. It has been implemented for the QAP and is subsequently known as HAS-QAP. It incorporates simple intensification and diversification processes into the algorithm. HAS-QAP begins each iteration with a complete solution (rather than constructing it). In diversification phases, both the pheromone matrix and the initial solution are reinitialised. In the case of the latter, the solution is a random solution (i.e., a random permutation for the QAP). Blum [48] uses a similar approach to the above by having a number of restart phases and resetting the pheromone values to random levels.

Hendtlass' [187] Ant Multi Tour System (AMTS) achieves a measure of diversification by allowing ants to retain a memory of TSP tours they constructed in previous generations. A weighted term in the solution component selection equations is used to discourage ants from choosing previously incorporated components. In a case study of a 14 city problem, it was shown that AMTS could outperform implementations of ACS and $\mathcal{MAX} - \mathcal{MIN}$ Ant System.

Randall and Tonkes [325] outline a scheme based on the ACO metaheuristic ACS in which the characteristic component selection equations (Equations 17.1 and 17.2 herein) are modified so that the level of pheromone, in relation to the heuristic information, is varied. The premise is that solution components having higher pheromone levels have shown in the past to be attractive and vice-versa. During a diversification phase, components with large amounts of pheromone are actively discouraged. There was however no statistically significant difference in solution quality between the intensification/diversification schemes and a control ACS strategy. Meyer [277] has extended this idea by introducing the concept of "α-annealing" based on simulated annealing. Like simulated annealing, diversity is highly encouraged at the beginning of the search process by controlling the relative weighting of the pheromone information. The paper reports encouraging initial results for small benchmark TSPs.

Instead of modifying trail strength importance, Randall [323] examines the frequency of incorporation of solution components. During diversification, frequently incorporated components are discouraged while less frequently occurring components are encouraged. This strategy was inspired by a classic tabu search intensification/diversification strategy. Intensification, diversification and normal phases in the search are triggered dynamically. The results showed that improved performances (over a control strategy) on large TSPs could be obtained.

Nakamichi and Arita [286] define a simple diversification strategy for the TSP. At each step of the ant algorithm, each ant has a probability of selecting a city at random, without regard to pheromone or heuristic (cost) information. This allows the search to diversify, however, the results were far from conclusive as only one relatively small problem instance (eil51, see TSPLIB [331]) was used.

Algorithm 17.1 This algorithm checks the tabu status of a component. Note that k is the current ant and p is the position of the solution component.

```
tabu = FALSE
for i = 1 to k - 1 do
    if x_ip = x_kp then
        if c_ik >= tabu_threshold then
            tabu = TRUE
        end if
    end if
end for
end
```

17.4 Maintaining Intra-Colony Diversity

Tabu search principles of the balance between the restriction and freedom of the incorporation of various solution components [146] can be extended to form the basis of an intra-colony diversification strategy. The overall strategy is referred to as *divers* herefrom. As previously mentioned, an undesirable property of ACO techniques is that a group of ants within a particular colony can take the same path through search space and subsequently produce the same solution. This is commonly referred to as *stagnation* behaviour [374].

The approach described below allows ants to achieve greater exploration within each iteration while still maintaining the ability to exploit good components via standard pheromone mechanisms. To illustrate this, consider the TSP. The exploration of solution space is impaired if two or more ants within a colony select the same sequence of cities within their tours. Therefore, it may be considered tabu for an ant to have a number of sequentially visited cities in common with another ant. If this occurs, one of the ants is forced to choose another city. This requires the creation of a separate memory, c, that stores the number of components (cities) that ants have in common with one another. Formally, c is a matrix in which c_{ij} contains the length of the common sequence that is shared between ants i and j $(1 \leq i, j \leq m)$. Note that this is a symmetric matrix such that $c_{ij} = c_{ji}$. The tabu status of a component is checked after it has been selected by either Equation 17.1 or 17.2. Algorithm 17.1 is used to determine the tabu status.

Algorithm 17.1 inspires two important questions, a) how is *tabu_threshold* set? and b) how can the tabu status of a component be lifted (i.e., aspiration)? In terms of the former, there are two simple ways in which this can achieved:

(1) *tabu_threshold* is made a constant (regardless of problem size) or
(2) *tabu_threshold* $= round(\alpha n)$, where α is a proportion ($0 \leq \alpha \leq 1$), n is the number of solution components (e.g., N, the number of cities in a TSP) and *round*() is a rounding function.

Implementation specific details of the setting of *tabu_threshold* are provided in Section 17.5. The application of aspiration can be achieved in either of the following ways:

(1) If all the non-tabu solution components are unavailable (as, for instance, they have already been incorporated into the solution), the elements of the tabu list are used instead.
(2) If by adding a tabu solution component to a partial solution, the cost of this partial solution is better than the best solution's partial cost (at that point within the constructive process), the component becomes non-tabu.

The c matrix is updated after every step of the ant algorithm. If ant i has the same component in the same position as ant j, then $c_{ij} = c_{ji} = c_{ij} + 1$. If it is different, then $c_{ij} = c_{ji} = 0$. The latter condition ensures that the value of each element of c is the number *sequential* solution components rather than just the number of common components. As the diversity scheme consists, in the main, of array lookups and updates, it should not significantly impact on the algorithm's computational requirements.

17.5 Computational Experience

The computing platform used to perform the experiments is a 2.6GHz Red Hat Linux (Pentium 4) PC with 512MB of RAM. The experimental programs are coded in the C language and compiled with `gcc`. The experimental work is designed to test the operation of the new strategy as well as to contrast it against a control ACS meta-heuristic implementation.

The following describes the design of the computational experiments; the implementation details for each problem; the benchmark problem instances and finally the results.

17.5.1 *Experimental Design*

The experiments are divided into two stages:

- **Stage 1**: In this stage, the mechanics and parameters of the diversification strategy are tested. The key parameter is *tabu_threshold*. In order to encourage diversity, these values will be kept small relative to the problem size. Constant values of 5, 10 and 15 (for TSP) and 3, 5, 7 (for QAP) will be trialed. For the linear proportional model, α will be set as 0.05, 0.1 and 0.15. Additionally, the effectiveness of the second aspiration criterion will be tested by turning it on and off. The combination of these settings will be tested on a subset of the problems, namely hk48, st70, nug30 and tai35a. The number of iterations per run is set as 1000. The Kruskal-Wallis statistical procedure will be used to determine if there is a significant difference between the combinations and if so, which combination performs the best overall. It is used as there are multiple factors to test and the data are non-normally distributed.
- **Stage 2**: The performances of three solvers (for both TSP and QAP) will be compared. These are a) the *divers* strategy (using the best combination of parameters as identified from Stage 1), b) the *divers* strategy (without the local pheromone updating rule (Equation 17.3)) which is referred to as *divers_nlp* and c) the control strategy (which is a standard ACS implementation as outlined in Section 17.2 and is referred to as *control*). *divers_nlp* is trialed as the *divers* strategy performs the same function as the local updating rule. Each run will constist of 3000 iterations, so as to give the solvers sufficient opportunity to find good solutions.

The standard ACS parameters have been set as $\{\beta = -2, \gamma = 0.1, \rho = 0.1, q_0 = 0.9, m = 10\}$ for all experiments as these values have been found to be robust by Dorigo and Gambardella [112]. The value of ρ is set to 0 to implement *diver_nlp*. Each problem instance is run across ten random seeds.

17.5.2 *Implementation Details*

Local search will not be used in the first stage as it alters individual solutions and will hence confound the results. However, it will be used for *divers*, *divers_nlp* and *control* in Stage 2. It is applied, for each ant, at each iteration, in the following manner. The transition operators used for the TSP and QAP are inversion and 2-opt respectively. These operators have been found by Randall [324] to provide good performance for these problems. For each operator, the entire neighbourhood is evaluated at each step of the local search phase. The phase is terminated when a better solution cannot be found, guaranteeing a local minimum. Thus it is a variable depth local search scheme.

For the visibility heuristic, the TSP uses the distance measure between the current city and the potential next city. For the QAP, there are a number of choices for this heuristic, such as the use of a range of approximation functions and not using them at all [373]. The definition used here is given by Equation 17.6.

$$\eta(w,j) = \sum_{i=1}^{w-1} a(i,w)b(x(i),j) \tag{17.6}$$

Where:

w is the current location,
$x(w)$ is the facility assigned to w,
j is the the potential facility to assign to $x(w)$,
$a(i,w)$ is the distance between locations i and w and
$b(i,j)$ is the flow between facilities i and j.

17.5.3 *Problem Instances*

Twelve TSP and QAP problem instances are used to test the effectiveness of the three solvers. These problems are from TSPLIB [331] and QAPLIB [61] respectively and are given in Table 17.1.

Table 17.1 Problem instances used in this study (TSPs on the left and QAPs on the right). "Size" for the TSP and QAP is recorded in terms of the number of cities and facilities/locations respectively.

Name	Size	Best-Known Cost	Name	Size	Best-Known Cost
hk48	48	11461	nug12	12	578
eil51	51	426	nug15	15	1150
st70	70	675	nug20	20	2570
eil76	76	538	tai25a	25	1167256
kroA100	100	21282	nug30	30	6124
bier127	127	118282	tai35a	35	2422002
d198	198	15780	ste36a	36	9526
ts225	225	126643	tho40	40	240516
gil262	262	2378	sko49	49	23386
pr299	299	48191	tai50a	50	4941410
lin318	318	42029	sko56	56	34458
pcb442	442	50778	sko64	64	48498

17.5.4 *Results*

Stage 1 tested for the best combinations of settings for *divers*. Using the Kruskal-Wallis procedure and a significance level of 0.05 revealed that having $\alpha = 0.15$ with the second aspiration criteria turned off produced the overall best overall results. As the results were not statistically significant, the combination was chosen on the Kruskal-Wallis rankings. Further testing showed that using the second aspiration criteria was significantly worse than not using it.

Using the best settings from Stage 1 (no second aspiration criteria and $\alpha = 0.15$), Tables 17.2 and 17.3 give the results of the new solvers (*divers* and *divers_nlp*) and the control strategies for TSP and QAP respectively. In order to describe the range of objective costs obtained by these experiments, the minimum (denoted "Min"), median (denoted "Med") and maximum (denoted "Max") are given. Non-parametric descriptive statistics are used as the data are highly non-normally distributed. The cost results are reported as the relative percentage deviation (RPD) from the best known solution cost. This is calculated as $\frac{E-F}{F} \times 100$ where E is the result cost and F is the best known cost. The runtime is recorded as the number of CPU seconds required to obtain the best solution within a particular run.

In terms of the runtime, there is no noteworthy difference between the control and diversity strategies (see Section 17.4). Considering each problem type separately for the objective cost results revealed that:

- In terms of the TSP, the diversity strategies are significantly better than *control*.
- In terms of the QAP, there is no significant difference between the three solvers.

A possible explanation of the latter can be formed by examining the trials/recorded time at which best solutions were found. Unlike the TSP, all the solvers regularly found the best solution later in the run.

If both problem types are considered together, there is a statistically significant difference between the three solvers. Post-hoc testing (using Scheffes's test) showed that both *divers* and *divers_nlp* are significantly better than *control*. No overall difference is recorded between *divers* and *divers_nlp*. This indicates that for both TSP and QAP the use of local pheromone (an implicit form of intra-colony diversification) is unnecessary provided that the explicit scheme is used.

In light of the success of the new strategy for TSPs, it was decided to test it on two larger problem instances. Solving the problems d657 (657 cities) and rat783 (783 cities) under the same conditions showed that the

Table 17.2 Results for all the solvers on the TSP.

Problem	Solver	Cost RPD			Time (seconds)		
		Min	Med	Max	Min	Med	Max
dives	hk48	0	0	0	0.02	0.12	0.16
	eil51	0	0.23	0.23	0.02	0.15	39.92
	st70	0	0	0	0.33	1.49	26.49
	eil76	0	0	0.74	0.13	2.21	8.17
	kroA100	0	0	0	0.3	0.99	4.7
	bier127	0	0	0.32	2.5	8.69	887.83
	d198	0	0.03	0.13	151.68	1157.75	3741.14
	ts225	0	0	0	12.46	23.51	62.68
	gil262	0	0.06	0.34	42.7	476.19	6735.24
	pr299	0	0.08	0.38	196.88	1304.07	10651.66
	lin318	0.13	0.32	1.33	328.09	2997.45	12286.94
	pcb442	0.32	0.69	1.14	2746.7	5570.13	11241.24
divers_nlp	hk48	0	0	0	0.04	0.19	1.47
	eil51	0	0.23	0.23	0.04	0.17	0.71
	st70	0	0	0	0.35	2.1	89.37
	eil76	0	0	0	0.3	6.29	130.47
	kroA100	0	0	0	0.45	3.4	6.41
	bier127	0	0	0.15	2.65	48.96	1020.8
	d198	0	0.03	0.09	235.39	622.34	4254.66
	ts225	0	0	0	31.56	82.97	264.4
	gil262	0	0.23	0.84	111.07	628.51	5008.39
	pr299	0	0.07	0.18	391.86	2523.08	9700.76
	lin318	0.15	0.3	0.43	660.86	3892.62	16584.53
	pcb442	0.36	0.69	0.77	4246.9	24747.65	42211.08
control	hk48	0	0	0.11	0.02	0.13	27.07
	eil51	0	0.23	0.47	0.03	0.21	0.38
	st70	0	0.37	1.33	0.28	0.6	2.22
	eil76	0	0.09	0.74	0.29	1.16	3.35
	kroA100	0	0.2	0.46	0.25	1.27	5.97
	bier127	0	0.29	0.75	2.44	7.97	665.57
	d198	0.04	0.17	0.8	14.56	62.86	1546.57
	ts225	0	0	0.13	6.24	17.87	71.15
	gil262	0.29	0.69	1.39	11.12	32.3	2769.19
	pr299	0.32	0.64	1.47	21.37	45.35	9202.87
	lin318	0.73	1.03	2.09	43.08	85.98	1141.83
	pcb442	0.95	1.68	2.51	81.63	161.71	6268.72

new diversification strategy again produced very good solutions. Typically these were within 0.5% of the best known cost.

17.6 Conclusions

This paper has described an explicit intra-colony diversification mechanism for ACO. The strategy ensures that ants within a colony produce different partial solution sequences. The results show that the strategies based on this diversification notion outperform a control ACO implementation. This is particularly evident for the TSP in which both *divers* and *divers_nlp* consistently find solutions within a percent of optimal - even for the largest

Table 17.3 Results for all the solvers on the QAP.

Problem	Solver	Cost RPD			Time (seconds)		
		Min	Med	Max	Min	Med	Max
divers	nug12	0	0	0	0	0.02	0.03
	nug15	0	0	0	0.01	0.02	0.23
	nug20	0	0	0	0.12	0.94	10.69
	tai25a	0.37	0.76	1.43	49.77	158.79	221.86
	nug30	0	0	0.07	7	21.26	761.47
	tai35a	1.12	1.54	1.87	3.24	245.76	1038.54
	ste36a	0	0.25	1.47	414.9	1013.62	1815.62
	tho40	0.01	0.43	0.61	279.6	2359.55	4199.35
	sko49	0.05	0.14	0.27	492.11	3472.31	7408.42
	tai50a	2	2.34	2.55	500.35	1836.43	5931.69
	sko56	0.02	0.13	0.34	2423.35	14095.43	22218.96
	sko64	0.02	0.23	0.31	20083.71	30350.97	47165.03
divers_nlp	nug12	0	0	0	0	0.02	0.17
	nug15	0	0	0	0.01	0.02	0.17
	nug20	0	0	0	0.11	0.79	20.47
	tai25a	0.37	0.87	1.22	16.31	121.32	214.7
	nug30	0	0	0.07	234.89	264.86	286.81
	tai35a	0.94	1.49	1.74	3.28	716.81	1187.37
	ste36a	0	0.25	1.47	232.26	502.46	2477.66
	tho40	0.01	0.21	0.34	733.01	2494.46	3770.59
	sko49	0.03	0.07	0.26	1039.54	2929.4	9371.47
	tai50a	2.08	2.36	2.55	773.71	4680.92	6705.23
	sko56	0.02	0.16	0.3	1315.99	6840.7	21409.46
	sko64	0	0.07	0.35	7148.14	33504.2	41416.52
control	nug12	0	0	0	0	0.02	0.33
	nug15	0	0	0	0.01	0.03	0.57
	nug20	0	0	0	0.15	0.74	5.38
	tai25a	0.71	1.14	1.24	11.99	89.71	230.97
	nug30	0	0	0.07	17.44	136.18	835.52
	tai35a	1.11	1.69	2	62.63	508.87	984.4
	ste36a	0	0.24	1.18	729.08	1452.94	2202.54
	tho40	0.05	0.37	0.66	10.41	1446.31	3795.78
	sko49	0	0.1	0.25	2237.63	9188.76	10967.64
	tai50a	2.1	2.24	2.6	184.94	3292.33	5049.98
	sko56	0.02	0.15	0.32	1731.5	9328.02	16282.59
	sko64	0	0.17	0.34	1781.22	17777.08	46135.64

problems. All strategies performed equally well on the QAP. The reason for this may be attributed to the fact the ACO generally found its best solution later in the run. Running the new solver across more problem types will help to verify this.

This work naturally leads to a range of possible extensions. The most important of which is allowing the diversification scheme to continue across iterations. An implementation of this would ensure that all the solutions generated within a run would better sample the solution space. Additionally, the mechanics of the *tabu_threshold* need to be refined and trialed across a range of combinatorial problems. Dynamic varying of this parameter is also an option that should be considered.

Chapter 18

Evolving Gene Regulatory Networks for Cellular Morphogenesis

T. Rudge and N. Geard
The ARC Centre for Complex Systems,
School of Information Technology and Electrical Engineering,
The University of Queensland, QLD 4072, Australia.
E-mail: {timrudge,nic}@itee.uq.edu.au

The generation of pattern and form in a developing organism results from a combination of interacting processes, guided by a programme encoded in its genome. The unfolding of this programme involves a complex interplay of gene regulation and intercellular signalling, as well as the mechanical processes of cell growth, division and movement.

In this study we present an integrated modeling framework for simulating multicellular morphogenesis that includes plausible models of both genetic and cellular processes, using leaf morphogenesis as an example. We present results of an experiment designed to investigate the contribution that genetic control of cell growth and division makes to the performance of a developing system.

18.1 Introduction

The generation of pattern and form in a developing organism results from a combination of interacting processes, guided by a programme encoded in its genome. The unfolding of this programme involves a complex interplay of gene regulation and intercellular signalling, as well as the morphogenetic processes of cell growth, division and movement [428]. Recently, computers have enabled these multi-scale developmental systems to be simulated, revealing new insights into the emergence of pattern and form [192; 343].

In this study we present an integrated modeling framework for simulating multicellular morphogenesis that includes plausible models of both genetic and cellular processes, using leaf morphogenesis as an example. Leaf forms display a wide variety of morphological features, the development of which provide excellent examples of robust control of shape formation. We focus here on the role played by genetic control of cell growth and division orientation in the generation of specific shapes, and how this interacts with the physical constraints on cell shape.

The format of this report is as follows: Section 18.2 presents background material on leaf morphogenesis and previous computational models of gene regulation and development. Section 18.3 describes our simulation framework, which consists of a gene network model, a physical cell model, a procedure for coupling these two components, and an evolutionary search algorithm for investigating model parameters. Section 18.4 presents initial results obtained using this simulation framework and Section 18.5 concludes with a discussion of future directions for this research.

18.2 Background

18.2.1 *Leaf Morphogenesis*

Morphogenesis, the formation of shapes and structures in plants and animals, occurs by three processes: 1. Tissue growth; 2. Cell movement; 3. Cell death (apoptosis) [83]. Active cell movement does not occur in plants and so morphogenesis is coordinated by tissue growth – determined by cell shape, growth, and proliferation – and cell death. Variation in these behaviours across tissues and over developmental time causes the development of specific forms. The term *patterning* is applied to the coordinated differential expression of genes over space and time. It is these gene expression patterns that give rise to the variation in cell behaviour that drives morphogenesis.

Patterning provides positional information that guides cell behaviour and although cell lineage also plays some role, it seems this positional information is of primary importance in plant development [109; 127]. The relationship between patterns of gene expression and the specification of tissue and organ shape is not well characterised. [83] cite the difficulty in measuring morphogenetic effects and the need for quantitative analysis as possible reasons for this gap. Another difficulty is understanding the tight coupling of morphogenesis and patterning: the patterns develop along with, and are embedded in the forms to which they give rise.

Formation of leaf shape is tightly regulated, and evidence exists for both cell-division dependent and cell-division independent regulatory mechanisms [127]. That is, final leaf form and size is to some extent independent of both cell number and cell size. There is also evidence that overall leaf shape is unaffected by cell division orientation [127]. This suggests regulation of cell behaviour depends on feedback of organ-level information [399]. The nature of this regulation is at present unknown. However, hormone (e.g. auxin) transport has been implicated in organ shape regulation [127; 222], and [221] suggest that hormones may also regulate organ shape by affecting cell expansion and/or by modulating the cell cycle. Noting that leaf size is dependent on whole-plant physiology, [399] suggests that a source-sink relationship within the plant (e.g. of nutrients) might limit leaf size.

[433] and [221] present results of experiments in which cell division rates, cell division orientation, and cell growth rates were perturbed both locally and across the leaf. These results provide evidence for the involvement of the above mentioned processes in regulating leaf morphogenesis, but how these multiple mechanisms interact and their relative importance are still unknown.

18.2.2 *Previous Models*

[313] identifies three categories of plant development models, focusing on plant architecture, individual organs, and the underlying mechanics of gene regulation, respectively. The first of these is well established, with L-systems being the dominant modelling framework; the latter two areas are still active areas of research – [313] and [83] provide an overview of recent developments. These categories occupy very different temporal and spatial scale ranges, and a full understanding of development requires the integration of multiple scales.

Developmental issues have been addressed by the Artificial Life research community [367]. An early attempt to integrate multiple scales of developmental mechanism into a single model included cells with complex internal dynamics that communicated with each other via chemical and electrical signals as well as physical interactions [126]. One of the findings of this study was that, while the multiple mechanisms enabled the robust production of interesting phenotypes, it also made the design of *specific* phenotypes more difficult. Later research demonstrated that this difficulty could be addressed by using a representation of the regulatory network that could be artificially evolved [119].

[192] also used a model that combined mechanisms at multiple scales of description – gene regulation, development and evolution – to investigate the interactions between evolutionary dynamics and morphology. Her evolutionary process was aimed at maximising cell type diversity, rather

than achieving a specific morphological shape. However, she found that some features of morphology, such as engulfing, budding and elongation, were relatively 'generic', that is, they appeared 'for free' in systems that satisfied certain prerequisites.

The formation of patterns across a fixed field of autonomous cells has also been studied in some detail [344; 359]. Here too, it was found that patterns were a common emergent feature of interacting gene networks, although again, selection was for 'pattern complexity' rather than a specific phenotypic target. Later research [343] focused on the issue of how pattern formation processes interact with growth processes, specifically with reference to the evolution of tooth development.

In general, the models mentioned above do not include individual cell morphology: Cells are represented as circles or squares, of equal or varying size. The model used by [192] does support anisotropic cell shapes; however, these result from the algorithm used to calculate cell boundaries, rather than reflecting anisotropy in the underlying growth process. In plants, cell behaviour is frequently anisotropic, with the axes of both growth and division under a degree of cellular and genetic control. As described above, control of cell morphology is intimately connected with the production of leaf form, therefore a detailed and flexible model of cell shape is of fundamental importance in any approach to modeling leaf morphogenesis.

18.3 The Simulation Framework

Our integrated model of plant morphogenesis brings together plausible representations of cell shape, genetic regulation, and cell-cell signalling. Cell shape is determined by growth and division activity as well as external physical forces, and the combination of the shapes of all cells determines the overall phenotypic form. Each cell is autonomous and its behaviour is regulated by its own copy of the organism's gene network, which also responds to signals received from neighbouring cells. The genetic network thus indirectly specifies phenotypic morphology.

As noted earlier, due to the complexity of cellular developmental systems, a search and optimisation approach is favoured when examining their properties *in silico*. We have chosen to use an evolutionary algorithm to search for systems with particular shape formation capabilities. Our approach is to decide on a target phenotypic shape, specify initial conditions, and then artificially evolve gene networks which come close to producing the desired shape.

In the following we describe the primary components of the model: the genetic component, consisting of a network embedded within each cell, and the spatial model, consisting of an arrangement of cells that constitutes the phenotype. Following that, the coupling between these two components is described. Finally, the evolutionary algorithm used to explore the parameter space of these networks is outlined.

18.3.1 *The Genetic Component*

In this study, we used a dynamic recurrent gene network (DRGN) model for the genetic component of the framework [141]. The DRGN model is based on a widely studied class of artificial neural network models known as recurrent neural networks [120], and has previously been used to investigate the generation of developmental cell lineages [141]. An advantage of a recurrent network representation is that it enables the model to express a complex range of gene interactions while abstracting away from the specific biological processes that underly those interactions.

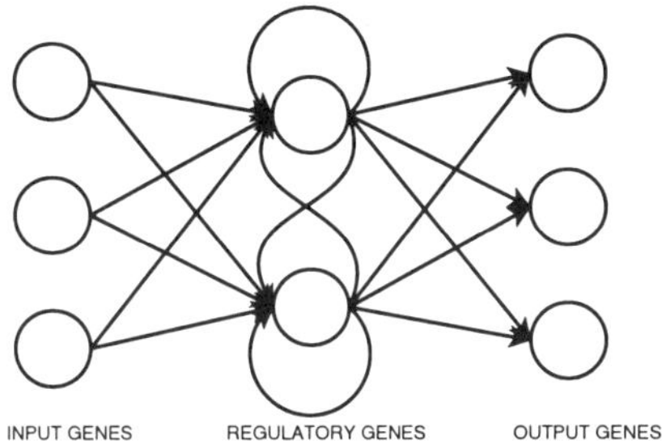

Fig. 18.1 The structure of the DRGN model. The network is partitioned into three categories: input genes that detect the presence of morphogens produced by other cells or the environment; regulatory genes that interact with one another to perform the computational tasks of the cell; and output genes producing morphogens that can be transmitted to other cells or that trigger events such as growth and division.

In the DRGN model, a genetic system is defined as a network of N interacting nodes (see Figure 18.1). Depending on the level of abstraction, each node can be considered to represent either a single gene, or a cluster of co-regulated genes. In this study we generally consider a node to be a equivalent to a single gene. The activation state of each node is a continuous variable in the range $[0, 1]$, where 0 represented a completely inactive gene and 1 a fully expressed gene. Nodes can be divided into three classes: input genes that detect the presence of morphogens; regulatory genes that interact with each other to carry out the computational task of the network; and output genes that produce morphogen signals.

The network is updated synchronously in discrete time steps. To capture the potential complexity of the interacting factors involved in gene expression, we have used a network in which each input gene is connected to each regulatory gene, all regulatory genes are connected to each other and themselves, and each regulatory gene is connected to each output gene. Thus an individual link in the network does not necessarily represent a direct physical interaction, but rather the degree of influence that the expression of the source gene at time t has on the expression of the target gene at time $t+1$.

These interactions can be summarised in a weight matrix, in which the entry at row i, column j specifies the influence that gene j has on gene i. These entries may be positive or negative, depending on whether the product of gene j is an activator or a repressor in the regulatory context of gene i. A zero entry indicates that there is no interaction between the two genes. The inclusion of self-connections (i.e. from node i to node i) allows for the possibility of genes influencing their own regulation.

The state of the network is updated synchronously, with the activation of node i at time $t+1$, $a_i(t+1)$, given by

$$a_i(t+1) = \sigma\Big(\sum_{j=1}^{N_r} w_{ij} a_j(t) - \theta_i\Big) \tag{18.1}$$

where N_r is the number of regulatory nodes, w_{ij} is the level of the interaction from node j to node i, θ_i is the activation threshold of node i, and $\sigma(.)$ is the sigmoid function, given by

$$\sigma(x) = \frac{1}{1+e^{-x}} \tag{18.2}$$

18.3.2 *The Cellular Component*

We use a 2-dimensional spatial model of the cellular arrangement. This is based on linear cell boundary elements (walls), which are modelled as elastic springs. The approach is similar to that of [198], however we also consider some more complex cell dynamics such as anisotropic growth. Cell-cell signalling is considered in the form of chemical diffusion, as in [126]. This approach has previously been used to examine rule-based control of plant morphogenesis [340].

Cell: The genome of our artificial organism is represented as a DRGN. Each cell is defined by its DRGN, a set of dynamic state parameters and a closed boundary. The DRGNs contained by each cell in a phenotype have identical structure and weights, reflecting the genetic homogeneity of an individual organism. The activation levels of the DRGN nodes in each cell, however, are independent and represent the variation in gene expression across the phenotype.

The cell state parameters include passively received information such as morphogen levels and cell volume, and behavioural states like growth rate and morphogen production rates. As part of their state, the cells also maintain polarity vectors that are used to direct anisotropic growth, to orient the division plane, and to asymmetrically divide the cells morphogens between its daughters, according to the behavioural state parameters.

The state of the cell determines its behaviour at any point in time. Cell dynamics are expressed as the transformation of cell state parameters to proceed to a new state. Behavioural states are transformed by the DRGN, with the inputs and outputs of the DRGN defined by a fixed mapping onto the cell state parameters. The passive state parameters are transformed by physical simulation of the cells' environment, including its own boundary shape and interactions with neighbours.

Spatio-Mechanical Model: The boundary of the cell describes its shape, and is decomposed into a set of walls. Each wall is the interface between two cells. Morphogens diffuse from one cell to the other via the wall, providing a cell-cell signalling mechanism. The walls are considered to be two linearly elastic elements (springs), one for each adjacent cell, bound together at the end points (vertices). Each of the adjacent cells influences the properties of only one of these springs. Each spring has stiffness K and natural length L_n determined from the state parameters of the appropriate cell [340].

Each cell exerts a turgor force perpendicular to each of its walls in an outward direction with respect to the cell, extending the springs, which then exert an opposing tension force. At each time-step these simulated forces are accumulated at the vertices, and the vertex positions are adjusted to find the equilibrium configuration.

Cell growth is achieved by increasing the natural lengths of each cells' springs to varying degrees (see [340] for details). Division consists of inserting a dividing wall across the centre of the cell, and redefining the daughter cell boundaries. When a cell divides, its DRGN (including current node activation levels) is copied into the two daughter cells.

18.3.3 *Genotype-Phenotype Coupling*

The system integrates multiple scales of model into a single framework. Figure 18.2 shows an overview of the way in which the levels of the model interact. Starting at the micro level, the DRGN transforms the cell state. The cell state is expressed as local behaviours such as growth, which then affect the entire phenotype via simulated mechanical forces and diffusion processes. This global effect is then transduced back into local information to each cell, and from there transformed into micro level input to the DRGN.

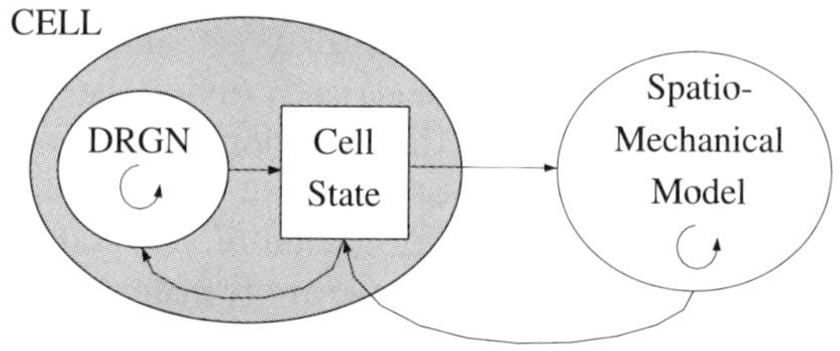

Fig. 18.2 Scheme of interactions between different levels in the model, from microscopic (left) to macroscopic (right). Circular arrows indicate faster time scale processes running multiple time steps between cell state updates.

The flow of control is therefore from micro to macro level, and the flow of information or feedback is from macro to micro. The nature of the coupling between information feedback and phenotypic output is ultimately determined by the structure and weights of the DRGN, as well as the mapping from DRGN to cell state – i.e. the genotype.

The dynamics of cell behaviour, such as growth and division, gene expression, and transmission of mechanical forces, occur on very different time scales. In general, variation in cell behaviour occurs most slowly and equilibration of forces occurs most quickly. We assume that mechanical equilibrium is reached instantaneously when relevant parameters such as growth rate change.

The DRGN can be used to model genetic regulation on several levels. Each node may represent a single gene or a cluster of genes, and each node update may represent one or many regulatory events. In order to incorporate this flexibility we allow the DRGN to update multiple times before affecting the cell state. The procedure that produces a cellular phenotype from the DRGN genotype is thus:

(1) Determine cell states from initial conditions
(2) Map DRGN inputs from cell states
(3) Update DRGN by some number of time steps
(4) Map cell states from DRGN outputs
(5) Compute cell shapes and morphogen diffusion
(6) Repeat from 2 until stopping condition met

The stopping condition may be chosen arbitrarily according to the experiment. We used a maximum number of time steps of 250 in our experiments.

18.3.4 *The Evolutionary Component*

The evolutionary component, which enables a population of DRGNs to be artificially "evolved" towards some particular target, serves two purposes. At a methodological level, it provides a useful machine learning technique for searching the parameter space of networks. At a theoretical level, it facilitates questions about the evolutionary dynamics of morphogenesis [192].

A simple evolutionary search strategy called the 1+1 ES was used [23]. Initially, a single DRGN was generated with weights randomly drawn from a Gaussian distribution with mean 0 and standard deviation 4. This DRGN was used to develop a phenotype, as described in Section 18.3.3. A fitness value for this phenotype was calculated as described in Section 18.4.1 below and stored. A new DRGN was derived from the existing DRGN by adding Gaussian noise (mean 0, standard deviation 0.01) to each of the node interactions. A new phenotype was developed and evaluated and the fitness value for the modified DRGN was compared to that of the original DRGN. The DRGN producing the phenotype with the greatest fitness was retained and used as the basis for the creation of a further new DRGN. This process was repeated until the stopping conditions were met. We used a maximum number of generations of 15,000 in our experiments.

18.4 Initial Experiments

To investigate the role of genetic control of growth and development in morphogenesis, we ran three sets of comparative evolutionary trials: random growth and division orientation, regular growth and division orientation and genetically controlled growth and division orientation. We set the DRGN the task of generating a circular shaped final phenotypic form.

(1) **Random orientation** In the first set of trials, there was no control of growth and division orientation – they were each chosen randomly at each time step. Only one output node was utilised, the morphogen controlling the decision to grow and divide.
(2) **Regular orientation** In the second set of trials, the orientation of growth and division of each cell was chosen to be opposite to that of its parent cell – giving alternating axial and lateral growth and division each generation. The DRGN was not able to change this sequence of

orientations; however, it was able to make use of the regularity of the predetermined sequence.

(3) **Controlled orientation** In the final set of trials, two additional output nodes were used, producing morphogens that controlled the orientation of growth and division respectively. Therefore the DRGNs had the capability to coordinate the two processes.

18.4.1 *Method*

DRGN Coupling: Three inputs were provided to the network. The first input responded to the concentration of a morphogen that was initialised to a concentration of 1.0 in the initial cell, and was not produced after that. Therefore, as the volume of the phenotype increased due to growth and division, the concentration of this morphogen decreased. The second and third inputs responded to morphogens related to the position of the cell. These were externally supplied as the (x, y) position of the cell centre. This may be considered as incorporating information supplied by underlying cell layers.

The phenotypes were initialised as a single unit square cell with unit morphogen concentration, and DRGN outputs p_j mapped to cell behaviour as follows:

- If $p_0 > 0.5$ then set growth rate to 0.2, and divide if volume > 2.
- If $p_0 \leq 0.5$ then set growth rate to 0 and do not divide.
- If $p_1 > 0.5$ set division orientation to *axial* otherwise set to *lateral.*
- If $p_2 > 0.5$ set growth orientation to *axial* otherwise set to *lateral.*

Fitness function: The task used for this study was to evolve a DRGN capable of generating a circular arrangement of cells of a given radius. Fitness was calculated for each phenotype at each time step based on the current cell arrangement – specifically, the absolute distance of each exterior (marginal) cell from the centre of mass of the phenotype, given by:

$$r_i = |\mathbf{x_i} - \mathbf{c}| \tag{18.3}$$

for cell i, where $\mathbf{x_i}$ is the cell's centre of mass, and $\mathbf{c}$ is the centre of mass of the whole phenotype. The error of the phenotype from a circle radius R at any time point, is calculated from the distance of each cell from the circle $dr_i = |r_i - R|$:

$$\epsilon(t) = \bar{dr}(t) + \sqrt{\frac{1}{N(t)} \sum_{i=1}^{N(t)} \left(dr_i(t) - \bar{dr}(t)\right)^2} \tag{18.4}$$

where the sum is over the N exterior cells. The first term is the mean distance error, and the second is the standard deviation in the distance error. We used a target radius of 5 units in our experiments. The overall phenotypic fitness was calculated from the cumulative error over all time steps $\{1, 2, 3...T\}$ and scaled by a constant factor such that the maximum possible fitness is approximately 1.0:

$$f = \frac{625}{1 + \sum_{t=1}^{T} \epsilon(t)} \tag{18.5}$$

Summary: In summary, three sets of evolutionary trials were run, each corresponding to one of the control conditions described in Section 18.4. Each evolutionary trial was run for up to 15,000 generations, with snapshots of the best phenotype being recorded at 500 generation intervals. In each generation, the DRGN was run for 250 developmental time steps, with it's fitness evaluated over this period as described above.

18.4.2 *Results*

The DRGNs that had explicit control of the growth and division orientation were able to generate considerably more accurate phenotypes than the those supplied with either a regular or random sequence of orientations (Table 18.1).

Table 18.1

Orientation	Maximum Fitness
Random	0.314
Regular	0.555
Controlled	0.642

The evolutionary history of the most successful evolutionary trial from the Controlled set displays a level of continual innovation typical of highly evolvable systems (Figure 18.3). By contrast, the most successful trials from the Regular and Random sets (not shown), reached their peak fitness early (around generation 4,000), and failed to improve any further. The phenotypes produced at different stages of evolution provide some clues to explain these differences (Figure 18.4).

(1) **Random orientation (Figures 18.4.a- 18.4.d):** Regulation of size appears relatively early, and is consistent throughout the course of evolution. However control of phenotypic shape has not evolved. The randomness of the sequence of orientations prevents the DRGN from

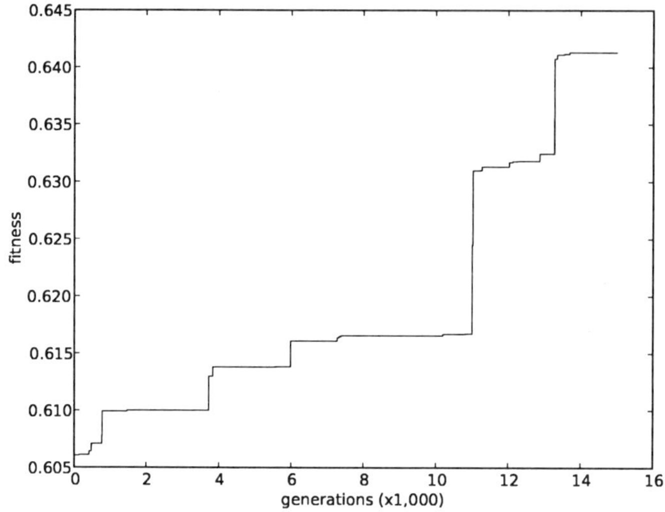

Fig. 18.3 The evolutionary history of the fittest system found in the Controlled set of trials, in which the DRGN had explicit control of the orientation of growth and division.

being able to successfully coordinate the development of a stable shape. The examples shown here represent only one instance of a set of possible outcomes for a given DRGN.

(2) **Regular orientation (Figures 18.4.e- 18.4.h):** When the DRGN is able to rely on a regular sequence of growth and division orientations, greater control of phenotypic shape is achieved. Very early in this evolutionary trial, a strategy emerged in which a group of growing cells is surrounded by non-growing cells. However, while ensuring a reasonably circular phenotype, this approach proves too strong a constraint, limiting any further improvement.

(3) **Controlled orientation (Figures 18.4i.- 18.4.l):** With full control over division and growth orientations, the evolutionary algorithm was able to explore a much broader range of developmental possibilities. In the example shown, the DRGNs found early in the evolutionary history developed by first growing and dividing laterally and then switching to axial division, resulting in the phenotype fanning out. The largest jump in fitness (Figure 18.3, around generation 11,000) occurred when a DRGN was discovered in which the fanning out process was inhibited by a cap of quiescent cells. The resulting "stem and bud" arrangement was refined in successive stages of evolution by increasing the roundness of the bud, and reducing the length of the stem.

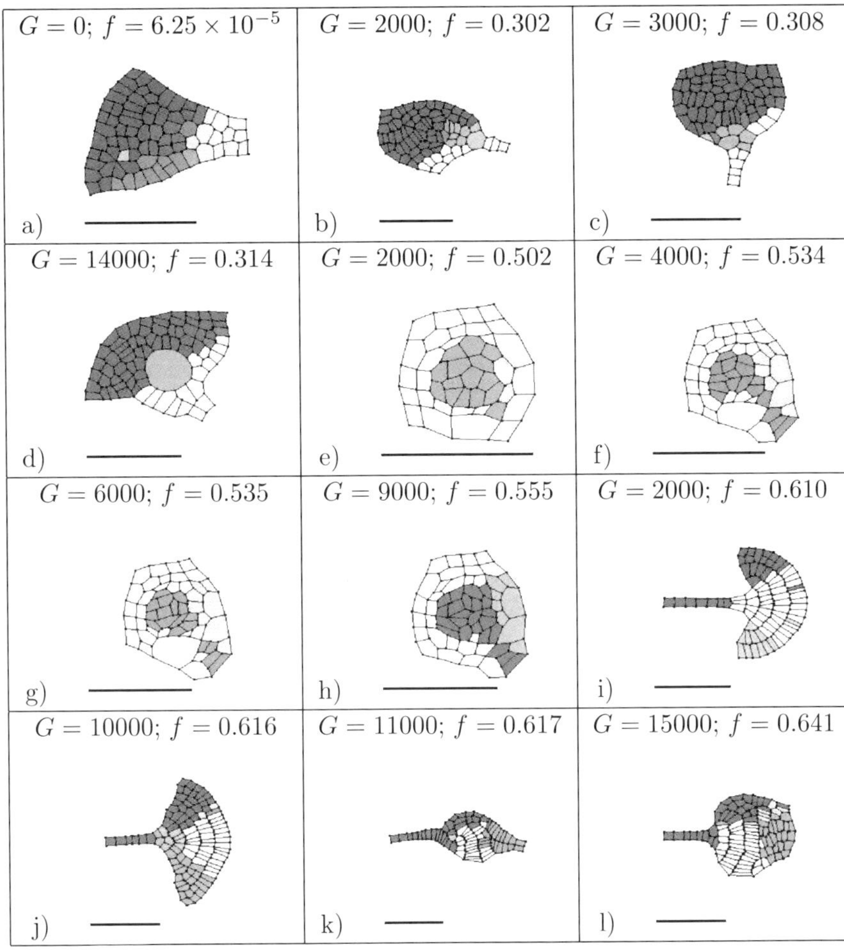

Fig. 18.4 Fittest phenotypes at key stages (generation G) in artificial evolution, where f is fitness: [a,d] Random growth/division, [e,h] Regular growth/division, [i,l] DRGN control. Shading shows DRGN output on a grey scale, white(0) to dark grey(1): [a,h] cell growth and division trigger, [i,l] division orientation. Scale bar is 10 units.

18.5 Discussion and Future Directions

In all three sets of trials, DRGN evolved that were capable of controlling phenotype size. With full DRGN control over development a significant degree of shape control evolved, using a variety of developmental approaches. With regular cell growth and division, DRGNs were able to control the phenotype shape to a similar extent, but the range of strategies for doing this was more limited. With random growth and division, little control of phenotypic shape emerged.

While preliminary, these results suggest that positional information only provides sufficient information to enable generation of stable phenotypic forms in the presence of predictable growth and division orientation. It would appear that the claim by [127], that leaf shape is, to a degree, independent of division orientation, requires the presence of more complex cell-cell signalling than simple positional cues. One strong possibility is that cell-cell communication plays a vital role in the robust development of form. Future work will involve the investigation of inductive interactions between cells and how this additional level of communication may facilitate more robust development.

Acknowledgments

This study was funded by the ARC Centre for Complex Systems (`http://www.accs.edu.au`). We thank Jim Hanan and Janet Wiles for stimulating discussions, helpful suggestions and guidance in carrying out this research.

Chapter 19

Complexity of Networks

R. K. Standish

Mathematics, University of New South Wales
E-mail:R.Standish@unsw.edu.au
http://parallel.hpc.unsw.edu.au/rks

Network or graph structures are ubiquitous in the study of complex systems. Often, we are interested in complexity trends of these system as it evolves under some dynamic. An example might be looking at the complexity of a food web as species enter an ecosystem via migration or speciation, and leave via extinction.

In this paper, a complexity measure of networks is proposed based on the *complexity is information content* paradigm. To apply this paradigm to any object, one must fix two things: a representation language, in which strings of symbols from some alphabet describe, or stand for the objects being considered; and a means of determining when two such descriptions refer to the same object. With these two things set, the information content of an object can be computed in principle from the number of equivalent descriptions describing a particular object.

I propose a simple representation language for undirected graphs that can be encoded as a bitstring, and equivalence is a topological equivalence. I also present an algorithm for computing the complexity of an arbitrary undirected network.

19.1 Introduction

In [363], I argue that *information content* provides an overarching complexity measure that connects the many and various complexity measures proposed (see [117] for a review). The idea is fairly simple. In most cases, there is an obvious *prefix-free* representation language within which descriptions of the objects of interest can be encoded. There is also a classifier of descriptions that can determine if two descriptions correspond to the same object. This classifier is commonly called the *observer*, denoted $O(x)$.

To compute the complexity of some object x, count the number of equivalent descriptions $\omega(\ell, x) =$ of length ℓ that map to the object x under the agreed classifier. Then the complexity of x is given in the limit as $\ell \to \infty$:

$$\mathcal{C}(x) = \lim_{\ell\to\infty} \ell \log N - \log \omega(\ell, x) \tag{19.1}$$

where N is the size of the alphabet used for the representation language.

Because the representation language is prefix-free, every description y in that language has a unique prefix of length $s(y)$. The classifier does not care what symbols appear after this unique prefix. Hence $\omega(\ell, O(y)) \geq N^{\ell-s(y)}$, $0 \leq C(O(y)) \leq s(y)$ and so equation (19.1) converges.

The relationship of this algorithmic complexity measure to more familiar measures such as Kolmogorov (KCS) complexity, is given by the coding theorem[242, Thm 4.3.3]. Equation (19.1) corresponds to the logarithm of the *universal a priori probability*. The difference between these measures is bounded by a constant independent of the complexity of x.

Many measures of network properties have been proposed, starting with node count and connectivity (no. of links), and passing in no particular order through cyclomatic number (no. of independent loops), spanning height (or width), no. of spanning trees, distribution of links per node and so on. Graphs tend to be classified using these measures — *small world* graphs tend to have small spanning height relative to the number of nodes and *scale free* networks exhibit a power law distribution of node link count.

Some of these measures are related to graph complexity, for example node count and connectivity can be argued to be lower and upper bounds of the network complexity respectively. However, none of the proposed measures gives a theoretically satisfactory complexity measure, which in any case is context dependent (ie dependent on the *observer* O, and the representation language).

In this paper we shall consider only undirected graphs, however the extension of this work to directed graphs should not pose too great a problem. In setting the classifier function, we assume that only the graph's topology counts — positions, and labels of nodes and links are not considered important. Clearly, this is not appropriate for all applications, for instance in food web theory, the interaction strengths (and signs) labeling each link is crucially important.

The issue of representation language, however is far more problematic. In some cases, eg with genetic regulatory networks, there may be a clear representation language, but for many cases there is no uniquely identifiable language. However, the *invariance theorem*[242, Thm 2.1.1] states that the difference in complexity determined by two different *Turing complete* representation languages (each of which is determined by a universal Turing machine) is at most a constant, independent of the objects being measured. Thus, in some sense it does not matter what representation one picks — one is free to pick a representation that is convenient, however one must take care with non Turing complete representations.

In the next section, I will present a concrete graph description language that can be represented as binary strings, and is amenable to analysis. The quantity ω in eq (19.1) can be simply computed from the size of the automorphism group, for which computationally feasible algorithms exist[271].

The notion of complexity presented in this paper naturally marries with thermodynamic entropy S[235]:

$$S_{\max} = \mathcal{C} + S \tag{19.2}$$

where $S_{\max}$ is called *potential entropy*, ie the largest possible value that entropy can assume under the specified conditions. The interest here is that a dynamical process updating network links can be viewed as a dissipative system, with links being made and broken corresponding to a thermodynamic flux. It would be interesting to see if such processes behave according the maximum entropy production principle[106] or the minimum entropy production principle[312].

In artificial life, the issue of complexity trend in evolution is extremely important[34]. I have explored the complexity of individual Tierran organisms[364; 365], which, if anything, shows a trend to simpler organisms. However, it is entirely plausible that complexity growth takes place in the network of ecological interactions between individuals. For example, in the evolution of the eukaryotic cell, mitochondria are simpler entities than the free-living bacteria they were supposedly descended. A computationally feasible measure of network complexity is an important prerequisite for further studies of evolutionary complexity trends.

19.2 Representation Language

One very simple implementation language for undirected graphs is to label the nodes $1..N$, and the links by the pair $(i,j), i<j$ of nodes that the links connect. The linklist can be represented simply by a $N(N-1)/2$ length bitstring, where the $\frac{1}{2}j(j-1)+i$th position is 1 if link (i,j) is present, and 0 otherwise. We also need to prepend the string with the value of N in order to make it prefix-free — the simplest approach is to interpret the

Network	Bitstring description
	1110100, 1110010, 1110001
	1110110, 1110101, 1110011
	11110110100, 11110101010, 11110011001, 11110000111

number of leading 1s as the number N, which adds a term $N + 1$ to the measured complexity.

Some example 3 and 4 node networks are shown in table 19.1. One can see how several descriptions correspond to the same topological network, but with different node numberings.

A few other properties are also apparent. A network A that has a link wherever B doesn't, and vice-versa might be called a complement of B. A bitstring for A can be found by inverting the 1s and 0s in the linklist part of the network description. Obviously, $\omega(A, L) = \omega(B, L)$.

The empty network, and the fully connected network have linklists that are all 0s or 1s. These networks are maximally complex at

$$\mathcal{C} = \frac{1}{2}N(N+1) + 1 \tag{19.3}$$

bits. This, perhaps surprising feature, is partly a consequence of the definition we're using for network equivalence. If instead we ignored unconnected nodes (say we had an infinite number of nodes, but a only a finite number of them connected into a network), then the empty network would have extremely low complexity, as one would need to sum up the ωs for $N = 0, 1, \ldots$. But in this case, there would no longer be any symmetry between a network and its complement.

It is also a consequence of not using a Turing complete representation language. Empty and full networks are highly compressible, therefore we'd expect a Turing complete representation language would be able to represent the network in a compressed form, lowering the measured complexity.

Networks of 3 nodes and 4 nodes are sufficiently simple that it is possible enumerate all possibilities by hand. It is possible to numerically enumerate larger networks using a computer, however one will rapidly run into diminishing returns, as the number of bitstrings to consider grows as $2^{\frac{1}{2}N(N-1)}$. I have done this up to 8 nodes, as shown in Fig. 19.1.

Network	Complement	ω	$\mathcal{C}$
		1	7
		3	5.42

Network	Complement	ω	$\mathcal{C}$
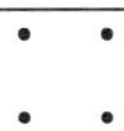		1	11
		6	8.42
		12	7.42
		3	9.42
	same	12	7.42
		4	9

19.3 Computing ω

The first problem to be solved is how to determine if two network descriptions in fact correspond to the same network. We borrow a trick from the field of symbolic computing, which is to say we arrange a canonical labeling of the nodes, and then compare the canonical forms of each description. Brendan McKay [271] has solved the problem of finding canonical labelings of arbitrary graphs, and supplies a convenient software library called *nauty*[1] that implements the algorithm.

The number of possible distinct descriptions is given by $N!$ (the number of possible renumberings of the nodes), *divided* by the number of such renumberings that reproduce the canonical form. As a stroke of good fortune, nauty reports this value as the order of the automorphism group, and

[1]Available from http://cs.anu.edu.au/~bdm/nauty.

is quite capable of computing this value for networks with 10s of thousands of nodes within seconds on modern CPUs. So the complexity value $\mathcal{C}$ in equation (19.1) is computationally feasible, with this particular choice of representation.

19.4 Compressed complexity and Offdiagonal complexity

I have already mentioned the issue of non Turing completeness of the proposed bitstring representation of a network. This has its most profound effect for regular networks, such as the empty or full networks, where $\mathcal{C}$ is at a maximum, yet contained a great deal of redundancy in the expression. To get a handle on how much difference this might make, we can try a compression algorithm of the all the equivalent bitstring representations, choosing the length most compressed representation as a new measure I call zcomplexity. Inspired by the brilliant use of standard compression programs (gzip, bzip2, Winzip etc.) to classify texts written in an unknown language[38], I initially thought to use one of these compression libraries. However, all of the usually open source compression libraries were optimised for compressing large computer files, and typically had around 100 bits of overhead. Since the complexities of all networks studied here are less than around 50 bits, this overhead precludes the use of standard techniques.

So I developed my own compression routine, based around run length encoding, one of the simplest compression techniques. The encoding is simple to explain: Firstly a "wordsize" w is chosen such that $\log_2 N \leq w \leq \log_2 N + \log_2(N-1) - 1$. Then the representation consists of w 1 bits, followed by a zero, then w bits encoding N, then the compressed sequence of links. Repeat sequences are represented by a pair of w bit words, which give the repeat count and length of a sequence, followed by the sequence to be repeated. As an example, the network:

$$1111110101010101010101$$

can be compressed to

$$\underbrace{111}_{w}0\underbrace{110}_{N}\underbrace{000}_{\text{rpt}}\underbrace{010}_{\text{len}}\underbrace{10}_{\text{seq}}.$$

Here 000 represents 8, not 0, as a zero repeat count makes no sense! Also, since the original representation is prefix free, the extra 0 that the compressed sequence adds to the original is ignored.

By analogy with equation (19.1) define zcomplexity as

$$\mathcal{C}_z = 1 + \log_2 \sum_b 2^{-\min\{\zeta(b), N(N+1)/2+1\}} \tag{19.4}$$

where b iterates over all bitstring representations of the network we're measuring, and $\zeta(b)$ is the compressed length of b, using the best w, by the aforementioned compression algorithm. The extra 1 takes into account a bit used to indicate whether the compressed or uncompressed sequence is used, so $\mathcal{C}_z \leq \mathcal{C} + 1$.

The optimal w for the empty (or full) network $w = \lceil \log_2 N \rceil$, and zcomplexity can be readily computed as

$$\begin{aligned} \mathcal{C}_z &= 2 + 3w + \lceil \frac{N(N-1)}{2^{w+1}} \rceil \\ &= \begin{cases} 2 + 3\log_2 N + \frac{N-1}{2} & \text{if} \quad n = 2^w \\ 5 + 3\log_2(N-s) + \lceil \frac{N(N-1)}{4(N-s)} \rceil & \text{if } n = 2^{w-1} + s \end{cases} \end{aligned} \tag{19.5}$$

Compared with equation (19.3), we can see that it makes a substantial difference. Already at $N = 5$, $\mathcal{C}_z = 13$ and $\mathcal{C} = 16$, with the difference increasing with N.

To compute the zcomplexity for an arbitrary graph, we need to iterate over all possible bit representations of a graph. There are two obvious ways to do this, since the number of nodes N, and number of links l are identical in all representations:

- Start with the bitstring with the initial l bits of the linkfield set to 1, and the remaining bits 0. Then iterate over all permutations, summing the right hand term into a bin indexed by the canonical representation of the network. This algorithm computes $\mathcal{C}_z$ for all networks of N nodes and l links. This algorithm has complexity $\binom{N(N-1)/2}{l} = N(N-1)\ldots(N-l)/l!$
- Take the network, and iterate over all permutations of the node labels. Some of these permutations will have identical bitstring representations as others — as each bitstring is found, store it in a set to avoid double counting. This algorithm has complexity $N!$

In my experiments, I calculate zcomplexity for all networks with linkcount l such that $\binom{N(N-1)/2}{l} < N!$, then sample randomly networks with greater link counts.

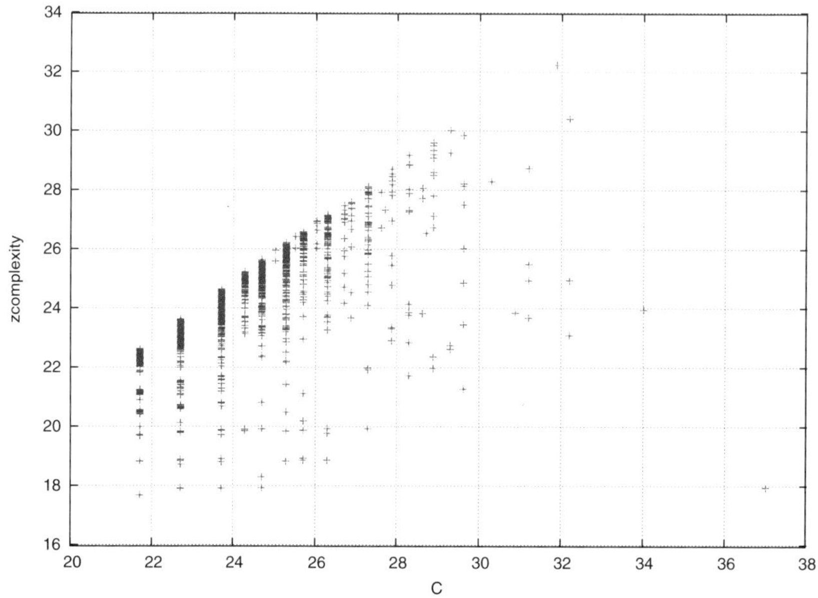

Fig. 19.1 $\mathcal{C}_z$ plotted against $\mathcal{C}$ for all networks of order 8. Note the empty/full network lying in the lower right hand corner

Fig. 19.1 shows $\mathcal{C}_z$ plotted against $\mathcal{C}$ for all networks of order 8, which is about the largest size network for which an exhaustive computation of $\mathcal{C}_z$ is feasible.

Unfortunately, without a smarter way of being able to iterate over equivalent bitstring representations, zcomplexity is not a feasible measure, even it more accurately represents complexity. The disparity between $\mathcal{C}_z$ and $\mathcal{C}$ is greatest for highly structured graphs, so it would be interest to know when we can use $\mathcal{C}$, and when a more detailed calculation is needed.

Claussen[81] introduced a measure he calls *offdiagonal complexity*, which measures the entropy of the distribution of links between different node degree. Regular graphs will have zero offdiagonal complexity, as the node degree distribution is sharply peaked, and takes on moderate values for random graphs (where node degree distribution is roughly exponential) and is extremal for scale-free graphs. Since the discrepancy between $\mathcal{C}$ and $\mathcal{C}_z$ was most pronounced with regular graphs, I looked at offdiagonal complexity as a predictor for this discrepancy.

Figure 19.2 shows the compression error (defined as $\frac{\mathcal{C}-\mathcal{C}_z}{\mathcal{C}}$) plotted as a function of offdiagonal complexity and $\mathcal{C}$. The dataset falls clearly into two

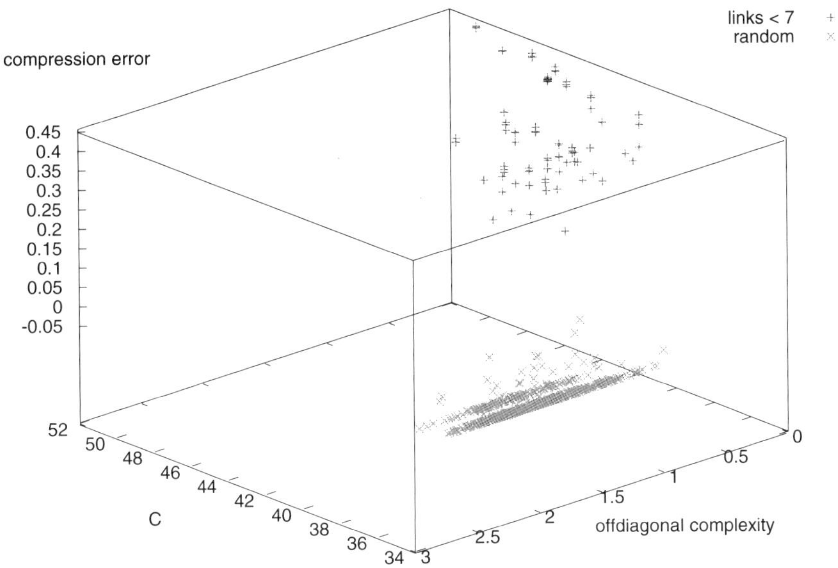

Fig. 19.2 Compression error as a function of C and offdiagonal complexity for networks with 10 nodes. All networks with link count less than 7 were evaluated by method 1, and 740 graphs with more than 7 links were selected at random, and computed using method 2. The separation between the two groups is due to compressibility of sparse networks.

groups — all sparse networks with link count less than 7, and those graphs sampled randomly, corresponding to the two different methods mentioned above. The sparse networks are expected to be fairly regular, hence have high compression error, whereas randomly selected networks are most likely to be incompressible, hence have low compression error.

Figure 19.3 shows the results of a linear regression analysis on offdiagonal complexity with compression error. The correlation coefficient is -0.87. So clearly offdiagonal complexity is correlated (negatively) with compression error, much as we expected, however it is not apparently a good test for indicating if the compression error is large. A better distinguishing characteristic is if C is greater than the mean random C (which can be feasibly calculated) by about 3–4 bits. What remains to be done is to look at networks generated by a dynamical process, for example Erdös-Rényi random graphs[122], or Barabási-Albert preferential attachment[26] to see if they fill in the gap between regular and algorithmically random graphs.

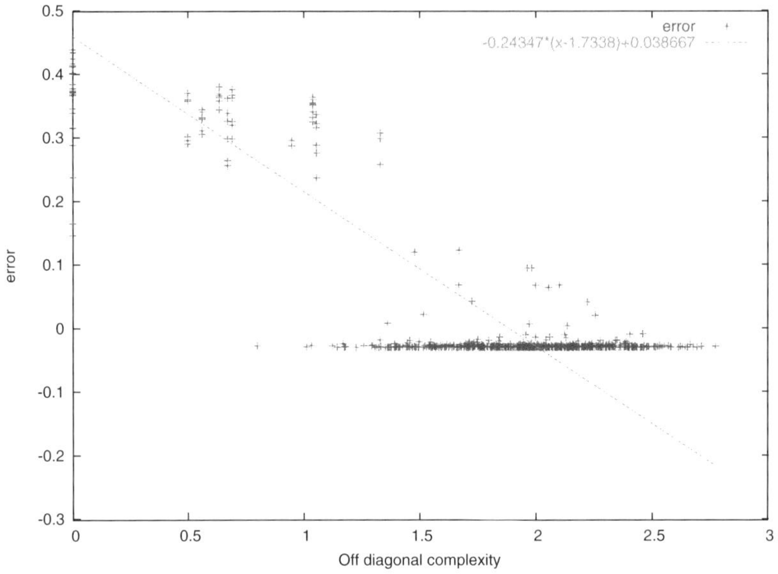

Fig. 19.3 Compression error as a function of offdiagonal complexity. A least squares linear fit is also shown

19.5 Conclusion

In this paper, a simple representation language for N-node undirected graphs is given. An algorithm is presented for computing the complexity of such graphs, and the difference between this measure, and one based on a Turing complete representation language is estimated. For most graphs, the measure presented here computes complexity correctly, only graphs with a great deal of regularity are overestimated.

A code implementing this algorithm is implemented in C++ library, and is available from version 4.D17 onwards as part of the EcoLab system, an open source modelling framework hosted at http://ecolab.sourceforge.net.

Obviously, undirected graphs is simply the start of this work — it can be readily generalised to directed graphs, and labeled graphs such as food webs (although if the edges a labeled by a real value, some form of discretisation of labels would be needed).

Furthermore, most interest is in complexities of networks generated by dynamical processes, particularly evolutionary processes. Some of the first processes that should be examined are the classic Erdös-Rényi random graphs and Barabási-Albert preferential attachment.

Acknowledgements

I wish to thank the *Australian Centre for Advanced Computing and Communications* for a grant of computer time, without which this study would not be possible.

Chapter 20

A Generalised Technique for Building 2D Structures with Robot Swarms

R.L. Stewart and R.A. Russell

Intelligent Robotics Research Centre,
Monash University, Clayton, VIC 3800, Australia
E-mail: {robert.stewart, andy.russell}@eng.monash.edu.au

Templates (or patterns) in the environment are often used by social insects to guide their building activities. Robot swarms can also use templates during construction tasks. This paper develops a generalised technique for generating templates that can vary in space and with time. It is proposed that such templates are theoretically sufficient to facilitate the loose construction of any desired planar structure. This claim is supported by a number of experimental trials in which structures are built by a real robot swarm.

20.1 Introduction

Collective construction is a relatively new problem domain for swarm-based robotics . Thus far, robot swarms have demonstrated the ability to build walls [409; 261; 274; 370], circular nests [300] and annular structures [423]. The question of how more complex structures might be constructed by swarms of minimalist robots is problematic. In addressing this question, the work in this paper is specifically directed at how complex structures, of a designer's choosing, might be constructed.

We define our generalised two-dimensional (2D) collective construction problem as follows:

> *How might a robot collective be designed such that, when provided with adequate building material, the robots are capable of loosely constructing any given planar structure?*

While partial answers to similar questions have recently been proposed [412], a real robot swarm capable of such a task has not (to the best of our knowledge) been demonstrated. In order to increase the potential generality of solutions to this problem, a number of restrictions are imposed. Namely, (i) robots should be designed with a minimalist philosophy, (ii) radio-communication between robots is not permitted, (iii) a global coordinate system is unavailable, and, (iv) an environment map is unavailable. In this paper, a solution to this generalised 2D collective construction problem is detailed. The solution relies on the use of spatio-temporal varying templates (patterns that change in space and with time) and a distributed feedback mechanism.

20.2 Background Information

The individuals of a social insect colony are relatively simple, yet together, through their combined actions they have the capability to solve difficult problems. One such problem is nest construction. A number of mechanisms are thought to be employed by social insects during construction. One of these is the template mechanism. A template is a pattern in the environment that is "used to construct another pattern" [50]. Templates are often in the form of chemical, temperature, humidity and light heterogeneities [382] and insects use these patterns to guide their building activities.

Previous work has investigated how templates might also be used to guide the building activities of robots. An early study revealed how a minimalist robot could build doughnut shaped structures using a template already present in the environment [369]. Following on from this, a robot swarm was designed and constructed with the ability to create templates of its own [370]. These templates were created by a light source mounted on a mobile organiser robot. The light source was set up so that it only projected light over an angle of 30 °. This situation is represented in Fig. 20.1. By generating and then using a spatio-temporal varying template, the robot swarm was able to construct a loose linear wall structure.

Equipped with light sensors, builder robots were programmed to deposit building blocks when they were in a certain window of light intensities (equivalent to a window of sensed voltages, $V_{min} \leq V \leq V_{max}$). This window of light intensities corresponded to a spatial region, $r_1 \leq r \leq r_2$, in the beam called the deposition window (the hatched region in Fig. 20.1). Here, the term beam refers to that region of the light beam defined by $V_{ambient} \leq V \leq V_{max}$.

The organiser robot drove in steps along a straight line, waiting a set time (called the latency time) at each stop. This resulted in a path being

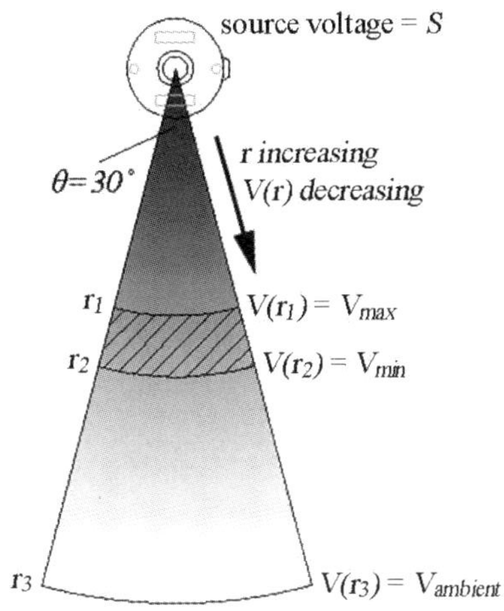

Fig. 20.1 A light source mounted on an organiser robot projects light over a fixed angle of 30 °. The deposition window is indicated by hatching.

traced out by the deposition windows that fell along a straight line. Because the builder robots deposited blocks when they were within the deposition window, a loose linear wall structure resulted. The latency time was chosen through trial and error to give a good trade-off between the number of blocks deposited and the total trial time.

In an environment, which may be changing, choosing an appropriate latency time becomes a difficult problem. To address this problem the robot system was then modified to incorporate a distributed feedback mechanism that allowed for an adaptable choice of latency time [371]. In the new system (shown in Fig. 20.2), builder robots assessed the current state of the deposition window using a cue. If they perceived the window to be full (with blocks) they became frustrated by their inability to make a deposit. After two consecutive failed attempts to deposit a block they displayed their frustration by producing a flash of high intensity light. This burst of light acted as a signal to the organiser robot. The organiser robot waited to receive a number of these signals, possibly from different robots, before moving to the next stop. Having moved to a new stop, the deposition window was then devoid of any blocks. This meant that builder robots were again able to make deposits and their frustration level decreased.

A moving organiser robot carrying a light source of fixed intensity created the spatio-temporal varying template. Because the window limits, V_{min} and V_{max}, were fixed, the distance of the deposition window from the

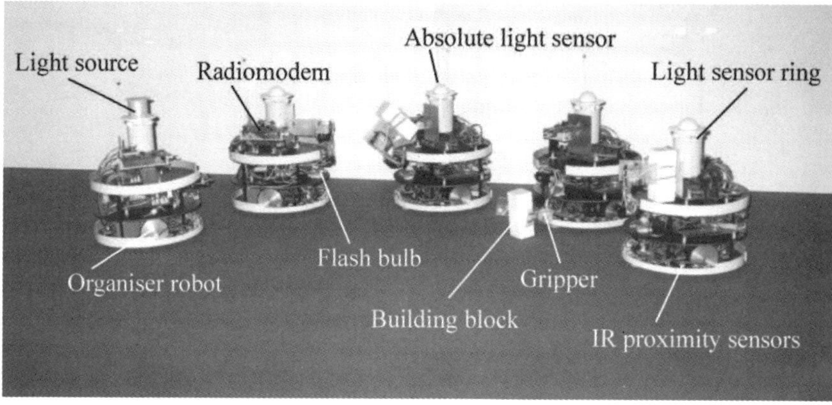

Fig. 20.2 The small swarm of robots, called *The Robotermites*, with important hardware features highlighted. The organiser robot is on the far left.

organiser robot was also fixed. In this paper a new technique is proposed, and verified, that allows for variable deposition window placement without the need for the organiser robot to employ complex motion patterns itself. This is achieved through a systematic intensity variation of the light source mounted on top of the organiser robot. Both techniques are then combined to allow more complex structures to be constructed.

20.3 A New Technique for Creating Spatio-temporal Varying Templates

20.3.1 *Calibration*

To understand how light intensity variation might lead to variable deposition window placement, a calibration routine was undertaken.

The calibration routine involved driving a builder robot away from the organiser robot's light source along a radial line. The starting position of the builder robot was such that the initial distance between the organiser robot's light source and the builder robot's absolute light sensor was $\sim 0.265m$. The builder robot was stopped every $\sim 0.1m$ and drove a total distance of $\sim 1.5m$. At each stop the intensity of the organiser robot's light bulb (light source) was varied by automatically adjusting the voltage applied to it. This light source voltage, S, was varied from 30 to 100 normalised units in increments of 5. For each different light source voltage, the light level sensed by the builder robot was recorded (in terms of the

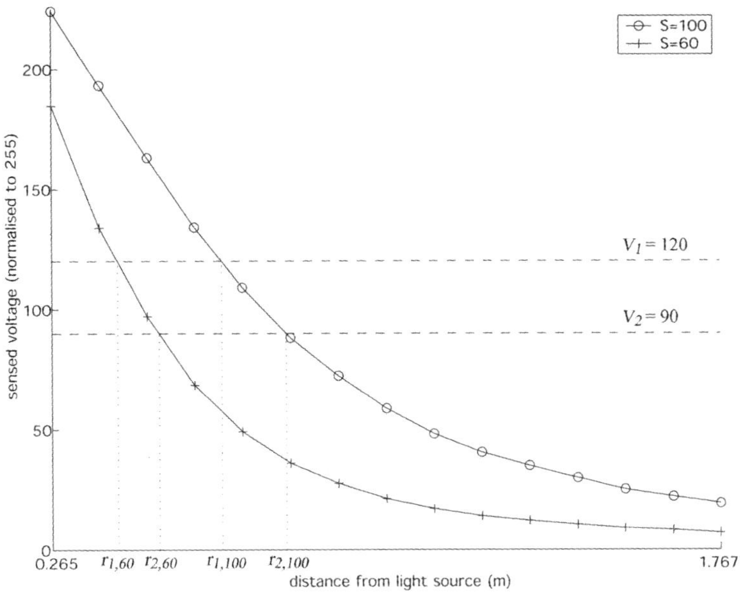

Fig. 20.3 Plots of two characteristic curves for different light source voltages (namely, $S = 60$ and $S = 100$). Also shown are two lines that define the deposition window, $90 \leq V \leq 120$.

sensed voltage, V). Note that the higher the value for S, the more intense the light source was. Two typical data curves obtained during this process are shown in Fig. 20.3. As can be seen, each curve is a monotonically decreasing function of distance as expected. Also evident from the entire set of data curves (not shown in Fig. 20.3) was that for any given distance from the light source, the sensed voltage was a monotonically increasing function of light source voltage.

To understand how the deposition window placement ($r_1 \leq r \leq r_2$) can be varied, consider the situation where builder robots are, as in previous trials, programmed to deposit building material when they are in a certain window of sensed voltages, $V_{min} \leq V \leq V_{max}$. Consider again Fig. 20.3 which shows the two example data curves that correspond to the light source voltages $S = 60$ and $S = 100$. Two lines for $V_1 = V_{max} = 120$ and $V_2 = V_{min} = 90$ are also shown on the same graph representing an arbitrarily chosen deposition window definition ($90 \leq V \leq 120$). The two lines are seen to intersect each curve at different locations. If the source intensity is low ($S = 60$) the deposition window placement is close to the light source ($r_{1,60} \leq r \leq r_{2,60}$). If on the other hand, the source intensity is high ($S = 100$) the deposition window placement is further away from the light source ($r_{1,100} \leq r \leq r_{2,100}$). It is therefore clear that by systematically varying the light source intensity the location of the deposition window can be controlled.

By re-expressing the obtained data, it was possible to determine appropriate values for V_{min}, V_{max} and $V_{ambient}$ (namely, 50, 70 and 15 respectively). A number of factors had to be considered when deciding upon values for these parameters. For brevity, details regarding this design process will be provided elsewhere. With the design parameters chosen, it is possible to determine the light source voltage required to give a deposition window placement that starts at some desired position (r_1), as well as the depth of the deposition window ($r_2 - r_1$). Due to the non-linear nature of the data curves (e.g. Fig. 20.3), the deposition window depth does increase somewhat with distance. This could be regarded as a loss of resolution with distance and is intrinsic to this system.

20.3.2 *Experimental Procedure*

The trials in this section (20.3) were conducted with a robot swarm consisting of one organiser robot and two builder robots. The two builder robots were known to possess similar absolute sensor responses to light intensity so that calibration only had to be performed once. As in previous experiments, the organiser robot had a light source mounted on top of it while builder robots possessed an absolute light sensor, a gripper and a flash bulb (for producing bursts of high intensity light). All robots had a light sensor ring, infra-red (IR) proximity sensors and a radiomodem (for data logging purposes). The complete robot swarm is shown in Fig. 20.2.

A rectangular test area (with inner dimensions of $2.56m \times 2.16m$) was used for the trials. The organiser robot was stationary for the duration of these trials and the builder robots roamed freely. As in previous experiments, two layers of building blocks were arranged outside the inner perimeter of the main test area. Blocks taken from the outermost layer during the trials were replaced manually by the observer. If blocks were pushed aside in the outermost layer so that a large gap developed, blocks were added to a third layer where needed.

The organiser and builder robots were programmed with the rule sets that are shown in Tables 20.1 and 20.2 respectively. The *If-Then* rules appearing first receive higher precedence. The feedback mechanism described earlier (Section 20.2) is evident in these rules (allowing for an adaptable choice of latency time).

Table 20.1 The rule set used by the organiser robot in Trials 1 and 2.

(1) **If** (number of deposition windows completed equals the total number of deposition windows required) **Then** (turn off light bulb)

(2) **If** (3 flashes have been detected for the current deposition window or the trial has just been started) **Then** (change light source voltage to the next value in the sequence (a_n))

(3) Count flash signals

NOTE: In Trial 1 the sequence $(a_n) = (S_1, S_2, S_3, S_4, S_5, S_6)$ and in Trial 2 the sequence $(a_n) = (S_1, S_2, S_6)$ where $S_1 = 41.5$, $S_2 = 48.2$, $S_3 = 56.4$, $S_4 = 67.4$, $S_5 = 81.6$ and $S_6 = 100.0$.

Table 20.2 The rule set used by the builder robots in all trials.

(1) **If** (not holding block and not in beam and object detected) **Then**(attempt to pick up object)

(2) **If** (holding block and in deposition window) **Then**(deposit block behind another block or at inner window limit and reset frustration counter)

(3) **If** (holding block and have just arrived in beam) **Then** (drive further into beam avoiding obstacles if detected)

(4) **If** (frustration counter equals 2) **Then** (flash light and reset frustration counter)

(5) **If** (holding block and in beam and obstacle detected) **Then** (increment frustration counter, turn randomly $\pm 135\,^\circ$ and then drive forwards)

(6) **If** (holding block and in beam) **Then** (drive up beam towards light)

(7) **If** (obstacle detected) **Then** avoid obstacle

(8) 8. Drive forwards

20.3.3 *Building a Radial Wall With and Without a Gap*

The first two trials undertaken demonstrate that the deposition window placement, and hence block placement, can indeed be varied by altering the light source intensity. From the calibration results and chosen design parameters, different light source voltages (equivalent to source intensities) were chosen to create a certain number of deposition windows that lined up with one another. This was achieved by making the inner limit placement of each deposition window equal to the outer limit placement of the proceeding deposition window. Figure 20.4 shows the placement of the different deposition windows and the corresponding light source voltages.

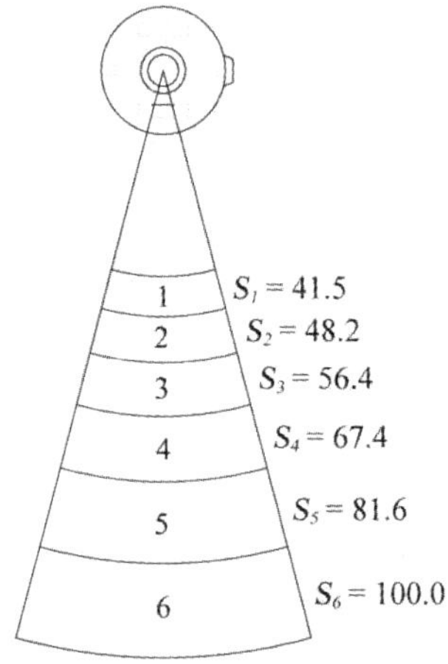

Fig. 20.4 Deposition window placements (numbered from 1 to 6) and the corresponding light source voltages used to create these placements.

In the first trial (Trial 1), the organiser robot was programmed to have an initial light source voltage of S_1. After waiting a certain latency time (determined by the feedback system) the organiser robot was programmed to then change its light source voltage to the next value, S_2, and wait again for a new latency time. This process continued until all light source voltages in the sequence $(a_n) = (S_1, S_2, S_3, S_4, S_5, S_6) = (41.5, 48.2, 56.4, 67.4, 81.6, 100.0)$ had been applied and the appropriate latency time waited for each. Note that the definition for latency time has now been broadened to mean the time spent waiting by the organiser robot for the current deposition window to be deemed full.

Because the deposition windows fall along a radial line from the light source (Fig. 20.4), it was expected that deposited blocks would also fall along a radial line. Figure 20.5.a shows the structure constructed by the robot swarm in this trial. Blocks were deposited in the expected manner allowing a radial wall structure to grow out from a starting point.

A natural extension to the first trial was to create a discontinuity in the radial wall structure. In a second trial (Trial 2) the organiser robot was given a reduced sequence of light source voltages $(a_n) = (S_1, S_2, S_6) = (41.5, 48.2, 100.0)$ so that the 3rd, 4th and 5th deposition windows in Fig. 20.4 would be absent. From this choice of light source voltages it was expected that a radial wall similar to that in Fig. 20.5.a would be constructed but that this time there would be a gap in the wall.

As expected, a radial wall with a gap was constructed by the robot

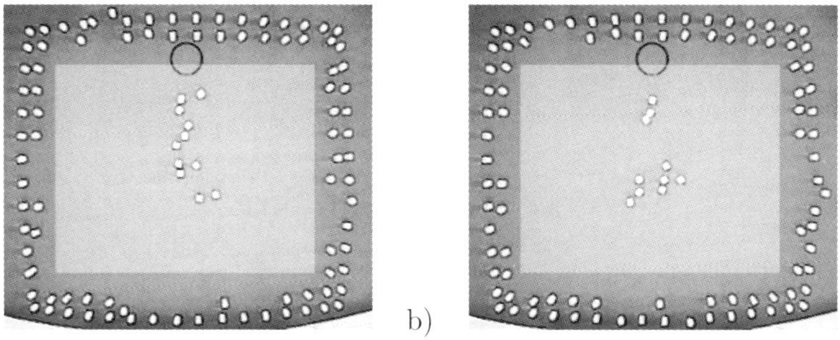

a) b)

Fig. 20.5 The configuration of the test area (with robots removed) after the completion of (a) Trial 1 and (b) Trial 2. The organiser robot's position during each trial is indicated by a circle and the structures are located in the graphically enhanced lighter areas. In the first trial a radial wall has been constructed and in the second trial a radial wall with a gap has been constructed.

swarm (see Fig. 20.5.b). For this to happen an appropriate jump in light source intensity was all that was required. Again the structure was observed to grow outwards away from the light source as was dictated by the sequence of light source voltages. The results from both trials confirm that the proposed technique for generating a spatio-temporal varying template is indeed viable.

20.4 Solving the Generalised 2D Collective Construction Problem

Thus far two techniques have been found for generating spatio-temporal varying templates. In general terms, these are (i) a moving source of fixed intensity [370] and (ii) a stationary source of variable intensity (Section 20.3). By combining these we arrive at the more general technique for generating spatio-temporal varying templates, namely, (iii) a moving source of variable intensity. It is through this generalised technique that it should be possible to realise any desired 2D structure without violating the problem restrictions.

By establishing a Cartesian coordinate system relative to the organiser robot's direction of movement (Fig. 20.6), it becomes clear that for any given deposition window placement centred on (x_i, y_i), there exists a corresponding source intensity (equivalent to a light source voltage, $S_i = f^{-1}(y_i)$) and organiser displacement (x_i). Any given 2D structure can then be decomposed into a building program that consists of a sequence of linear displacement values and light source voltages that govern the spatial location of deposition windows. If each deposition window is filled with

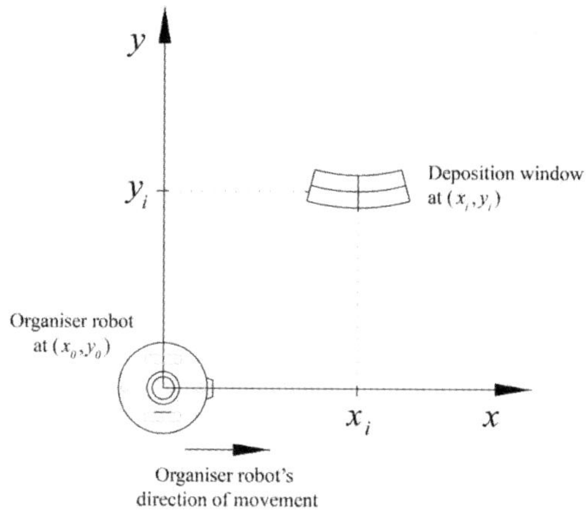

Fig. 20.6 A conceptual coordinate system established relative to the organiser robot's initial position and its direction of intended movement.

blocks by the builder robots then we would expect the final built structure to be a loosely constructed realisation of the intended structure design. To verify this generalised technique, a number of trials were proposed and undertaken.

20.4.1 *Experimental Procedure*

The entire 5-member robot swarm was used in the remaining trials (Fig. 20.2). Calibration was performed for all builder robots. The design parameters chosen for each builder robot ensured they all had a consistent quantifiable response to light. The physical dimensions for the test area set-up were kept the same (as in Section 20.3) as was the rule set followed by the builder robots (see Table 20.2). The organiser robot's rule set (Table 20.1) was modified somewhat to allow for the possibility of movement after the completion of each deposition window (Table 20.3). This gave the organiser robot the ability to reach all required displacement values (x_i in Fig. 20.6).

20.4.2 *Building Structures of Greater Complexity*

Three trials were undertaken to verify the generalised technique for building structures using spatio-temporal varying templates. Sequences of light source voltages (a_n) and movement distances (b_n) were programmed into

Table 20.3 The modified rule set used by the organiser robot in Trials 3, 4 and 5.

(1) **If** (number of deposition windows completed equals the total number of deposition windows required) **Then** (turn off light bulb)

(2) **If** (3 flashes have been detected for the current deposition window or the trial has just been started) **Then** (move/rotate a distance given by the next value in the sequence b_n and change the light source voltage to the next value from the sequence a_n)

(3) Count flash signals

the organiser robot. These sequences specified the placement of deposition windows relative to the organiser robot. Deposition window placements were ordered so that previous construction did not inhibit future construction. This was achieved by ensuring structural growth was always away from the organiser robot. In each trial the feedback system enabled the transition between deposition windows.

Figure 20.7 gives a representation of three different spatio-temporal varying template designs and the resulting structures for each of these that were actually built by the robot swarm. In the first trial (Trial 3, 20.7.a and 20.7.b), a diagonal wall was constructed to show how linear motion and variable source intensity can readily be combined. In the second trial (Trial 4, 20.7.c and 20.7.d), a cross shape structure was built to demonstrate the capacity for the swarm to build intersecting walls.

In the third trial (Trial 5, 20.7.e and 20.7.f), the organiser robot rotated incrementally and varied its intensity so that two concentric arcs were constructed. This trial showed that rotational motion can be used instead of linear motion.

20.5 General Discussion

In each of the trials detailed in this paper, the builder robots did not need to discern changes made to the light source intensity by the organiser robot. Instead, they wandered about searching for the beam and the fixed conditions (namely, a sensed voltage in the range $V_{min} \leq V \leq V_{max}$) under which they would make a deposit. That is, there was no need for the builder robots to be informed directly that the spatial location of the deposition window was being varied. It was enough for the organiser robot to make changes to a pattern in the environment that indirectly affected the actions undertaken by builder robots.

The generalised technique for building 2D structures has a number

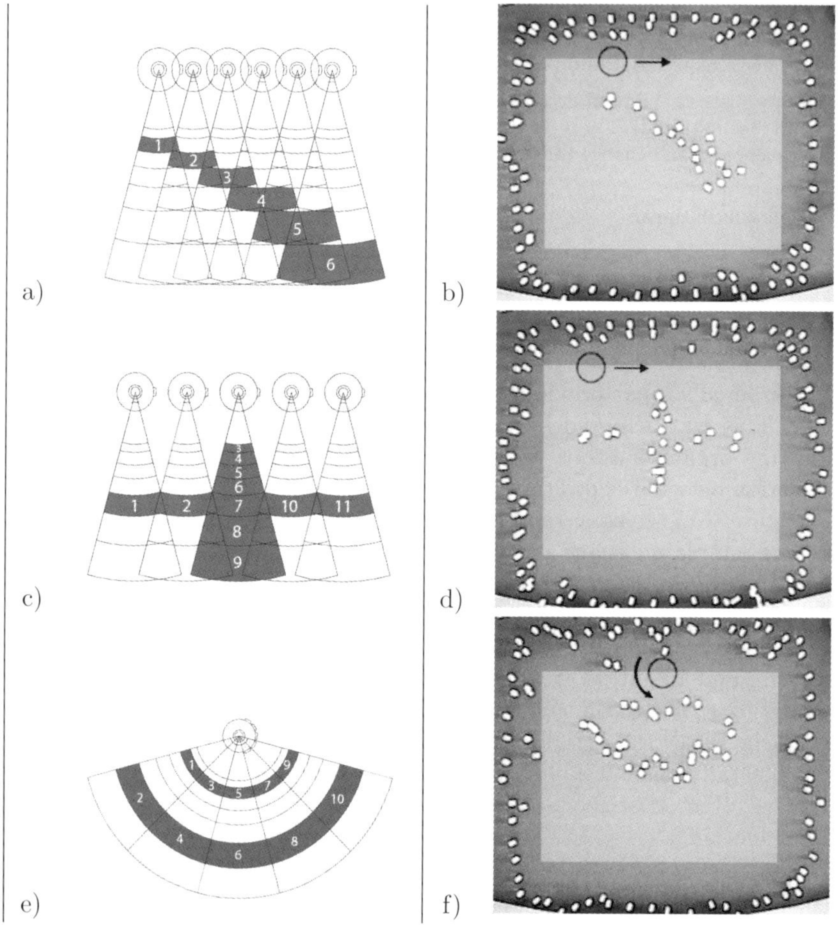

Fig. 20.7 Representations, (a), (c) and (e), of the spatio-temporal varying templates used by the robot swarm to construct, (b) a diagonal wall, (d) a cross shape, and, (f) concentric arcs, respectively. For the template representations, the different deposition windows used for the building of each structure are shaded and labelled to indicate their order. For each of the resulting structures, a circle indicates the organiser robot's initial position and an arrow shows its direction of movement. The structures are located in the graphically enhanced lighter areas.

of benefits. (i) Robots do not require sophisticated sensing or computational requirements, instead, they can be almost reactive. (ii) Global coordinates are not required. (iii) A map is not required. (iv) Different structures can be constructed without the need to update the rule sets of all the robots (only the organiser robot's rule set needs updating). (v) Radio-communications are not required. Instead robots use indirect communication through the physical environment and flash signals to regulate construction activities (see [371]).

20.6 Conclusion

A second technique for creating a spatio-temporal varying template has been introduced in this paper. The technique involves the intensity variation of a light source carried by an organiser robot. Builder robots, programmed to make deposits under certain fixed lighting conditions, can be made to vary the placement of their building material by changes in the organiser's light source intensity. The proposed technique was verified in two experimental trials where a robot swarm constructed a radial wall and a radial wall with a gap.

Both techniques for the creation of spatio-temporal varying templates were then combined to form a generalised technique that gives a robot swarm the capacity, theoretically, to build any given planar structure. This claim was supported by a number of experimental trials highlighting various aspects of the system.

Acknowledgement

This work is supported by the Australian Research Council funded Centre for Perceptive and Intelligent Machines in Complex Environments. The work is also supported by a grant from Monash University's Faculty of Engineering Small Grants Scheme. Robert Stewart would like to acknowledge the support of an Australian Postgraduate Award.

Chapter 21

H-ABC: A Scalable Dynamic Routing Algorithm

B. Tatomir and L.J.M. Rothkrantz

COMBINED Project, DECIS Lab,
Delftechpark 24, 2628 XH Delft, The Netherlands,
E-mail: b.tatomir@ewi.tudelft.nl

Man-Machine Interaction,
Faculty of Computer Science and Electrical Engineering,
Delft University of Technology,
Mekelweg 4, 2628 CD Delft, The Netherlands
E-mail: l.j.m.rothkrantz@ewi.tudelft.nl

In small networks, ant based algorithms proved to perform better than the conventional routing algorithms. Their performance decreases by increasing the number of nodes in the network. The scalability of the algorithms is affected by the increasing number of agents used. In this paper we present a scalable Hierarchical Ant Based Control algorithm (*H-ABC*) for dynamic routing. The network is split into several smaller and less complex networks called sectors. The agents are divided in two categories: local ants and exploring ants. Only the nodes situated at the border between sectors can generate exploring ants, the ones used to maintain the paths between different sectors. They are carrying no stack which reduces the overhead in the network. The algorithm was implemented and its performance compared with the well known *AntNet*.

21.1 Introduction

The *AntNet* adaptive agent-based routing algorithm [68; 69], is the best-known routing algorithm for packet-switched communications networks, which is inspired from the ants life. Besides the probability tables, at each node the average trip time, the best trip time, and the variance of the trip times for each destination are saved. Routing is determined through complex interactions of network exploration agents. These agents (ants) are divided into two classes, the forward ants and the backward ants. The idea behind this sub-division of agents is to allow the backward ants to utilize the useful information gathered by the forward ants on their trip from source to destination. Based on this principle, no node routing updates are performed by the forward ants, whose only purpose in life is to report network delay conditions to the backward ants. This information appears in the form of trip times between each network node. The backward ants inherit this raw data and use it to update the routing tables of the nodes. In [27] *AntNet* was improved with an intelligent routing table initialization, a restriction on the number of ants in the network and a special pheromone update after node failures. An increased adaptivity of ants [415] and reduced size of the routing tables [243] was achieved by combining *AntNet* with genetic algorithms.

Ant-based control (*ABC*) was the first successful swarm based routing algorithm and designed for telephone networks [349]. This algorithm uses only one class of forward ants, which travel along the network gathering information. They use this information for updating the routing table rows corresponding to their source node. In [51] *ABC* was extended and the pheromone updating was introduced also for the other intermediate nodes visited by the agent. Two other ramifications of *ABC* for packet-switched networks were presented in [376] and [189].

Although proved to perform better than the best classic algorithms like *RIP* (Routing Information Protocol) and *OSPF* (Open Shortest Path First), *AntNet* and *ABC* have scalability problems [215]. Since each node has to send an ant to all the other nodes of the network, for large networks, the amount of traffic generated by the ants would be prohibitive. Furthermore, for distant destinations there is a larger likelihood of the ants being lost. Moreover, the large traveling times of the ant render the information they carry outdated.

One way to solve this load problem and attain scalability is by using hierarchical routing. Adaptive-*SDR* [215] groups nodes into clusters and directs data packets from a source node to a destination by using intra and inter-cluster routing. Two types of agents, are introduced into the network. The first type is colony ants and the second type is local ants. The task of the colony ants is to find routes from one cluster to the other, while local ants are confined within a colony and are responsible for finding routes within their colonies. The colony ants are launched at every node. This keeps the overhead high.

BeeHive [410] is a novel routing algorithm, which has been inspired by the communicative and evaluative methods and procedures of honey bees. In this algorithm, bee agents travel through network regions called foraging zones. The model requires only forward moving agents and they utilize an estimation model to calculate the trip time from their source to a given node. The agents are sent to other nodes by broadcasting. Each node keeps a copy of each agent. Although they are really small the overhead is higher than in case of *AntNet*. Another real disadvantage is the higher memory use for storing every agent to distinguish between different replicas received at the same node.

In the next section we introduce a new hierarchical routing algorithm (*H-ABC*) which combines features of the algorithms presented before. We tested and compared its performance with *AntNet* in a network with 114 nodes and 172 bidirectional links. The simulation environment and the experimental settings are presented in section 3. The tests and results are described in section 4. The last section contains conclusions and future work.

21.2 The Hierarchical Ant Based Control algorithm

21.2.1 *Network model*

We decided to split the network in sectors. The nodes situated at the border of a sector and which have connection with other sectors are called *routing nodes*. As we will see later these nodes will play a special role, their activity being different than the one of an inner sector node. An example of such a network is shown in Figure 21.1 representing the Japanese Backbone NTTNet divided in 3 sectors. The *routing nodes* are circled.

A common feature of all the routing algorithms is the presence in every network node of data structure, called routing table, holding all the information used by the algorithm to make the local forwarding decisions. The routing table is both a local database and a local model of the global network status. Each node i in the network has a probability table for every possible final destination d. The tables have entries for each next neighbouring node n, P_{dn}. This expresses the goodness of choosing node n as its next node from the current node i if the packet has to go to the destination node d.

In our case this data structure has to be modified. For every sector of the network a virtual node is introduced. This can be understood as an abstraction for all the nodes of the sector. Each virtual node will have an entry in the data structures of every node. They will be used to route the

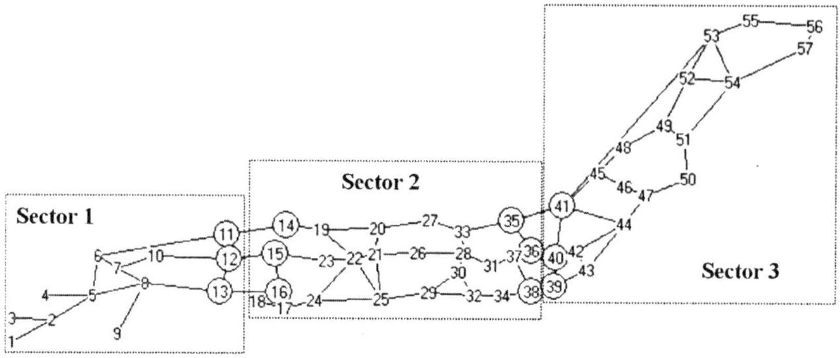

Fig. 21.1 Japanese NTTNet

data between different sectors. Considering the network with 3 sectors in Figure 21.1, an example of a routing table in node 12 of a sector will look like this.

Table 21.1 Routing table in node 12

	Neighbour			
Destination	**N10**	**N11**	**N13**	**N15**
D1	**0.28**	**0.45**	**0.28**	**0.00**
D2	**0.42**	**0.46**	**0.12**	**0.00**
D3	**0.21**	**0.52**	**0.27**	**0.00**
...	...	...	...	**0.00**
D11	**0.02**	**0.95**	**0.02**	**0.00**
D13	**0.02**	**0.02**	**0.95**	**0.00**
V2	**0.02**	**0.02**	**0.02**	**0.93**
V3	**0.05**	**0.05**	**0.05**	**0.85**

Besides the probability table in every node i we have the following additional data structure:

- μ_d: an array storing the mean value of the delay encountered for the destination d
- $S[d]$: an array which maps every node in the network to the corresponding sector.
- $U[d]$: an array of flags which mention if a node d is 'up' or 'down'

The algorithm makes use of 3 types of ants: local, backward and exploring ants. Next we will describe the behaviour of each entity present in the network.

21.2.2 *Local ants*

The purpose of the local ants is to maintain the routes between nodes of the same sector or to the closest routing node for another sector.

Each node s inside a sector periodically generates a local ant F_{sd}. The destination d can be another node in the sector or a virtual node:

$$d = \begin{cases} \text{a node in the same sector}(S[d] = S[s]) \text{ with probability of } 0.9 \\ \\ \text{a virtual node}(S[d] \neq S[s]) \text{ with probability of } 0.1 \end{cases} \tag{21.1}$$

The *routing nodes* have a different behaviour. In this case:

$$F_{sd} = \begin{cases} \text{local ant}(S[d] = S[s]) \text{ with probability of } 0.1 \\ \\ \text{exploring ant } (S[d] \neq S[s]) \text{ with probability of } 0.9 \end{cases} \tag{21.2}$$

A local ant behaves similar with the forward ant in *AntNet*. It keeps a memory about the visited nodes and the estimated time to reach them. At each node i, before going to the next neighbour n, the ant memorizes the identifier of the next node n and a virtual delay T_{in}. This delay represents the estimative time necessary for the ant to travel from node i to n using the same queues as the data packets.

$$T_{in} = \frac{q_n + Size(F_{sd})}{B_{in}} + D_{in} \tag{21.3}$$

- q_n [bits] is the length of the packets buffer queue towards the link connecting node i and its neighbour n;
- B_{in} is the bandwidth of the link between i and n in [bit/s];
- $Size(F_{sd})$ [bits] is the size of the local ant;
- D_{in} is the propagation delay of the link.

The selection of the next node n, to move to, is done according with the probabilities P_d and the traffic load in the node i.

$$P'_{dn} = \frac{P_{dn} + \alpha \cdot l_n}{1 + \alpha \cdot (|N_i| - 1)}, \alpha = 0.4 \in [0, 1] \tag{21.4}$$

$l_n \in [0, 1]$ is a normalized value proportional to the amount q_n (in bits waiting to be sent) on the link connecting the node i with its neighbour n:

$$l_n = 1 - \frac{q_n}{\sum_{j=1}^{|N_i|} q_j} \tag{21.5}$$

If a cycle is detected, the cycle's nodes are popped from the ant's stack and all the memory about them is destroyed. If the cycle is greater than half the ant's age, the complete ant is destroyed.

A local ant is not allowed to leave his sector. In this way in all the probability tables of the routing nodes we have $P_{dn} = 0$ (see Table 21.1 - N15) if:

- $S[d] = S[i]$: the destination is a node inside the sector
- $S[n] \neq S[i]$: the neighbour n is in another sector

For a local ant there are two possibilities to reach its destination. One is of course when it arrives in the node d. But when d is a virtual node it stops at the first encountered routing node. In this case it pushes on the stack the node d identifier and μ_d, the average time to go from node i to the sector d. At this moment the agent F_{sd} finishes its trip. It transfers all of its memory to a new generated backward ant B_{ds} and dies.

21.2.3 *Backward ants*

A backward ant takes the same path as that of its corresponding local ant, but in the opposite direction. At each node i along the path it pops its stack to know the next hop node. It updates the routing tables for the node d but also for all the subpaths from $i \rightarrow d$. The time T_{id} to reach d is the sum of all the segments T_{jk} on the path:

$$T_{id} = \sum T_{jk}, \text{ where } j,\, k \in i \rightarrow d \tag{21.6}$$

First it modifies the value of μ_d.

$$\mu_d = \mu_d + \eta(T_{id} - \mu_d),\ \eta = 0.1 \tag{21.7}$$

After this it updates the probability table with a reinforcement value r. This is a function of the time T_{id} and its mean value μ_d.

$$r = \begin{cases} \frac{T_{id}}{c\mu_d};\ \frac{T_{id}}{c\mu_d} < 1,\ c = 1.1 > 1 \\ \\ r = 1 \text{ otherwhise} \end{cases} \tag{21.8}$$

$$P'_{dn} = P_{dn} + (1-r)(1-P_{dn}) \text{ for } n \text{ the node chosen by the ant} \quad (21.9)$$

$$P'_{dj} = P_{dj} - (1-r)P_{dj}, \text{ for } j \neq n \quad (21.10)$$

When the source node s is reached again, the agent B_{ds} dies.

21.2.4 *Exploring ants*

The purpose of exploring ants is to find and maintain the routes between different sectors. They are 'light' and keep no track about the path they followed. The only information they register is an estimate time to reach their source sector T_s.

The exploring ants are generated only by the *routing nodes* of each sector s. They receive as destination d a virtual node representing another sector. In this case we will refer by s not to the source node but to the sector.

They are forwarded to destination using the same mechanism as the local ants. If a cycle is detected it is removed, and in case it is bigger than half of the ant age, the ant is killed.

As the local ants which are not allowed to get out of the current sector, the exploring ants are not allowed to move to a node inside their source sector. Once they left the home sector they can't return. If this still happens the exploring ant is killed. This is because there are other routing nodes in the same sector which are closer to the destination sector d of the ant. The ants generated there will be more efficient with that destination.

When an exploring ant arrives in a node i coming from a node p, it adds to its trip time T_s the trip time necessary for ant to travel from i to p.

$$T_s = T_s + T_{ip} \quad (21.11)$$

$$T_{ip} = \frac{q_p + Size(F_{sd})}{B_{ip}} + D_{ip} \quad (21.12)$$

The new computed T_s value is used to update the routing table at node i. The changes are are similar with the ones made by the backward ants, but they are done only for the source sector s.

$$\mu_s = \mu_s + \eta(T_{is} - \mu_s) \quad (21.13)$$

$$r = \begin{cases} \frac{T_s}{c\mu_s}; \frac{T_s}{c\mu_s} < 1 \\ \\ r = 1 \text{ otherwhise} \end{cases} \tag{21.14}$$

The reinforcement is given to the link $i \rightarrow p$:

$$P'_{sp} = P_{sp} + (1 - r)(1 - P_{sp}) \tag{21.15}$$

$$P'_{sj} = P_{sj} - (1 - r)P_{sj}, \text{ for } j \neq p \tag{21.16}$$

An exploring ant ends its trip when arrives at a routing node of the destination sector d.

21.2.5 *Data packets*

A packet P_{sd} generated at a node s can have as destination d any node in the network.

Arriving at a node i they are routed according with the probabilities P_d. If the destination node d can not be found in the same sector with the node i then the packet is routed to the destination sector $S[d]$ using the correspondent virtual node entry in the routing table.

21.2.6 *Flag packets*

When a neighbouring node n is detected as inactive, a special packet Y_n is broadcasted in the network to inform the other nodes. It is carrying the identifier of the node n and a flag set to *down* in this case. When the crashed node becomes active again it broadcast the same type of packet but with the flag set on *up*. When a packet Y_n is received at a node i it compare the U_n value with his flag value:

- if $U_n = Y_n$, then the change was already done by another packet and the current packet is destroyed.
- if $U_n \neq Y_n$, then set $U_n = Y_n$ and send Y_n to all the neighbours using the same priority queues as the ants.

Next we present the pseudocode of the algorithm

Algorithm 21.1 H-ABC part-0

```
{i - current node, d - destination node, s - source node}
{n - successor node of i, p - predecessor node of i}
for all Nodes {concurrent activity over the network} do
    if time to generate an agent at node s then
        Create(F_sd)
    end if
    Call H-ABC part-1
end for
```

21.3 Simulation environment

We implemented and tested our new *H-ABC* algorithm in a new developed simulation environment. For comparison we choose the *AntNet* algorithm which we already had implemented. The common parameters of the algorithms were set to the same values as for *AntNet* in [69]. In the literature, the most complex network instance that was mostly used in simulations is the Japanese Internet Backbone (NTTNET)(see Figure 21.1). It is a 57 node, 81 bidirectional links network. The link bandwidth is 6 Mbits/sec and propagation delay rages from 0.3 to 12 milliseconds. In order to create a more challenging environment we linked two copies of the NTTNet network and added 10 extra links (see 21.2). The result is a network with 114 nodes and 172 bidirectional links. Traffic is defined in terms of open sessions between two different nodes. At each node the traffic destination is randomly chosen between the active nodes in the network and remains fixed until a certain number of packets (50) have been sent in that direction. The mean size of data packet is 4 Kbytes, the size of an agent is 192 bytes. The queue size is limited to 1 Gb in all experiments. We studied the *H-ABC* performance related to the number of sectors the network was divided. 3 versions of the *H-ABC* algorithm were tested:

- *H-ABC*2: network with 2 sectors
- *H-ABC*4: network with 4 sectors
- *H-ABC*6: network with 6 sectors

Actually the version of the algorithm for only one network (*H-ABC*1) is exactly *AntNet*.

Algorithm 21.2 H-ABC part-1

```
for all forward ants F_sd received at node i do
    if big cycle detected then
        kill F_sd
    else
        if F_sd is local then
            dest_reached ← false
            if d is a virtual sector and i is a routing node then
                F_sd ← (d, μ_d) {add the virtual node on the stack}
                dest_reached ← true
            else
                if d = i then
                    dest_reached ← true
                end if
            end if
            if dest_reached then
                create B_is
                send B_is to p using the priority queue
                kill F_sd
            else
                n ← GetNext(F_sd)
                T_in ← GetVirtualtime(n, q_n) {compute the virtual time
                to get to n}
                F_sd ← (n, T_in) {add the new information on the stack}
                send F_sd to n using priority queue
            end if
        else
            T_ip ← GetVirtualtime(p, q_p)
            T_is ← T_ps + T_ip
            UpdateProbabilities(i, p, T_is)
            if i is routing node and S[i] = d {destination was reached}
            then
                kill F_sd
            else
                n ← GetNext(F_sd)
                if S[n] = S[s] and S[i] ≠ S[s] then
                    kill F_sd {a forward agent should not return to his
                    source sector}
                else
                    send F_sd to n using priority queue
                end if
            end if
        end if
    end if
end for
Call H-ABC part-2
```

Algorithm 21.3 H-ABC part-2

```
for all backward ants B_sd received at node i do
    UpdateProbabilities(i, d, T_id) {do updates to the node information}
    if d ≠ i {destination not reached} then
        n ← GetNext(B_sd) select next node to go
        send B_sd to n using high priority queues
    else
        kill B_sd
    end if
end for
for all data packet P_sd received at node i do
    if d ≠ i {destination not reached} then
        if S[i] = S[d] {the destination d is in the current sector} then
            n ← GetNext(P_sd)
        else
            n ← GetNext(P_sS[d])
        end if
        send P_sd to n using normal queue
    end if
end for
for all flag packet Y_n received at node i do
    if U_n = Y_n then
        kill Y_n
    else
        U_n ← Y_n
        broadcast Y_n using high priority queues
    end if
end for
```

21.4 Test and results

During the simulations, we focused on 4 metrics for evaluation: the trough-put, the delivered packets rate, the arrived packets delay and the overhead. Each test lasted 300s with a training period (just ants were generated) of 30s.

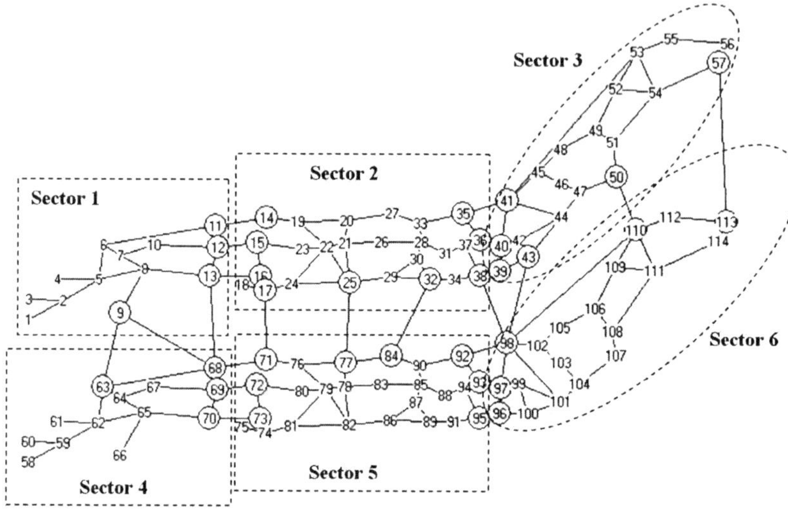

Fig. 21.2 Traffic network with three cities

21.4.1 *Low traffic load*

A routing algorithm should perform well not only under heavy traffic load but also under low load over the network. To achieve such a traffic we set the mean packet generation period(PGP) to 0.3s. Figure 21.3 shows the average packet delay. The average packet delay for *H-ABC*6 and *H-ABC*4 is below 250ms, around 250ms for *H-ABC*2 and above 300ms for *AntNet*. All the algorithms delivered similar throughput and more than 99% of the packets.

21.4.2 *High traffic load*

For this test we decreased the mean packet generation period to 2ms. Again *H-ABC*4 and *H-ABC*6 scored the best. They delivered 99% of the packets with an average delay below 1s. *H-ABC*2 delivered 98% of the packets and after an increasing slope the average delay went down to 1s. The *AntNet* had not so good performance. Just 83% of the packets reached the destination with an average delay of more than 5s.

As expected from the difference between the packets delivered ratios, the throughput of *H-ABC* algorithms $\simeq 2250Kb/s^{-1}$ is higher than the average throughput of $AntNet \simeq 1900Kb/s^{-1}$.

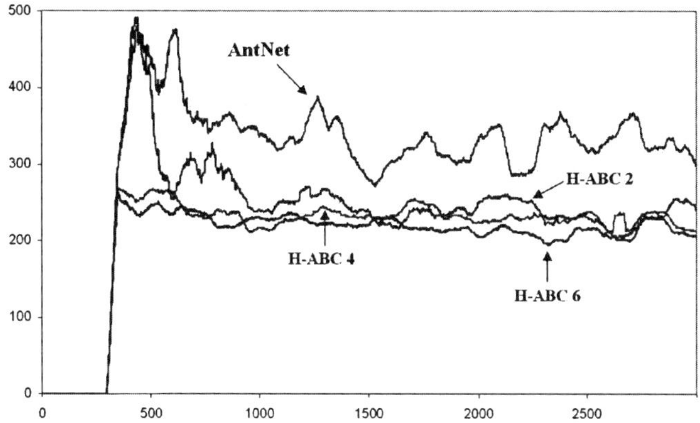

Fig. 21.3 Average packet delay under low traffic load

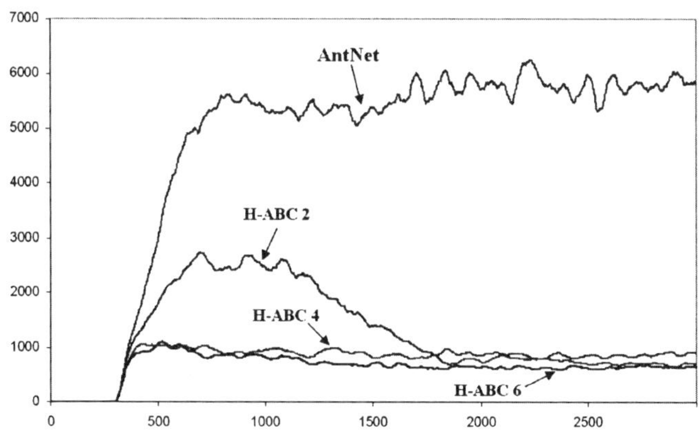

Fig. 21.4 Average packet delay under high traffic load

21.4.3 *Hot spot*

We started the simulation with PGP=100ms. After 100s the node 54 became a 'hot spot': all the nodes started to send the packets to it with PGP=75ms. The hot spot was active for 100s. All 4 algorithms behaved well under this test delivering 99% of the packets. The only difference was again the average packet delay (Figure 21.6).

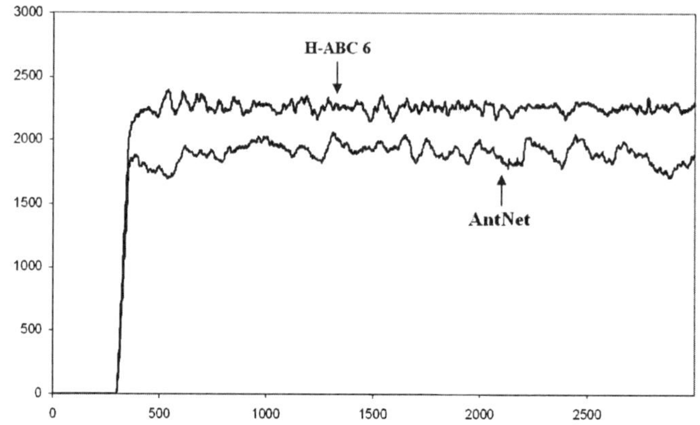

Fig. 21.5 Average throughput under high traffic load

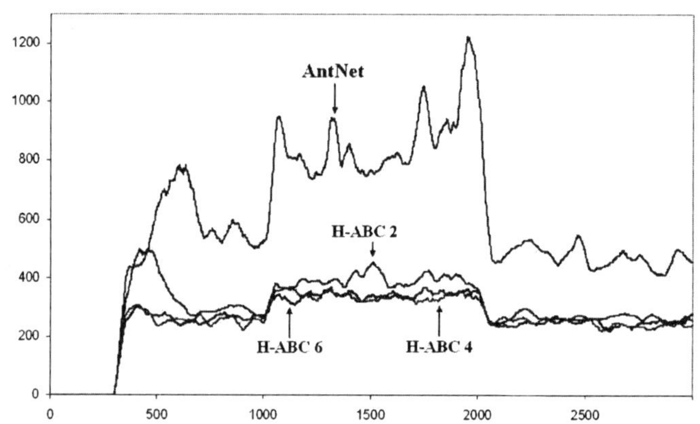

Fig. 21.6 Average packet delay under hot spot state

21.4.4 *Overhead*

Finally we measured the overhead of the algorithms. For this we run a short 50s test with only agents running in the network and measured bandwidth capacity usage. The overhead decrease with the number of sectors. *AntNet* agents used about 0.375%, *H-ABC*2 $\simeq$ 0.175%, *H-ABC*4 $\simeq$ 0.095% and *H-ABC*6 $\simeq$ 0.065% of the network capacity.

21.5 Conclusions and future work

We combined features from several ant based algorithms and obtained a highly scalable and robust algorithm. Its performance was tested in a simulation environment and compared with *AntNet*. *H-ABC* performed better both in low and high traffic load, but also in case of transient overloads over the network. With a low overhead it delivered faster the packets to destination, decreasing also the number of lost ones. Dividing the network in sectors was very effective. The results of the *H-ABC*4 and *H-ABC*6 are similar so a big fragmentation of the network is not necessary.

Our research is focused on developing a Routing System for cars in traffic networks. At TUDelft we have a city traffic environment which uses a version of *AntNet* for routing the vehicles. The scalability problem because of the big size of the city maps made us to look for a better scalable algorithm like the one presented in this paper. We expect it to be effective in our case because of the hierarchical distribution already existent in the current traffic network (cities linked by motorways).

Chapter 22

Describing DNA Automata Using an Artificial Chemistry Based on Pattern Matching and Recombination

T. Watanabe, K. Kobayashi, M. Nakamura, K. Kishi, M. Kazuno and K. Tominaga

School of Computer Science, Tokyo University of Technology, 1404–1 Katakura, Hachioji, Tokyo 192–0982 Japan
E-mail: tomi@acm.org

Artificial chemistries are promising candidates for formalisms to support designing molecular computation. One of the most significant works in the area of molecular computing is the biomolecular implementation of finite automata by Benenson *et al.*, in which finite automata with two states were constructed using DNA and enzyme. In this study, we described their implementation using our simple artificial chemistry to model the mechanism of the computation. The rules intuitively describe the interactions among DNA molecules and enzyme. We executed the description on our simulator to confirm the correctness of the description. The capability of modelling and simulating molecular reaction will be beneficial not only in designing molecular computation but also in modelling natural living systems.

22.1 Introduction

In order to understand complex phenomena found in natural living systems, it is beneficial to make an abstract model of such a system. Among a number of formal models for this purpose employed in the research area of artificial life, artificial chemistries are establishing a large area of study and have been used to model various interesting phenomena such as self-replication [108].

Artificial chemistries are abstract models of chemical reactions. They are used to model the behaviour of natural systems as well as creating systems that have similar properties to those of living systems. In modelling natural systems, artificial chemistries can be regarded as promising candidates for formalisms not only to help understanding living systems but also to support designing molecular computing systems.

A main trend of molecular computing is utilizing DNA [174], and significant results have been obtained in this area [5; 39; 245; 342; 425]. The first approach, called the Adleman-Lipton paradigm, which was named after Adleman's experiment [5] and Lipton's method [245], uses the parallelism of molecular reactions to solve combinatorial problems. But since computing in this approach requires manual processing in laboratory, another research direction is also being explored; that is to utilize a DNA molecule as an autonomous computing machine [39; 342], which does not need human intervention in performing computation.

We modelled Adleman's experiment and Lipton's method using our artificial chemistry [389], and successfully simulated them on small scale [390]. The description in the artificial chemistry simply expresses a property of DNA with which two single-stranded DNA molecules spontaneously join to form a double-stranded DNA (dsDNA) molecule if they have complementary sequences of bases. Although this is the principal interaction in DNA computing, interactions of other types, such as one between DNA and enzyme, are also indispensable in implementing molecular computation. Modelling these types of interactions will also help designing one.

In this study, we described in our artificial chemistry the way of implementing finite automata as autonomous biomolecular machines achieved by Benenson *et al.* [39]; the implementation includes an enzyme that cleaves a specific part of DNA molecules. We also ran the descriptions on our simulator to confirm the correctness of them.

22.2 An Artificial Chemistry for Stacks of Character Strings

This section briefly explains a simple artificial chemistry based on pattern matching and recombination of stacks of strings [389], which we are going to use in the following discussion.

22.2.1 *Elements and objects*

An *element* is a character; it corresponds to an atom or a group of atoms in nature. An *object* (corresponds to a molecule) is a line of character string or a stack of lines, such as those depicted below.

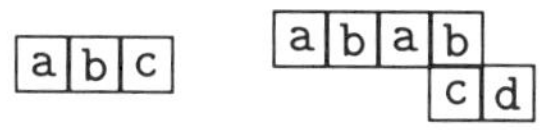

These objects are denoted by the string notations `0#abc/` and `0#abab/3#cd/`, respectively. One line is expressed as a string that starts with an integer number and ends with a slash (/); the number, which is followed by a pound sign (#), represents the displacement of the line relative to the first line.

22.2.2 *Patterns*

A *pattern* matches (or does not match) an object, and it can utilize two kinds of *wildcards*. An *element wildcard*, which is denoted by a number such as `1`, matches any element. A *sequence wildcard*, denoted by a number and an asterisk such as `2*` and `*2`, matches any sequence of zero or more elements; the position of an asterisk represents the direction in which the sequence can extend.

A pattern is denoted in a similar way to an object. For example, the pattern `0#a1c/` matches any of the objects `0#aac/`, `0#abc/`, `0#acc/`, *etc.*; the pattern `0#*1ab/1#cd/` matches `0#ab/1#cd/`, `0#aab/2#cd/`, `0#abab/3#cd/`, *etc.* Note that the displacements are meaningful and that the length of a sequence wildcard is treated as zero in the notation for patterns. A pattern without wildcards, such as `0#abc/` is literal, i.e., it matches only the object `0#abc/`.

22.2.3 *Recombination rules*

A *recombination rule* transforms a group of objects into a group of objects, conserving elements just like a chemical reaction does. It is expressed in a manner similar to chemical formulae, but in terms of patterns. An example rule is

$$\texttt{0\#*1ab/1\#cd/} + \texttt{0\#ab2*/} \rightarrow \texttt{0\#*1abab2*/1\#cd/}$$

which is illustrated as follows.

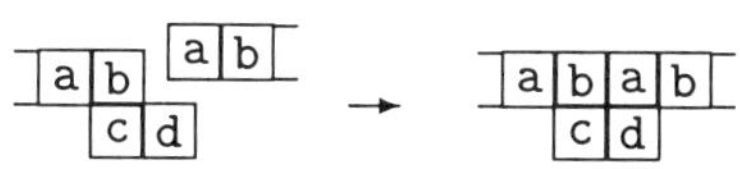

If this rule is applied to the objects `0#abab/3#cd/` and `0#abc/`, the object `0#abababc/3#cd/` is produced and the reactants disappear.

22.2.4 *Sources and drains*

Objects are kept in the *working multiset.* A *source* is defined as an object, and it supplies objects of the specified form to the multiset one at a time without limit. A *drain* is defined as a pattern, and it eliminates objects matched by the pattern, one at a time, from the multiset.

22.2.5 *Dynamics*

A *system* is a construct $\langle \Sigma, S, D, R, P_0 \rangle$ where Σ is a set of elements, S is a multiset of sources, D is a multiset of drains, R is a set of recombination rules, and P_0 is the *initial working multiset*, which specifies objects in the working multiset at the initial state. The system is interpreted nondeterministically as follows.

(1) Initialize the working multiset to be P_0.
(2) Do one of the following operations.

- Apply one recombination rule to a collection of objects.
- Operate one source.
- Operate one drain.

(3) Go to Step 2.

22.3 Implementation of Finite Automata with DNA

We are modelling an implementation of finite automata using actual DNA molecules, which was achieved by Benenson *et al* [39]. A brief overview of the implementation is given below.

An example automaton implemented is shown as a diagram in Fig. 22.1. It has two states S_0 and S_1 and processes its input composed of two symbols a and b. The initial state is S_0 (designated by the straight incoming arrow from nowhere) and the accepting state is S_0 (the double circle). This automaton accepts a sequence with an even number of the symbol b.

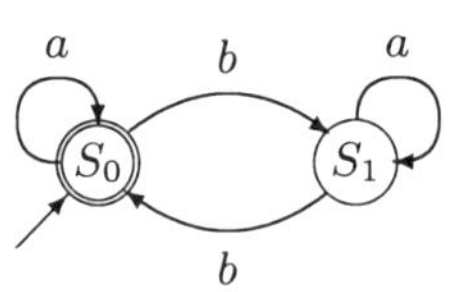

Fig. 22.1 An example finite automaton.

The biomolecular implementation of the automaton comprises the following components.[1]

- *Fok* I restriction nuclease. This enzyme recognizes the base sequence 5′-GGATG-3′ and its complementary 3′-CCTAC-5′ in a dsDNA molecule (we will denote this by GGATG/CCTAC hereafter) and cleaves it at a specific position, as shown in Fig. 22.2 (a blank box represents any base).
- Four kinds of transition molecules. These are dsDNA molecules, each of which has an *Fok* I recognition sequence (GGATG/CCTAC) and a cohesive (*i.e.*, single-stranded) end, and corresponds to one of the four transitions (the curving arrows in Fig. 22.1).
- Output-detection molecules. These are two kinds of dsDNA molecules with cohesive ends; one corresponds to S_0, the other to S_1.

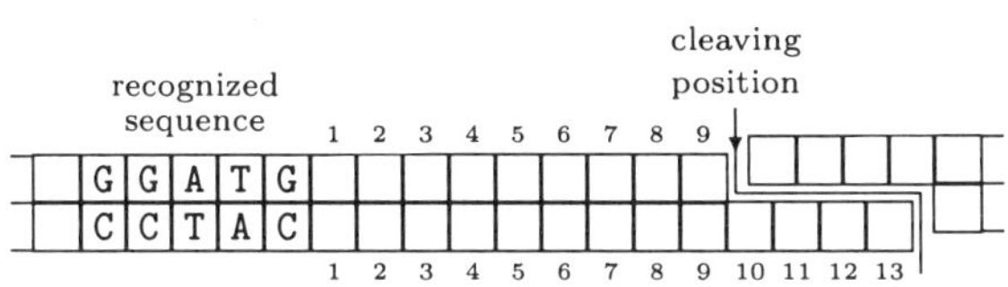

Fig. 22.2 Cleaving a double-stranded DNA molecule by *Fok* I.

An input, which is a series of a and/or b, is encoded as dsDNA; the base sequence CTGGCT/GACCGA represents the symbol a and CGCAGC/GCGTCG represents b. A special sequence TGTCGC/ACAGCG, called terminator, indicates the end of input. A whole input DNA molecule comprises a heading part including an *Fok* I recognition sequence, the series of sequences that represents symbols, the terminator, and a trailing part.

When the automaton is given an input, the following steps of reactions occur.

(1) *Fok* I cleaves the input DNA molecule according to the recognition sequence in the heading part; this exposes a cohesive part within the base sequence for the first remaining symbol, or the terminator if all the input symbols have been consumed.

  - If a sequence for a symbol is cleaved, go to Step 2.
  - If the terminator is cleaved, go to Step 3.

(2) The cohesive part sticks uniquely to one of the transition molecules according to base complementarity (they are so designed), and makes

[1]Although the implementation includes ligase (enzyme to connect ends of DNA molecules) and ATP (energy source of enzyme) in addition to those listed, we omit this detail in our modelling.

a dsDNA molecule. For example, if the current state is S_0 and the first remaining symbol is a, the transition molecule that represents $S_0 \xrightarrow{a} X$ where X is the next state (X is S_0 for this automaton) is selected. A transition molecule has an *Fok* I recognition site in its heading part, which is to be used in the next cleaving. Go to Step 1.

(3) One of the two detectors sticks to the cohesive part of the input molecule to make dsDNA; this molecule represents the final state, and is called reporter.

After enough time for the molecular reactions to generate reporters, they are detected by gel electrophoresis to decide whether the input is accepted by the automaton.

22.4 Describing the Implementation with the Artificial Chemistry

In this section, we first describe the implementation of the example automaton with two states and two symbols, which was shown in Fig. 22.1, as a system in our artificial chemistry. Then we extend the description to have three states.

22.4.1 *Describing the automaton with two states*

The initial working multiset of the system comprises the following types of objects corresponding to the molecules used in the original biomolecular implementation. The number of objects of a type is shown in brackets.

- *Fok* I [100]: `0#FFFFFFFFF/`
- Transition molecules [20 for each]

  $S_0 \xrightarrow{a} S_0$: `0#GGATGTAC/0#CCTACATGCCGA/`
  $S_0 \xrightarrow{b} S_1$: `0#GGATGACGAC/0#CCTACTGCTGGTCG/`
  $S_1 \xrightarrow{a} S_1$: `0#GGATGTCG/0#CCTACAGCGACC/`
  $S_1 \xrightarrow{b} S_0$: `0#GGATGG/0#CCTACCGCGT/`

- Detector [20]: `0#X/0#XAGCG/`
- Input molecules [10 for each] (folded to fit the page)

  abb: (`0#GGATGXXXXXXXCTGGCTCGCAGCCGCAGCTGTCGCX/`
  `0#CCTACXXXXXXXGACCGAGCGTCGGCGTCGACAGCGX/`)
  aba: (`0#GGATGXXXXXXXCTGGCTCGCAGCCTGGCTTGTCGCXX/`
  `0#CCTACXXXXXXXGACCGAGCGTCGGACCGAACAGCGXX/`)

The objects except for *Fok* I are composed of elements that represent DNA bases: the elements T, C, A and G represent thymine, cytosine, adenine and guanine, respectively. Elements X in the input molecules represent arbitrary bases; however, any part containing X is assumed not to be recognized by *Fok* I. The DNA molecules used in the original implementation have additional parts that do not directly relate to the mechanism of implemented automaton, such as those for electrophoresis; we omit them in our modelling to focus on the principle of the mechanism. .

We provide the objects for the two different input molecules *abb* and *aba* in the initial working multiset at the same time. They have the trailing parts of different lengths (X/X for *abb* and XX/XX for *aba*) so that they can be easily seen in the simulator's window.

We give the system the following recombination rules.

- Joining of *Fok* I and the input molecule

$$\texttt{0\#F0*/} + \texttt{0\#GGATG1*/0\#2*/} \rightarrow \texttt{0\#F0*/-5\#GGATG1*/-5\#2*/} \quad (22.1)$$

- Separation of *Fok* I and the input molecule

$$\texttt{0\#F0*/-5\#GGATG1*/-5\#2*/} \rightarrow \texttt{0\#F0*/} + \texttt{0\#GGATG1*/0\#2*/} \quad (22.2)$$

- Cleaving of the input molecule by *Fok* I

$$\texttt{0\#*0F/1\#*123*/1\#*456789*/} \rightarrow \texttt{0\#*0F/1\#*1/1\#*45678/} + \texttt{0\#23*/4\#9*/} \quad (22.3)$$

- Joining of state transition molecules and the input molecule

$$\texttt{0\#*0/0\#*1CCGA/} + \texttt{0\#GGCT2*/4\#3*/} \rightarrow \texttt{0\#*0GGCT2*/0\#*1CCGA3*/} \quad (22.4)$$

$$\texttt{0\#*0/0\#*1GTCG/} + \texttt{0\#CAGC2*/4\#3*/} \rightarrow \texttt{0\#*0CAGC2*/0\#*1GTCG3*/} \quad (22.5)$$

$$\texttt{0\#*0/0\#*1GACC/} + \texttt{0\#CTGG2*/4\#3*/} \rightarrow \texttt{0\#*0CTGG2*/0\#*1GACC3*/} \quad (22.6)$$

$$\texttt{0\#*0/0\#*1GCGT/} + \texttt{0\#CGCA2*/4\#3*/} \rightarrow \texttt{0\#*0CGCA2*/0\#*1GCGT3*/} \quad (22.7)$$

- Generation of reporters

$$\texttt{0\#*0/0\#*1AGCG/} + \texttt{0\#TCGC2*/4\#3*/} \rightarrow \texttt{0\#*0TCGC2*/0\#*1AGCG3*/} \quad (22.8)$$

$$\texttt{0\#*0/0\#*1ACAG/} + \texttt{0\#TGTC2*/4\#3*/} \rightarrow \texttt{0\#*0TGTC2*/0\#*1ACAG3*/} \quad (22.9)$$

The applications of recombination rules are illustrated below in the order of applications. First, by Rule 22.1, *Fok* I attaches to the input molecule at the specified position by the recognition sequence.

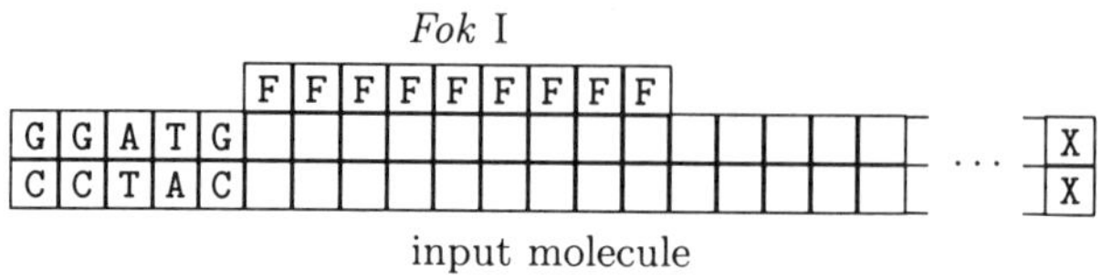

Product of Rule 22.1

This compound is cleaved by Rule 22.3, and the input molecule exposes a cohesive end, which is a part of the base sequence representing the first symbol.

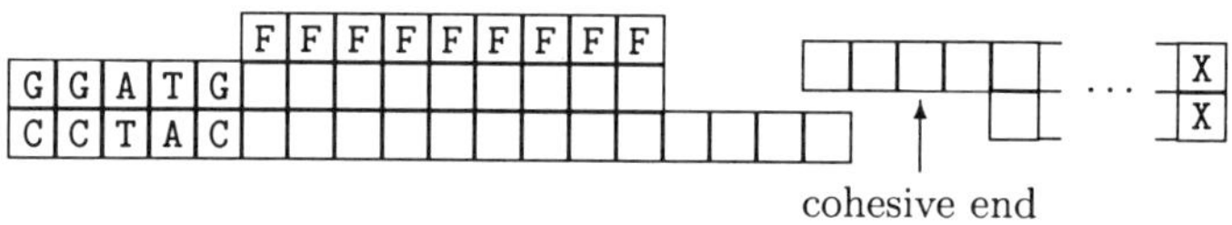

Cleaving by Rule 22.3

To the cohesive end attaches one of the transition molecule by an appropriate transition rule from among Rules 22.4–22.7; the below example is the application of Rule 22.4.

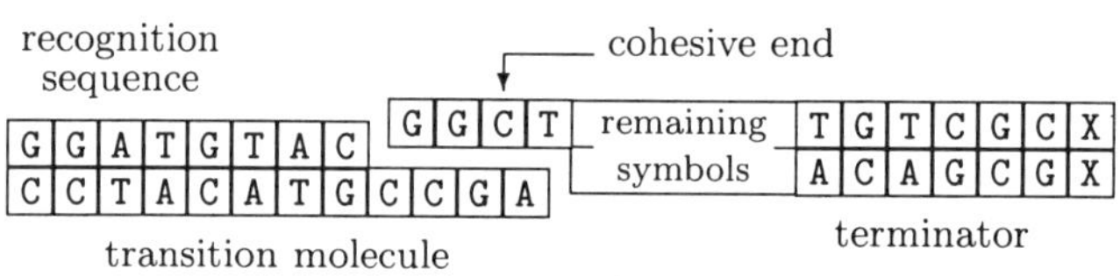

Transition by Rule 22.4

Since a transition molecule has the *Fok* I recognition sequence (GGATG/CCTAC) at its left end, the product of the recombination can be again cleaved by *Fok* I using Rule 22.1 and Rule 22.3.

After the repetition of the above steps, the terminator (TGTCGC/ACAGCG) in the input molecule is finally cleaved. Then Rule 22.8 or 22.9 combines it with a detector and generates a reporter. The example below is the application of Rule 22.9.

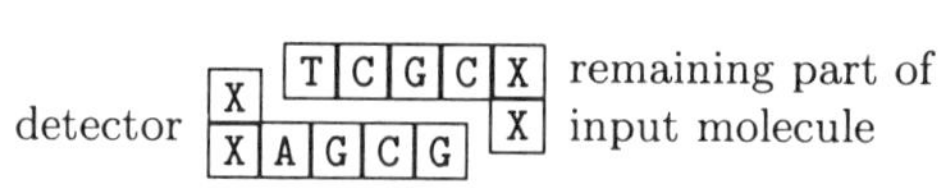

Generation of a reporter by Rule 22.8

The compound of *Fok* I and a part of the input molecule resulted by the application of Rule 22.3 is decomposed by Rule 22.2 into *Fok* I and a partial DNA molecule. *Fok* I is reused in another reaction; the partial DNA is unnecessary in the further execution of the automaton.

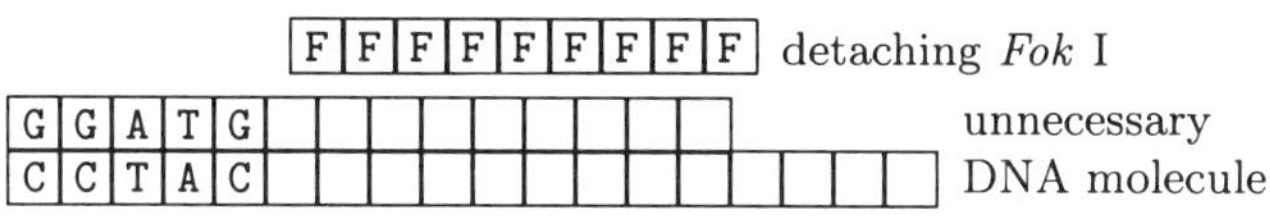

Separation of *Fok* I by Rule 22.2

Fok I can attach to transition molecules or unnecessary partial DNA (instead of the input molecule) by Rule 22.1. In such cases, however, Rule 22.3 cannot be applied to the product, and Rule 22.2, which is the only applicable rule, separates them again. Rule 22.2 can also separate the compound of *Fok* I and the input molecule.

We implemented a simulator for this artificial chemistry using Objective-C and the Cocoa Framework on Mac OS X. Figure 22.3 illustrates an example execution of the description given above; it shows the contents of the working multiset after an approximately 30-second run starting from the initial state on PowerPC G5 1.8GHz. The table shows objects in the right column and their numbers in the left column (not all objects are shown); there are generated reporters among the objects (`0#XTCGCX/0#XAGCGX/` shown at the bottom line, shaded; also depicted in the Molecule Viewer window). They have the terminating part of the input *abb* (having `X/X` at the right end); this means that the automaton accepts *abb*, which is the expected behaviour.

22.4.2 *Describing an automaton with three states*

Benenson *et al.* showed the construction of automata with two states and two symbols [39]. In this section, we extend the description given in Sec. 22.4.1 and give a description for a three-state automaton using the same principle. The described automaton is shown in Fig. 22.4; it accepts an input with $3n$ a's ($n \geq 0$).

The following types of objects are given in the initial working multiset.

- *Fok* I [100]: `0#FFFFFFFFF/`
- Transition molecules [20 for each]

$S_0 \xrightarrow{a} S_1$: `0#GGATGACGA/0#CCTACTGCTCCGA/`

$S_0 \xrightarrow{b} S_0$: `0#GGATGACG/0#CCTACTGCGTCG/`

$S_1 \xrightarrow{a} S_2$: `0#GGATGACGA/0#CCTACTGCTACCG/`

$S_1 \xrightarrow{b} S_1$: `0#GGATGACG/0#CCTACTGCCGTC/`

$S_2 \xrightarrow{a} S_0$: `0#GGATGA/0#CCTACTGACC/`

$S_2 \xrightarrow{b} S_2$: `0#GGATGACG/0#CCTACTGCGCGT/`

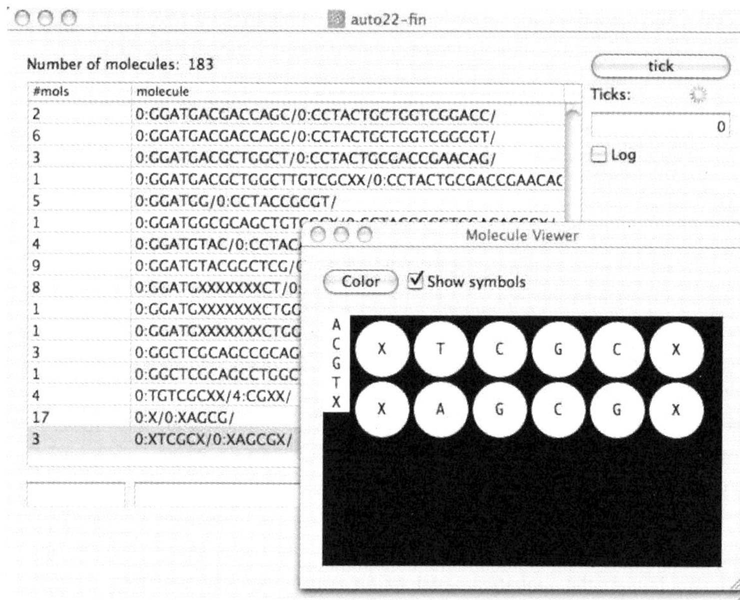

Fig. 22.3 A sample execution of the automaton with two states.

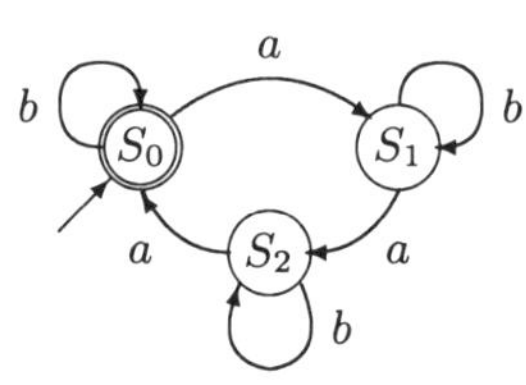

Fig. 22.4 A three-state automaton.

- Detector for S_0 [20]: `0#X/0#XAGCG/`
- Input molecules [10 for each] (folded to fit the page)

aaab:
```
0#GGATGXXXXXXXCTGGCTCTGGCTCTGGCTCGCAGCTGTCGCX/
0#CCTACXXXXXXXGACCGAGACCGAGACCGAGCGTCGACAGCGX/
```

abab:
```
0#GGATGXXXXXXXCTGGCTCGCAGCCTGGCTCGCAGCTGTCGCXX/
0#CCTACXXXXXXXGACCGAGCGTCGGACCGAGCGTCGACAGCGXX/
```

In this description, we used the base sequences shown in Tbl. 22.1. A pair $\langle s, S_i \rangle$ in the left column means that an input symbol s is consumed when a transition from the state S_i occurs, and the right column shows the cohesive end (underlined) exposed by the transition. The end of input is

Tbl. 1.1 The base sequences for the three-state automaton.

⟨*symbol, state*⟩	*base sequence*
$\langle a, S_0\rangle$	CTGGCT
$\langle a, S_1\rangle$	CTGGCT
$\langle a, S_2\rangle$	CTGGCT
$\langle b, S_0\rangle$	CGCAGC
$\langle b, S_1\rangle$	CGCAGC
$\langle b, S_2\rangle$	CGCAGC
$\langle t, S_0\rangle$	TGTCGC
$\langle t, S_1\rangle$	TGTCGC
$\langle t, S_2\rangle$	TGTCGC

represented by the symbol t. Note that this description uses the cohesive sequences for S_1 in the automaton shown in Sec. 22.4.1 as those for S_2.

The system uses the following rules in addition to the rules given in Sec. 22.4.1. Rules 22.10 and 22.11 are used for transitions from the state S_1, and Rule 22.12 generates reporters for the state S_1.

- Joining of state transition molecules and the input molecule

`0#*0/0#*1ACCG/` + `0#TGGC2*/4#3*/` → `0#*0TGGC2*/0#*1ACCG3*/` (22.10)

`0#*0/0#*1CGTC/` + `0#GCAG2*/4#3*/` → `0#*0GCAG2*/0#*1CGTC3*/` (22.11)

- Generation of reporters

`0#*0/0#*1CAGC/` + `0#GTCG2*/4#3*/` → `0#*0GTCG2*/0#*1CAGC3*/` (22.12)

The execution took approximately 50 seconds on PowerPC G5 1.8GHz to generate reporters of the form `0#XTCGCX/0#XAGCGX/` for the input *aaab*. The reporters indicate that the automaton has accepted this input as expected.

22.5 Discussion

The systems, described by the artificial chemistry, modelled the biomolecular implementations of automata without details. For example, detectors have long base sequences (161 and 251 bases) in the original implementation [39]. The difference in length is used in electrophoresis to distinguish which reporter is generated. This setting is necessary for processing in laboratory, but it is not directly related to the principle of the mechanism with which the automaton works. In contrast, our systems omitted the details (detectors have the same heading part `X/X`) and have given the succinct descriptions. We think this is a good way of abstraction that makes the principle easily understood.

The descriptions are executable by the simulator. Simulation will be useful when a new molecular computation is designed, tested and/or debugged. For example, if one is to construct the three-state automaton (Fig. 22.4), he/she can first design it using the artificial chemistry and test it on the simulator to find any fatal flaw in the design, before he/she actually builds and runs a biomolecular system, which is far more costly than simulation.

We gave the rules that express attaching, detaching and cleaving of molecules by the enzyme, as well as those for joining complementary DNA strands. In the descriptions, *Fok* I was represented as an object that has a sequence of nine Fs. It does not denote that the actual enzyme has that length relative to DNA; rather, it indicates the relative positions of the recognition site and the cleaving point. We think the representation is simple and clear. We described joining of complementary DNA strands in a different way from the one we used in our previous work [390]: we employed abstract bases previously; in this study, we used individual bases.

22.6 Comparison with Related Works

Several models for DNA computation have been presented [6; 329; 339; 440; 441]. Most of them are specific only to DNA computing, and to the authors' knowledge, their descriptions are not directly executable on simulators. This artificial chemistry, on the other hand, is general and not limited to modelling DNA computation. For example, we modelled an imaginary protein synthesis using it [391].

Several simulators for DNA computing have also been constructed such as [292]. While some of them adopt techniques from artificial chemistries [352], one could be used only for a specific type of DNA computation. This artificial chemistry modelled a computation in the Adleman-Lipton paradigm [390] as well as the biomolecular machine discussed in this paper. Although simulation with this artificial chemistry would not have quantitative accuracy compared to results from DNA-specific simulators, the generality will be helpful in modelling a new type of computation.

Not many artificial chemistries have been used to directly model the behaviour of natural molecules. We think that the capability of modelling natural molecular phenomena is a beneficial property of an artificial chemistry, which will help understanding natural living systems.

22.7 Concluding Remarks

We modelled the implementation of biomolecular automata with two states by Benenson *et al.* and its extension having three states as systems in our artificial chemistry. Each rule in the descriptions naturally corresponds to a step of molecular interaction such as attaching an enzyme to substrate and hybridization of two DNA molecules. The descriptions were executed on our simulator and we observed they correctly modelled the automata.

The artificial chemistry is general and applicable to different types of molecular computations, and descriptions are executable. This property will be beneficial in designing, testing and debugging a new molecular computation, as well as in modelling natural living systems at molecular level.

Chapter 23

Towards a Network Pattern Language for Complex Systems

J. Watson[1,2], J. Hawkins[1,2], D. Bradley[3,4], D. Dassanayake[5], J. Wiles[1,2] and J. Hanan[1,4,6]

[1] *ARC Centre for Complex Systems,*
The University of Queensland, Brisbane QLD 4072 Australia
http://www.accs.edu.au/patterns/

[2] *School of Information Technology and Electrical Engineering,*
The University of Queensland, Brisbane QLD 4072 Australia
Email: {jwatson, jhawkins, j.wiles}@itee.uq.edu.au

[3] *Institute for Molecular Bioscience,*
The University of Queensland, Brisbane QLD 4072 Australia
Email: d.bradley@imb.uq.edu.au

[4] *ARC Centre in Bioinformatics,*
The University of Queensland, Brisbane QLD 4072 Australia

[5] *CSIRO Land & Water,*
School of Science and Technology,
Charles Sturt University, Wagga Wagga NSW 2678 Australia
Email: dharma.dassanayake@csiro.au

[6] *ARC Centre of Excellence for Integrative Legume Research,*
The University of Queensland, Brisbane QLD 4072 Australia
Email: jim@maths.uq.edu.au

The development and use of complex systems models can involve many common problems, problems that are solved again and again by different researchers with various backgrounds and experience. The application of the software engineering patterns paradigm to complex systems modeling will enable capture of the wisdom of the network modeling community in such a way that proven solutions to recurring challenges can be identified and tailored to the specific problem at hand. The use of networks for simulation and analysis are two areas of complex systems modeling that stand to benefit from the patterns approach. The use of networks to guide thinking, as analytic tools and as visualizations is ubiquitous in the field of complex systems. However, various methods of using networks (e.g. design, updating functions, visualization, analysis) are not always obvious to a newcomer and are often assumed as general knowledge in the literature. This paper is a first step towards a pattern language addressing these issues for the complex systems community.

23.1 Introduction

The field of complex systems is concerned with systems whose global behaviour cannot be predicted simply by examining components in isolation. Such studies of emergent behaviour have provided insights into disparate disciplines such as economics [128], air traffic control, and genetic regulatory networks [176]. Since complex systems are comprised of numerous entities, each interacting with some subset of the total, the use of networks is fundamental to the field. Networks can be used to analyze the entities and their relationships, and to investigate state transitions [430].

An often overlooked fact is how networks guide the way an investigator may think about a complex system, and the types of analysis and insights that may be drawn from theorizing about the system. Therefore, the way to use networks in complex systems modeling is not always obvious. Also, the finer details of a methodology – such as why a technique was or was not used, limitations and trade-offs of methods, the reasoning for discarding designs, etc. – often do not make it into the published literature. Consequently, this knowledge is often continuously rediscovered, or passed on only through informal discussion.

Patterns are a proven technique developed to deal with similar issues of reuse and knowledge capture in the fields of software engineering and architecture [138]. Far from being just documented techniques or heuristics, patterns include not only the benefits of a method, but also the inherent trade-offs and consequences. They are organized into classifications so that, for a given situation, they can be easily found and used in collaboration with other patterns as appropriate. Patterns span the range of complex systems methodology, from the abstract design of a system down to an implementation of a particular update function. Importantly, they describe proven solutions that have been discovered multiple times.

The capture of commonly recurring problems, along with the details and consequences of their solutions, can have multiple benefits. Practitioners, both novice and experienced, can benefit from a library of tested ideas. After undergoing over 40 years of development in a variety of fields, from architecture to software engineering [138], existing pattern formats facilitate the ongoing community development of a pattern library by (i) asking the right questions to appropriately derive the pattern, and (ii) by providing a shared language to rapidly communicate ideas agreed upon by the community. In addition to saving time otherwise spent reinventing solutions, using patterns in the design, development and analysis of a complex systems model can build confidence in the quality of the model produced. For more information on the use of patterns in complex systems modeling, see Wiles and Watson [417].

The goal of this paper is to introduce three network-centric prototype patterns (or proto-patterns [17]), intended to form the basis of a pattern language. A pattern language is a collection of related patterns [17] covering the use of networks in complex systems modeling. It is intended that input from the complex systems community will refine these to the state of completed patterns. The first is an abstract Network Diagram pattern, which describes the general ways a network can be used to solve complex systems problems. The second is a more concrete Synchronous System State Update pattern, which describes how a current network state may be used to produce the next simulation state. Finally, a Discrete State-Space Trajectory Diagram pattern is outlined, which offers a visualization of system state dynamics.

23.2 Methodology

The three proto-patterns were refined after two days of development at the first Complex Systems Patterns Workshop hosted by the ARC Centre for Complex Systems at the University of Queensland. The workshop attendees discussed commonalities among their problems and solutions, and then focused on areas in need of complex systems patterns.

Using an appropriate, agreed-upon format is critical to pattern development. By asking the right questions, a good format facilitates communication and the extraction of modeler experience at an appropriate level of detail. The pattern format used to develop the complex systems patterns is adapted from the object-oriented design patterns format used by the software engineering community. The adapted format is shown in Table 23.1.

In keeping with the patterns development process used in software engineering, the three proto-patterns described in this paper were developed

Table 23.1 Complex systems patterns format

Section	Description
Submitted by.	Includes name, affiliation and date.
1a. Pattern name.	The pattern's name conveys the essence of the pattern succinctly. A good name is vital, because it will become part of the design vocabulary.
1b. Classification.	The pattern's classification should reflect it's scope and use
2. Intent.	A short statement that answers the following questions: What does the pattern do? What is its rationale and intent? What particular issue or problem does it address?
3. Also known as.	Other well-known names for the pattern, if any.
4. Motivation.	A scenario that illustrates a problem and how the pattern solves the problem. The scenario aids understanding the more abstract description of the pattern that follows.
5. Applicability.	What are the situations in which the Complex Systems pattern can be applied? What are examples of Complex Systems problems that the pattern can address? How can you recognize these situations?
6. Structure.	A graphical representation of the components of the pattern if possible (leave blank if not applicable).
7. Participants.	The components of the pattern and their responsibilities (leave blank if not applicable).
8. Collaborations.	How the participants collaborate to carry out their responsibilities (leave blank if not applicable).
9. Consequences.	How does the pattern support its objectives? What are the trade-offs and results of using the pattern?
10. Implementation.	What pitfalls, hints or techniques should one be aware of when implementing the pattern? Are there platform-specific issues?
11. Sample code.	Code fragments or pseudo-code that illustrate how you might implement the pattern.
12. Known uses.	Examples of the pattern found in real systems. Preferably at least two examples from different domains. Give references.
13. Related patterns.	Which patterns are closely related to this one? What are the important differences? With which other patterns should this one be used?

(i) with a focus on recurring problems and their proven solutions, (ii) through development in workshop discussions instead of solo presentations, and (iii) not necessarily by the original inventor of the solution.

23.3 Results

The following sections describe how the three proto-patterns were developed (see Appendices A, Network Diagram, B, Synchronous System State Update, and C, Discrete State-Space Trajectory Diagram).

23.3.1 *Development of the Network Diagram pattern*

The discussion leading to the Network Diagram pattern initially centred on issues of visualizing and manipulating models of networks in complex systems simulations. However, the discussion soon elucidated a second major theme: the issue of knowing when to use a network representation in the first place. Even though in many instances a system can be naturally seen as a network due to its physical structure, in some applications the network is a significant abstraction from the system being studied. Furthermore, there are often numerous ways in which one can abstract a network structure from a representation of a physical system. For example the relationships between academic authors can be studied using either co-authorship or citations as connections between authors.

The first point of contention in this discussion was whether we were talking about the general issue of when one should use a network to visualize a problem, or the issue of how one visualizes something as a network when a network approach has already been decided on. The two issues exist at different levels of complex systems modeling methodology. The first is an issue of deciding how to approach a problem; the second is an issue of how to implement a solution. Both issues are equally deserving of a complex systems pattern; however, focus was given to the visualization of a network structure.

This emergence of two competing notions of what the pattern should communicate illustrates an important lesson about the process of pattern development. Even though patterns are by their nature abstract, they can still exist at differing levels of abstraction. They can be relatively specific, dealing with structuring code to solve a particular design problem, or alternatively they can address conceptual aspects of the modeling process.

Two tensions emerged in the discussion of this pattern. The first was a natural consequence of the different domain knowledge among participants. These differences lead to a discussion that gradually drew out the commonalities across domains. Secondly, it was apparent during the discussion that each of the participants had a natural level of abstraction to which they would return. This resulted in contention regarding the exact level of abstraction at which the pattern exists.

The first of these issues was somewhat obvious and quickly resolved, the details of differing fields of application were put aside and the essential modeling problem distilled. However, the second issue was not readily apparent and consequently often remained unresolved until well into the discussions. While it remained unresolved it led to people talking at cross purposes, and failing to find consensus. In the end it was solved by a specific discussion of abstraction, which operated to create a balance between generality and usefulness.

It is important to recognise that the strength of a workshop approach is the drawing out of commonality from the experiences of different individuals. However, while issues remain unresolved, productive discussion can stall. We would recommend that an early part of the pattern development process involve agreeing specifically about the level of abstraction. Concrete topics such as identifying the target audience and the words that would be used to describe the particular problem solved by the pattern can help in elucidating this issue. Without deliberate targeting, the nature of the problem can remain vague and change continuously throughout the discussion.

For several reasons the ultimate content of the pattern converged upon the more abstract issue of recognizing when to apply a network diagram. Firstly, we had originally planned to make at least one other concrete network pattern to complement the general network diagram. Hence, it seemed appropriate to begin the pattern language with the most abstract aspect of networks in modeling, 'When and why to use them?' Secondly, in developing a pattern repository, people should ideally be able to come to the repository with only a set of questions and few preconceptions about the solution. If the network pattern language began with the assumption that users already knew about and intended to use a network representation, then the repository would exclude a potentially large group if users who have problems and no inkling that they can be solved with a network representation.

23.3.2 *Development of the Synchronous System State Update pattern*

The discussion for the Synchronous System Update pattern began with the observation that some newcomers to network modeling choose naive implementations for sets of synchronously interacting entities that can be all updated simultaneously (i.e., node state updates neglect to account for the interactions of all other nodes before the updates are committed). Even though the solution to this problem may seem trivial to an experienced modeler, it is one that many people solve independently, and almost certainly causes problems for novice computer modelers. The fact that people tend to converge to one of a small set of solutions makes this problem an ideal case for an implementation level pattern.

Further discussion of the exact nature of the problem and the structure of the solution illustrated that it is a problem much broader than network models, and even complex systems. It applies to any simulated system that consists of numerous entities that must all be updated synchronously where changes of entities are affected by the current state of some of the others. For example, this situation occurs in the construction of computer games, where it is called the central game loop. Many game programming books devote an entire chapter to the subject, since it is the fundamental controller whereby the discrete temporal nature of the system is enforced.

23.3.3 *Development of the Discrete State-Space Trajectory pattern*

Much of the field of complex systems requires some understanding of the overall behavioural characteristics of a system. Researchers require methods of capturing essential behavioural properties that filter out superfluous information. This is the purpose of the Discrete State-Space Trajectory Diagram. In this pattern the states of a system are depicted as nodes of a network, with edges representing the transitions between states.

The discussion of this pattern began with its application in abstract genetic regulatory networks (GRNs). The use of state space diagrams has been extensive in this area, beginning with Kaufmann's random Boolean networks [216][217]. It has been particularly successful due to the layout approach known as the Wuensche diagram [430]. In this visualization all the nodes involved in the central attractor of the system are placed in a circle in the centre of the page. Nodes that feed into this attractor are placed at ever increasing distances from the central circle, so that the outermost nodes are those from which the system must pass through the most transient states in order to reach the attractor. Any other smaller attractors are arranged similarly on other parts of the page. This layout approach renders the overall system dynamics comprehensible to visual inspection.

Discussion of this pattern soon veered away from the particulars of its application in GRN modeling to defining the essentials of circumstances in which it can be applied. We discussed the fact that although it is most naturally applied to models with discrete states, it can be applied to continuous systems by quantising system variables. In fact such an approach could potentially lead to valuable insights into a continuous system; by investigating changes in the system dynamics as the scale of quantisation is varied.

Discusssions also dealt with the issue of state space size. If the number of system states prohibits exhaustive enumeration, it was agreed that sampling a small number of start states can still yield insight into the state space. In many instances this will illuminate the dominant attractors of the system. Secondly, one may also apply the pattern to investigate particular subsections of the state space that are of interest.

Numerous quantitative results can be produced in the application of this pattern. One may measure the number of attractors, the distribution of their sizes, the number of transients, the distribution of their lengths, the number of unreachable states and stationary points. Furthermore, this pattern leads on to the more detailed analysis of issues such as the stability of the attractors via perturbation analysis [142].

23.4 Conclusions

Patterns are a means of capturing the collective experience of the complex systems community. Using the proven patterns format of software engineering to guide the development of complex systems patterns offers two main benefits. First, by posing suitable questions, the knowledge gained by experienced practitioners is extracted at an appropriate level of detail. Second, a consistent and shared language is provided, which facilitates discussion of this captured knowledge.

We have presented the development of three patterns related to the use of networks in complex systems modeling. Some of the typical issues that occur in defining patterns were highlighted through a discussion of the process used to develop these patterns.

23.5 Future Work

The patterns presented in this paper are progress towards a pattern language that captures essential applications of networks to complex systems. During the workshop and subsequent refinement process we identified two prominent areas for future work. First, analytical techniques for studying network structure form an important area of network modeling that hasn't been touched upon in this paper. There are numerous metrics for comparing networks, which would be served well by being organized as patterns that explain how and when to use them. Second, in our discussion we have focused on networks with static structures. There exists an extensive amount of work done on network rewiring algorithms, and on complex systems in which the network structure changes endogenously over time. We expect that formalizing the content of these two broad areas of research would be the next step along the path to a network pattern language.

Further work including a complex systems patterns repository is available online at `http://www.accs.edu.au/patterns/`.

Acknowledgements

The authors would like to thank the participants of the First Complex Systems Patterns Workshop for illuminating discussions and insights. In particular, thanks to Scott Heckbert, David Carrington and Andrew Hockey. This workshop, held 6-7 June 2005, was supported by the ARC Centre for Complex Systems (ACCS), the ARC Complex Open Systems Network (COSNet), and CSIRO's Complex Systems Science Area. The Complex Systems Patterns Project is funded by the ARC Centre for Complex Systems. Daniel Bradley was supported by Australian Research Council grant CE0348221.

Chapter 24

The Evolution of Aging

O. G. Woodberry, K. B. Korb and A. E. Nicholson

Clayton School of Information Technology
Monash University, Victoria 3800, Australia
Email {owenw,korb,annn}@csse.monash.edu.au

Inclusive fitness theory [177], better known as kin selection, has often been cited as an alternative to group selection as a way of explaining the evolution of altruistic behavior. However, an evolving understanding of inclusive fitness has seen it redefined, by its creator, in terms of levels of selection, leading to a blurring of the distinctions between the two. Hamilton [178] suggests that if a distinction is to be made between group and kin selection, the term 'group selection' should only be used when there is no reliance on kin associations.

Based on the early group selection model of Gilpin [145] for the evolution of predatory restraint, Mitteldorf [282] designed an ALife simulation that models the evolution of aging and population regulation. Mitteldorf sees the evolution of aging as a case of 'extreme' altruism "...in the sense that the cost to the individual is high and direct, while the benefit to the population is far too diffuse to be accounted for by kin selection" [282, p. 346].

We demonstrate that Mitteldorf's simulation is dependent on kin selection, by reproducing his ALife simulations and then introducing a mechanism to remove all and only the effects of kin selection within it. The result is the collapse of group selection in the simulation, suggesting a new understanding of the relation between group and kin selection is needed.

24.1 Introduction

The evolution of an aging trait would appear to be in direct conflict with the individual selection concept of natural selection. Individual selection refers to selection of the organism with the greatest individual fitness, measured as ability to survive and reproduce, or simply the individual's expected number of descendants. It is easy to see that an organism exhibiting a trait, such as aging, which by definition reduces its own survivability, will leave fewer descendants than a competing organism without the trait. The benefits of an aging trait, if there are any, could only be received by organisms other than the organism exhibiting the trait. For this reason such a trait, if beneficial, must be altruistic. In order to give an explanation of such an altruistic adaptation, one must call upon a mechanism of selection which incorporates such altruistic benefits, or otherwise deny that it is an adaptation.

There have been two notable attempts at explaining such a mechanism, these are: group selection, proposed by Wynne-Edwards [432], and inclusive fitness theory, proposed by Hamilton [177]. Group selection differs from individual selection in that it is the *group* rather the *individual* organism that selection acts upon. Group fitness is measured as the group's ability to prolong the period before extinction (group survival) and to produce emigrants and pioneer new groups (group reproduction). Maynard Smith suggested that this could be measured simply as the probability of the group producing a successful pioneering emigrant before extinction [268]. Group selection would sometimes act in opposition to individual selection, especially with selfish traits which may indirectly cause the group's extinction, such as unsustainable resource usage. Inclusive fitness theory gives yet another definition for the term 'fitness'. It differs from individual fitness in a further augmentation by the benefits, and harms, caused to the fitness of neighbours, weighted by their relatedness. This *gene* selection mechanism is often labelled kin selection, as neighbours will often be related to the individual holding the gene, inheriting the same genes, and hence an altruistic gene will increase its own inclusive fitness by benefiting copies of itself in kin. Inclusive fitness theory, often cited as an alternative to group selection in explaining altruistic adaptations, has since been redefined by Hamilton & Price into levels of selective forces [310; 178]. That is, they use within and between group levels of selection force which are paralleled to individual and group selection respectively.

Alternative explanations for the presence of an aging trait include the idea that it is not an adaptation at all but rather the side effect of another beneficial adaption or that it is the manifestation of mutational load. Such alternative explanations are appealing as it is difficult to conceive of an altruistic benefit which could outweigh the obvious direct cost to the organism's individual fitness. Mitteldorf [281] argues that although popular opinion is that aging has a non-adaptive explanation, the experimental evidence suggests otherwise. Mitteldorf [282] provides a group selection simulation which ascribes the benefit of aging to demographic homeostasis. This is, populations which live longer lives will exhibit chaotic population dynamics and will be more likely to become extinct. Individual selection will result in the selection of organisms which live longer and longer lives, eventually causing chaotic population dynamics, leading to the group's extinction. Mitteldorf claims this "tragedy of the commons" can never be addressed by individual selection and that the differential extinction of groups outweighs individual selection to enforce growth restraint through birth restraint and aging. Mitteldorf claims that these traits "...constitute 'extreme' altruism in the sense that the cost to the individual is high and direct, while the benefit to the population is far too diffuse to be accounted for by kin selection" [282, p. 346].

We argue that Mitteldorf's simulation is, despite Mitteldorf claims, reliant on kin associations and is therefore also a kin selection model. In Sections 24.2.1 – 24.2.3 we review the group selection debate in order to better understand the relationship between group and kin selection, and the mechanisms behind Mitteldorf's simulation experiments. In Section 24.2.4 we review Mitteldorf's group selection model and in Section 24.3 devise a method of removing kin selection from his model. In Sections 24.4.1 & 24.4.2 we replicate Mitteldorf's results with and without kin selection using our proposed method. Finally, we conclude with an analysis of the relationship between kin and group selection.

24.2 Background

24.2.1 *Group Selection*

The early generation of working group selection models, first proposed by Wynne-Edwards [432] and later adopted by Gilpin [145] among others, ascribed to group selection a major role in the evolution of population regulation. In these models it is the differing viability of the groups, together with their fecundity, that drives selection. Groups that contain selfish genes are more likely to become extinct and have less opportunity to produce emigrants to pioneer new groups. These types of models are reviewed by Maynard Smith [268] who describes a simplified version of Gilpin's predator-prey model (Figure 24.1).

Maynard Smith's model is divided into a number of discrete patches,

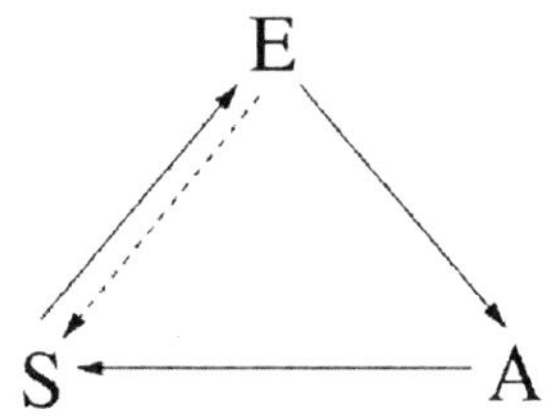

Fig. 24.1 States and transitions of the early group selection models.

each capable of supporting a single group. There are three different states each patch can take at a given time: empty (E), where the group population has become extinct; selfish (S), holding a group that contains *at least some* selfish individuals; or altruistic (A), holding a group that contains *only* altruistic individuals. Transitions occur between states due to extinction ($S \rightarrow E$) and migration (or mutation), either by re-population ($E \rightarrow S|A$) or by a selfish gene takeover ($A \rightarrow S$). In these models the main question is how the cooperation of the altruistic individuals will affect the factors of extinction and migration.

Maynard Smith observes that the fate of these models, when viewed in this manner, is dependent on a single parameter M, which is "the expected number of successful 'selfish' emigrants from an S patch during the lifetime of the patch" [268, p. 281]. A successful selfish emigrant is one that establishes itself and leaves descendants in a neighbouring E or A patch. If the expected number of emigrants from S patches is greater than one ($M > 1$) then the S patches will increase in frequency. Otherwise, the S patches will become extinct faster than they can found new groups and will therefore be selected out of the system. These models demonstrate that the mechanism of group selection is a logical possibility. However, it is debated whether or not the stringent conditions required for the evolution of an altruistic gene could be realised in nature [268; 155].

24.2.2 *Kin Selection*

Shortly after Wynne-Edwards' group selection mechanism was proposed, Hamilton [177] introduced inclusive fitness theory, which was initially seen as an alternative method of explaining the evolution of altruism. Inclusive fitness theory shifts the selection emphasis from the individual to the gene, be it held by the individual, or as a replica in another — it is the organism's individual fitness, augmented by the harms and benefits caused to the fitness of neighbours, weighted by their relatedness [177]. As the neighbours of the individual are also likely to be kin, the selection of the gene with the greatest inclusive fitness is often termed kin selection. It is represented by Hamilton's rule [177], which holds that the criterion for the positive selection of a gene is:

$$\sum_i (b_i - c_i) r_i > 0 \tag{24.1}$$

where the subscript i denotes the ith member of the species; r_i is the relatedness between actor and individual i; b_i is the benefit to the fitness of the individual i; and c_i is the cost to the fitness of the individual i. The relatedness r approximates the chance that a copy of the same allele at a given locus will be held by both the donor and recipient. For example, siblings have an equal chance of inheriting the alleles of either parent at a particular locus and hence have a relatedness of $\frac{1}{2}$. Considering the simple case of a gene which only bestows benefit on a sibling a at the cost of the gene carrier b, applying Equation 24.1, we can see that for a gene to be positively selected, $b_a \times \frac{1}{2} - c_b > 0$. That is, the benefit to the receiving sibling must be greater than twice the cost to the donor. The consequences of the theory are summed up by Hamilton [177] in two points,

(1) for a gene to receive positive selection it is not necessarily enough that it should increase the fitness of its bearer above the average if this tends to be done at the heavy expense of related individuals; and
(2) conversely, that a gene may receive positive selection even though disadvantageous to its bearers if it causes them to confer sufficiently large advantages on relatives.

24.2.3 *Price Equation*

Inclusive fitness theory was often cited as an alternative to group selection in explaining the evolution of altruistic behaviour until Price [310] and Hamilton [178] reformulated it into equations which identified different levels of selection: within- and between-group levels. Hamilton defines his groups unusually, as sharing *equally* the benefits of altruism, creating an important difference between his groups and real groups. As there is no preference for holders of the same genes, such as kin, altruistic genes are always selected against within the group, as any selfish free riders receive the same benefits as everybody else without paying the cost.

This assumption is also made implicitly in the group selection models discussed earlier (Section 24.2.1). For an altruistic gene to be positively selected, the magnitude of between-group selection must be greater than within-group selection. That is, groups with a higher frequency of altruists can perform better by increasing group fitness and hence increasing the relative size of altruistic groups in the population. If this increase outweighs the decrease in frequency of altruists within each group, the gene will increase in global frequency. The more varied the frequency of altruists across the groups, and the more benefit bestowed by the altruists on the group, the greater this between-group selection will be. Figure 24.2 (adapted from [356]) illustrates this effect in a population divided, for a period, into two groups with a varied frequency of altruists, represented by slices of pies. Within-group selection causes the altruistic "pie slice" to shrink in both groups. However, between-group selection causes an increase in the altruistic "pie size" resulting in an increase in the overall frequency of altruist genes.

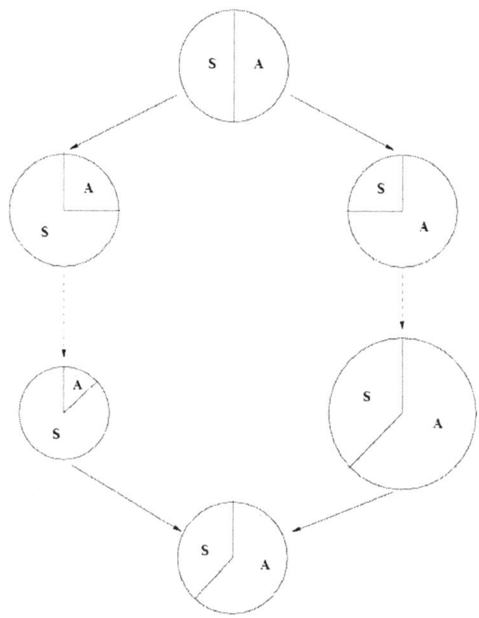

Fig. 24.2 Within- and between-group selection.

The relationship between the Price equation and group selection is considered by some as simply one of mathematical convenience [268]. On the other hand, Wilson [421] uses the equation as the basis of his 'multilevel selection', justifying redefining groups as 'trait groups' which might only exist for a part of their life cycle. As for conventional, partially isolated groups, Hamilton [178] points out that relatedness, due to complex kin associations, eventually builds up to:

$$r_g = \frac{1}{2E+1} \tag{24.2}$$

independent of group size, where r_g is the mean intra-group relatedness; E is the absolute number of emigrants per group, per generation. That is, virtually closed groups become highly related units, regardless of size.

24.2.4 *Mitteldorf's Aging Simulation*

Mitteldorf's simulation [282] is based on the early group selection models of Wynne-Edwards [432] & Gilpin [145], in which group extinction is the driving force (see Section 24.2.1). The simulation experiments demonstrate the evolution of population regulation through aging and birth restraint, which Mitteldorf claims "... constitute 'extreme' altruism in the sense that the cost to the individual is high and direct, while the benefit to the population is far too diffuse to be accounted for by kin selection" [282, p. 346].

Mitteldorf's model is composed of a grid of 16x16 cells, each capable of holding a single group of approximately 100 individuals, which migrate to neighbouring cells at a rate of 10^{-5} individuals per cycle. Mitteldorf employs a logistic equation to model death by overcrowding:

$$\frac{dx}{dt} = bx(1 - \frac{x}{K}) \tag{24.3}$$

where $\frac{dx}{dt}$ is the population growth rate; x is the population size; b is the birth rate; and K is the steady state population level. When the population size is less than the steady state level ($x < K$) the population exhibits exponential growth, whereas, when $x > K$ the population exhibits exponential decline. Populations governed by this equation are normally well-behaved, approaching K asymptotically either from above or below. When a small delay is introduced into these equations, instead of approaching K asymptotically x will overshoot and oscillate. If this delay is further increased, the behaviour of the group undergoes a transition into dynamic chaos, resulting in fluctuations that cause extinction. The solution of population regulation, either by restraining birth rate or increasing aging rate, is permitted to evolve, by asexual reproduction, in Mitteldorf's simulations. If the individuals fail to regulate population growth, the group will experience chaotic fluctuations, causing it, and all its members, to become extinct.

In Mitteldorf's paper, he describes three simulations: the first is a calibration run, determining, in the absence of aging, the maximum sustainable birth rates; the second run permitted aging rates to evolve whilst the birth rate was kept constant; and in the third run both aging and birth rates where permitted to evolve independently. In this paper we review only experiments concerning the evolution of an aging rate, holding birth rates fixed. Mitteldorf's simulation works because, below a certain threshold rate of aging, groups become extinct faster than they can export their members, as can be seen from Maynard Smith's analysis, $M < 1$ (see Section 24.2.1). This threshold rate of aging is determined from the migration rate — at higher migration rates the threshold rate will be lower, as groups require a shorter "lifetime" to export their members.

24.3 Methods

Our Mitteldorf simulation replica uses a 16x16 grid of cells (patches) with a steady state population level (K) of 100. Each cycle every agent had a fixed chance of reproducing asexually (b) with a probability of 0.045; and migrating to a neighbouring cell with a probability of 10^{-5}. The agent also had a chance of dying either by cell crowding or old age. The chance of a death by crowding is given by a probability proportional to the population sharing the site after a time delay of 50 cycles was applied (see Equation 24.3). Otherwise the agent would die of aging once it exceeded its genetically determined natural age of death. Each agent had: a position; an age incremented each time step; and an age of natural death, which was determined from a chromosome holding an evolving aging rate, a Gompertz function [149]. The Gompertz function is used in actuarial science to determine the probability a newborn will survive to an age, t. It is given by the function:

$$S(t) = e^{\frac{I(1-e^{Gt})}{G}} \tag{24.4}$$

where I is the intrinsic vulnerability, fixed at 0.001 in our simulations; and G is the Gompertz value, which is permitted to evolve. A lower Gompertz aging rate equates to a longer life. This aging rate was inherited and mutated with a probability of 10^{-3} by a normal distribution with variance of 0.01.

In order to test the reliance of Mitteldorf's model on kin selection, it required the identification and removal of kin associations to determine the effects on the model. Maynard Smith states that "...kin selection can operate whenever relatives live close to one another, and hence can influence one another's chances of survival and reproduction" [268, p. 279]. This suggests that to remove kin selection it is simply enough to ensure that there is no correlation between the locations of the parent and child. This could be done simply by selecting a location at random to spawn each new child. However, in the case of the Mitteldorf model (Figure 24.1) randomly spawning children to empty patches would negate the effects of migration founding new groups. To address this problem we adopt the mechanism of a compulsory adoption queue. As new children are born, they are placed at the end of the adoption queue and in their place a child is taken from the front of the queue and placed in the same cell as its new adopting parent. This way we can switch kin selection "on" or "off" and ensure that all other factors, such as group density, remain unaffected.

24.4 Results

In order to demonstrate Mitteldorf's simulation's reliance on kin selection, we replicated his simulation experiments (see Section 24.2.4), first without aging and then evolving an aging rate. Next, we repeated the simulation runs, implementing an adoption queue (see Section 24.3) in order to discern the impact of kin associations and kin selection. In our simulations we concern ourselves exclusively with the evolution of aging rate, while holding the birth rate constant.

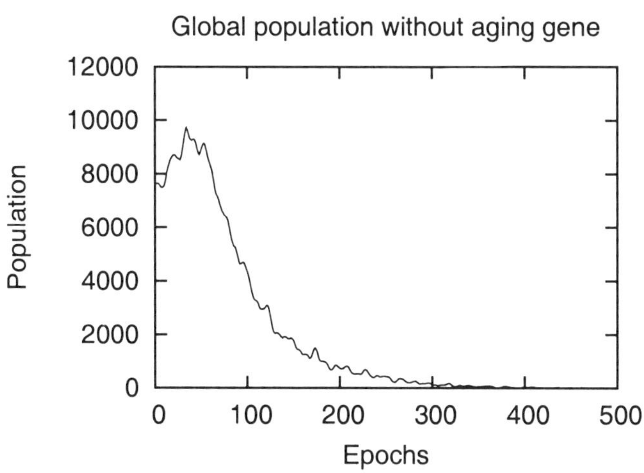

Fig. 24.3 Without aging the global population is quickly driven extinct through chaotic population dynamics within groups.

24.4.1 *Simulations Replicating Mitteldorf's Results*

We initially replicated Mitteldorf's simulation experiments (see Section 24.2.4): first simulating the group chaotic population dynamics without aging genes, and the expected result of eventual global extinction; and second with aging genes, and the expected evolution of a steady aging rate.

24.4.1.1 *Without Aging:*

Figure 24.3 was generated by averaging the results of 20 simulation runs, each lasting until all groups had become extinct, less than 500 epochs (×10 cycles). In these runs the simulation described in Section 24.3 was used, with all deaths by aging switched "off", removing all effects of the aging gene. This means that all agent deaths are only attributable to group crowding. Figure 24.3 shows that the groups, in agreement with Mitteldorf's findings, eventually drive themselves, through chaotic population fluctuations, into extinction.

24.4.1.2 *With Aging:*

Figures 24.4, 24.6 and 24.8 (i.e. the left side of the full page of figures) were generated by averaging the results of 20 simulation runs, each lasting 5000 epochs (×10 cycles). In these runs the simulation described in Section 24.3 was used. Figure 24.4 shows that the global population evolves to a steady state at approximately 6000 agents, indicating that, on average, approximately $\frac{2}{3}$ of patches are empty at any point in time. Figure 24.6 shows the percentage of all deaths that are attributable to aging, approximately 20%. Figure 24.8 shows the evolving rate of aging, the population evolves a Gompertz value, G, of 0.2 which equates to an expected age of natural death of approximately 50 cycles. These results are in accord with Mitteldorf's findings.

24.4.2 *Simulation Without Kin Selection*

To test the importance of kin selection on the model, we performed runs using the simulation described in Section 24.3 with kin selection turned "off", generating the Figures 24.5, 24.7, and 24.9 (i.e. the right side of the full page of figures), by averaging the results of 20 simulation runs, each lasting until all groups had become extinct, less than 1500 epochs (×10 cycles). As can be seen from Figure 24.9 the aging rate is quickly selected against, resulting in the increase of crowding deaths seen in Figure 24.7. Figure 24.5 shows the population eventually dying out, as in the runs without aging, unable to sustain itself with the high growth rates and consequent chaotic population fluctuations.

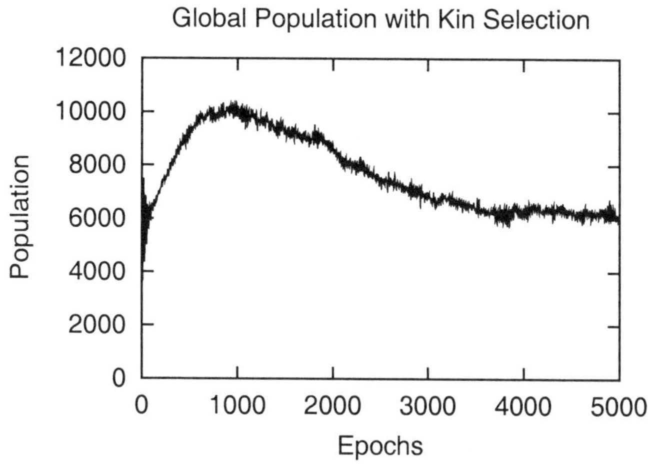

Fig. 24.4 With aging and kin selection the global population evolves to a steady state of approximately 6000 agents.

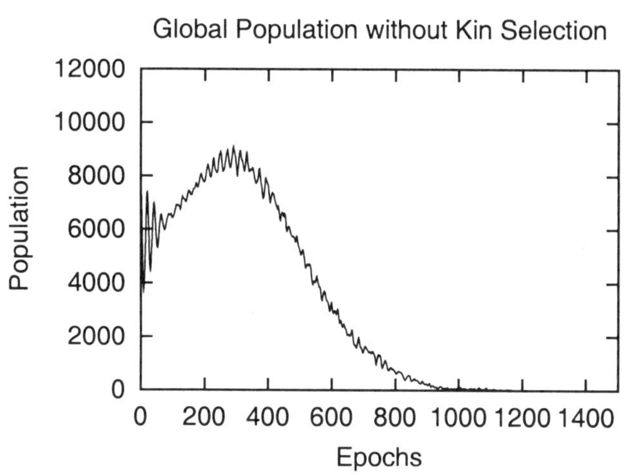

Fig. 24.5 Without kin selection the global population is quickly driven extinct through chaotic population dynamics within groups after the aging gene has been selected out of the system.

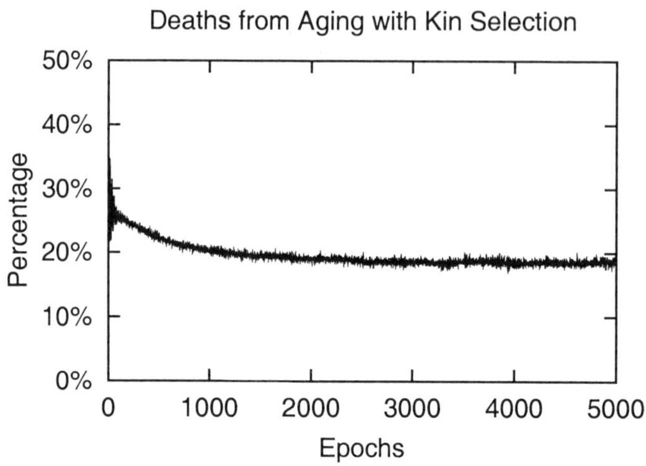

Fig. 24.6 With kin selection the type of deaths evolves to a steady state with approximately 20% of deaths attributable to aging.

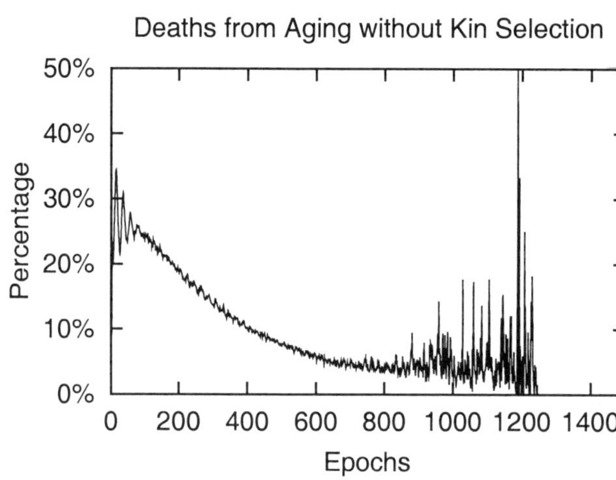

Fig. 24.7 Without kin selection the aging gene is selected against, resulting in the increase of crowding death as cause of agent death.

When kin selection is "on", we would expect the members of the same group to be more genetically related to each other than to members of different groups. Conversely, when kin selection is "off", we would expect

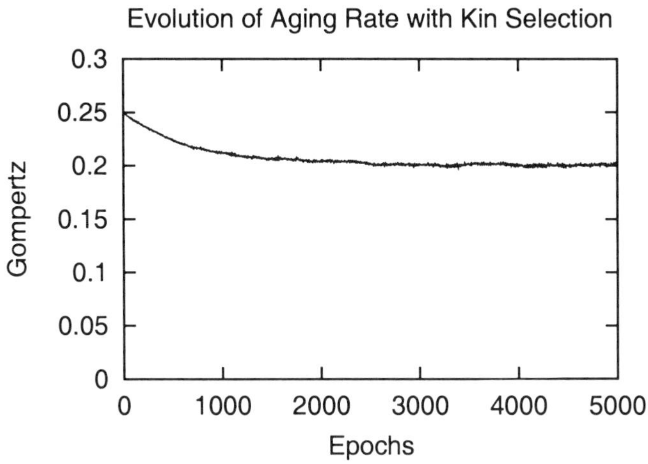

Fig. 24.8 With kin selection an aging rate evolves to a steady Gompertz value of 0.2 (approximately 50 cycle lifetime).

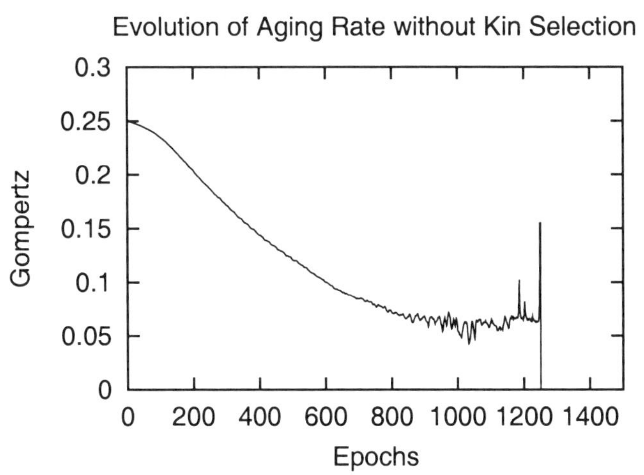

Fig. 24.9 Without kin selection the aging gene is quickly selected out of system.

that members of the same group will be just as closely related to members of different groups as to each other. In order to test this we measured the genetic relatedness between agents as the difference between their evolved

Gompertz values, G, performing a t-test to compare the means between groups. When kin selection was "on", the means of genetic relatedness of members of the same group and members of different groups were found to differ significantly ($t(59.4) = -28.26, p \ll .05$). When kin selection was "off", while there was some difference in the mean genetic relatedness of members of the same group and members of different groups, the difference was not statistically significant ($t(785) = 1.90, p > .05$).

24.5 Conclusion

Of the controversy over group selection and the importance of kin selection for the evolution of altruism Hamilton remarked:

> Because of the way that it was first explained, the approach using inclusive fitness has often been identified with 'kin selection' and presented as an alternative to 'group selection' as a way of establishing altruistic social behavior by natural selection.... Kinship should be considered just one way of getting positive regression of genotype in the recipient, and that it is positive regression that is vitally necessary for altruism [178].

The idea is that inclusive fitness is more general than kin selection and might arise by other mechanisms than kin selection. What is needed for group selection of altruistic behavior is the positive association between group fitness and the altruistic genes required by the Price Equation. Such an association can arise by kin selection or by other means. But however it arises, it will result in differential group fitness leading to a spreading of altruism.

Our results are certainly consistent with this view. We would say that group selection is supervenient upon kin selection in the simulations we have conducted; that is, there are multiple possible ways of realizing group selection, kin selection being one [223]. Kin selection can lead to within group selection for altruistic behavior, so long as the groups are *not* Hamilton's groups which miraculously share the benefits of altruistic behavior identically across all group members (see [263] for an example). And kin selection can lead to between group selection for altruistic behavior, as our current study demonstrates. But the association between longevity of groups and altruism could in principle happen otherwise, for example, by chance or (at least in the case of artificial simulations) by intervention of a Designer. In our simulations, and so also in Mitteldorf's simulations, neither of these alternatives apply; indeed, there was no alternative to kin selection in making group selection operative. We believe that, although kin selection is not logically necessary for group selection, it is practically necessary: nature also provides no alternative basis for differential group survival and reproduction. Hamilton's other suggestion that we reserve "group selection" for situations where kin selection is inoperative [178] would empty the term of all practical application.

Thus, and as the results of Section 24.4.2 show, Mitteldorf's model is dependent on the kin associations within its groups and therefore is a kin selection model. The operations of kin selection may be too diffuse to be seen by the naked eye, but they can easily be seen in t-tests over simulation runs, which we consider the more reliable instrument.

We can see that Mitteldorf's groups are in fact highly related, as new groups are founded by small groups: indeed, in Mitteldorf's asexually reproducing, sparse and low migrating rate model, most groups will be founded by a single individual and have little or no contact with other groups throughout its lifetime. This results in groups of individuals which are practically clones of each other. We can also see from Hamilton's observation (see Section 24.2.3) that the genetic relatedness builds up in virtually closed groups to approximately $\frac{1}{2E+1}$. Both these points, separately and combined, argue strongly that Mitteldorf's groups are highly related units.

Future work. As a follow-on to this study, we are developing a simulation which tests the Weismann hypothesis [411], an adaptive explanation of the evolution of aging. The Weismann hypothesis attributes the benefits of aging to "making room for the young". That is, a population that ages will turnover faster, promoting genetic diversity and adapting more nimbly to a changing environment. In our simulation a host population co-evolves with a disease population. Individual selection within host groups results in the evolution of longer living hosts and consequently groups with less diversity. Groups of hosts with less diversity are more easily exploited by the disease population and are less successful at producing emigrants. This differential success of groups counterbalances the selection against aging within groups, resulting in the evolution of a steady aging rate.

Chapter 25

Evolving Capability Requirements in WISDOM–II

A. Yang[1], H.A. Abbass[1], M. Barlow[1], R. Sarker[1], and N. Curtis[2]

[1] *The Artificial Life and Adaptive Robotics Laboratory, School of Information Technology and Electrical Engineering, UNSW@ADFA, NorthCott Drive, Canberra, ACT 2600, Australia*
E-mail: {ang.yang,m.barlow,h.abbass,r.sarker}@adfa.edu.au

[2] *Land Operation Division*
Defence Science & Technology Organisation
Department of Defence
Edinburgh, SA 5111, Australia
E-mail: neville.curtis@dsto.defence.gov.au

Over the last decade, it has widely been accepted that warfare is a complex adaptive system (CAS). The multi-agent technology is a promising tool to study CASs. The warfare intelligent system for dynamic optimization of missions (WISDOM) is an agent based distillation (ABD) system for warfare. WISDOM–II, the second version of WISDOM is based on a novel network-centric multi-agent architecture (NCMAA). WISDOM-II allows analysts to study the new theory of network centric warfare (NCW) easily and effectively. In this paper, we use multi-objective optimization to evolve capability requirements for the blue (friendly) force. We examine four setups, where either the blue or the red (adversary) force adapts NCW. We show that capability requirements are different when the red force switches from a platform centric warfare (PCW) to a NCW. For a platform–based blue force, an increase in cost is required to meet the same mission when compared to a network-centric blue force.

25.1 Introduction

Network centric warfare (NCW) is a recent concept in defence, where forces are able to communicate better and increase the accuracy of their common operation picture (COP) [13]. In 1999, Alberts *et al.* [13] proposed the following definition of NCW.

> "We define NCW as an information superiority-enabled concept of operations that generates increased combat power by networking sensors, decision makers, and shooters to achieve shared awareness, increased speed of command, higher tempo of operations, greater lethality, increased survivability, and a degree of self-synchronization. In essence, NCW translates information superiority into combat power by effectively linking knowledgeable entities in the battlespace." [13]

Since its emergence in 1983 [422], the debate between proponents and opponents continues. The proponents advocate that networked entities may produce information superiority, which in turn dramatically increases combat power. The theory that power is increasingly derived from information sharing, knowledge sharing and more rapid command has been supported by results of recent military operational experience [140; 420]. The advantages of NCW have been recognized as:

- Small-size networked forces can perform missions effectively at a lower cost;
- New tactics can be adopted in combat, e.g. "swarm" tactics. A larger force may be separated into several small swarms. These swarms do not need to keep communicating with each other at all time. Through networking they are aware of each other. If one swarm gets into trouble, other swarms will detect this and give help to it immediately.
- The mechanism for decision making on the battle field has changed. It is easy to obtain help from geographically dispersed experts with the aid of networking;
- The sensor-to-shooter time is reduced. The soldiers in the field may take actions based on their own raw intelligence from sensor's displays instead of waiting for orders.

Certainly, a few challenges and limitations with NCW still exists [219; 73; 53; 420], e.g. information overloaded, bandwidth limitations, difficulty of applying theory to practice and the theory remaining un-demonstrated and unproven to date because the opponent in recent real wars is far from formidable. It is not possible to verify and better understand NCW in real engagements. Fortunately red teaming which is the process of assessing the risk evolving in a system or a plan by anticipating adversary behaviors, agent based distillation (ABD) where only essential aspects are modeled and simulated for a CAS, complex system theory, network theory, evolutionary computation and other modern information technologies may help us understand NCW [435].

Recognizing warfare as a complex adaptive system [205; 233; 348; 207] opened the doors for a number of ABDs to emerge. These include ISAAC [205; 206] and EINSTein [206; 207] from the US Marine Corps Combat Development Command, MANA [233; 234; 136; 135] from New Zealand's Defence Technology Agency, BactoWars [414] from the Defence Science and Technology Organisation (DSTO), Australia, CROCADILE [28] and WISDOM [436; 437; 438] developed at the University of New South Wales at the Australian Defence Force Academy (UNSW@ADFA). These systems have facilitated the analysis and understanding of combat, for example, using MANA to explore factors for success in conflict [54].

In this paper, WISDOM–II is used as the simulation platform. We use multi-objective optimization techniques, compare the performance of a networked force and a traditional force in several scenarios, and try to gain a better understanding of NCW. Evolutionary techniques are used to derive the blue force structure most suitable for each scenario. In the rest of the paper, we first briefly review WISDOM–II followed with a description of the scenarios and experiments, then the analysis of the results. Finally conclusion and future work are discussed.

25.2 WISDOM–II

WISDOM–II was inspired by existing ABD combat systems. Our recent research [435; 438] shows that most existing ABDs, including WISDOM version I, were developed mainly on platform centric (traditional) warfare and current agent architectures, which limit their ability to study NCW. WISDOM–II was re-designed and re-developed based on a new agent architecture, called Network centric multi-agent architecture (NCMAA) [438]. NCMAA is purely based on network theory. The system is designed on the concept of networks, where each operational entity in the system is either a network or a part of a network. The engine of the simulation is also designed around the concept of networks.

Generally speaking, there are five distinguishing components in WISDOM–II. The first three components are used to model the internal behavior of warfare, while the last two are used for analysis.

(1) the C3 component – including both command and control (C2), and communication.
(2) the sensor component – retrieving information from environment
(3) the engagement component – including firing and movement activities
(4) the visualization component – presenting various information with graphs
(5) the reasoning component – interpreting the results in natural language during the simulation process

Five types of networks are defined in WISDOM–II; these are the command & Control (C2), vision, communication, information fusion, and engagement networks. Four types of agents are supported in WISDOM–II: combatant agent, group leader, team leader and swarm leader. Agents are defined by their characteristics and personalities. Each agent has nine types of characteristics: health, skill, probability to follow command, visibility, vision, communication, movement and engagement. Initial levels of health, skill, visibility and vision are predefined by the user. They may be different for different agent types. The swarm leader can also build plans and give orders to combat groups. The personality in WISDOM-II is a defined by two values: a magnitude and a direction vector representing the attraction-repulsion direction and weight for each agent.

The movement of each agent is determined by its situation awareness and personality vector. In each time step, the agent can only move to its neighbor cells based on the overall influence of all perceived agents. A strategic decision is made by the swarm leader of each force based on the common operating picture (COP), which is the global view of the battle field for that force. Decision making on the force level utilizes the same environment the agents are embedded in, but on a coarser resolution.

WISDOM–II collects information for each entity as well as for the interaction between entities. In this way, a large number of statistics are collected. Then these statistics are fed to a reasoning engine, where natural language reasoning is provided to the user. WISDOM–II also provide capabilities such as interactive simulation. For more details of WISDOM-II, please refer to [438].

25.3 Experimental setup

The objective of these experiments is to investigate the differences in force capability requirements when using PCW or NCW. The red force has fixed capability resources and will be tested under PCW and NCW. The blue force has less than half the number of agents of the red force and will be tested under PCW and NCW. The blue force needs to decide on its future capability requirements. Consequently, four scenarios are designed as shown in Table 25.1.

Table 25.1 Four Scenarios

	Blue Force	Red Force
PCB – PCR	Platform Centric Blue	Platform Centric Red
NCB – PCR	Network Centric Blue	Platform Centric Red
PCB – NCR	Platform Centric Blue	Network Centric Red
NCB – NCR	Network Centric Blue	Network Centric Red

In the NCW mode of operation, we assume that all combatants have the ability to transmit their information to the swarm leader, based on which the swarm leader develops a COP for the force. Once this COP is filtered by the swarm leader, it is sent back to the combatants to act upon them. In the PCW mode of operation, each combatant has its own local operating picture (LOP) based on information collected from vision and communication. The cost of a force is based on the infrastructure cost and operational cost.

Eight different types of communication and weapons are predefined as shown in Table 25.2 and 25.3. Each agent can only be equipped with one type of communication and weapons. The groups in both the blue and red forces are composed of a set of homogenous agents, but the force as a whole consists of heterogeneous groups.

Table 25.2 Communication Type

ID	0	**1**	2	3	4	5	6	7
Range	Long+	**Short**	Medium	Long	Short	Medium	Long	Long+
Lost Prob	0	**0**	0	0	10%	10%	10%	0
Lantency	2	**0**	0	0	2	2	2	0

Table 25.3 Weapon Type

ID	0	**1**	2	3	4	5	**6**	7
Range	Long	**Short**	Long	Short	Long	Short	**Long**	Long+
Strength	Light	**Light**	Light	Light	Heavy	Heavy	**Heavy**	Heavy+
Radius	Heavy	**Point**	Light	Heavy	Point	Light	**Heavy**	Heavy+

The capabilities used by the red force is indicated in bold in Tables 25.2 and 25.3. The total number of red agents are 30 combatants and 2 surveillance agents. The 30 combatant agents are divided into three groups of eight agents each and one group of 6 agents. All red agents use communication type 1. The 8-agent groups use weapon type 1, while the 6-agent group uses weapon type 6. The surveillance agents in both forces do not use weapons as they do not engage nor they can be destroyed. The blue force consists of 15 agents in total in a maximum of 5 groups. The fifth group is reserved for surveillance. The evolution decides on the number of agents per each group, where a group is allowed to be empty, and the type of communication and weapon used in each group.

25.3.1 *Chromosome representation*

We evolve the capability requirements of the blue force. A chromosome is divided into 5 blocks, each block corresponds to one group of the blue force. Each block consists of 4 variables encoded using binary representation. The variables are: the communication id (3 bits), weapon type id (3 bits), amount of ammunitions (8 bits) and the proportion of agents belonging to this group (3 bits) as indicated in Figure 25.1. The total number of bits in the chromosome is 85.

COM (3 bits)	ART (3 bits)	ATT (8 bits)	# W (3 bits)	X 5

Fig. 25.1 Chromosome Representation

The number of agents per group is calculated as follows: first, we distinguish between the first four blocks representing combatant agents and the last one corresponding to the surveillance group. For the first four blocks, the 3 bits representing the proportion of agents belonging to this group are decoded into the corresponding integer number (0-7). The following equation is then used to decide on the number of agents n_i in group $i = 1 \ldots 4$, given that the total number of agents available N is 15.

$$n_i = \lfloor N \times \frac{w_i}{\sum_{j=0}^{4} w_j} \rfloor \qquad 0 \leq i \leq 3 \tag{25.1}$$

where $\lfloor \quad \rfloor$ is the floor function, w_i is the weight as specified in the chromosome (0-7). Once the number of agents is calculated for each of the combatant groups, we sum to find the total number of combatant agents. The number of surveillance agents is then calculated as 15 - the total number of combatant agents. This guarantees that the total number of blue agents is always 15.

25.3.2 Objectives and evolutionary computation setup

Yang *et al.* [436; 437] argued that taking a simple linear combination of the objectives may hide some information which is crucial in understanding the dynamics within a warfare simulation. Therefore, we explicitly represent the three objectives in the problem and use the non-dominated Sorting Genetic Algorithm – II (NSGA–II) [100], a multi-objective optimization algorithm. Three objectives are defined:

(1) **minimizing** the cost of the blue force $\Downarrow$
(2) **minimizing** the casualty of the blue force $\Downarrow$
(3) **maximizing** the casualty of the red force $\Uparrow$

The cost of the blue force includes the capital cost and the operational cost. The capital cost is defined as in Equation 25.2, 25.3, 25.4 and 25.5:

$$C_c^t = \sum_i^N C_i^c + \sum_j^n C_j^a \tag{25.2}$$

$$C_i^c = K_i^c \times \frac{(R_i^c)^\alpha \times (1 - P_i^c)^\beta}{(1 + L_i^c)^\theta} \tag{25.3}$$

$$C_i^a = K_i^a \times \overline{R_i^a}^\gamma \times (D_i^a)^\delta \tag{25.4}$$

$$D_i^a = S_i^a + \sum_{r=0}^{r_i^a} (8 \times r \times (S_i^a - r)) \tag{25.5}$$

where
C_c^t denotes the total capital cost of the blue force;
N is the total number of blue agents;
C^c denotes the communication cost of agent i;
n is the number of blue agents except the blue surveillance agents because the surveillance agent always does not have any weapon;
C_j^a denotes the weapon cost of blue agent j;
R_i^c denotes the communication range of blue agent i;
P_i^c denotes the probability of a message loss in the communication channel used by blue agent i;
L_i^c denotes the latency of the communication channel used by blue agent i;
C_i^a denotes the weapon cost of blue agent i;

$\overline{R_i^a}$ denotes the average firing range of the weapon used by blue agent i;
D_i^a denotes the total damage of one shot by blue agent i;
S_i^a denotes the strength of the weapon used by blue agent i;
r_i^a denotes the radius of the weapon used by blue agent i;

All other parameters are constants and are set to 1. If the weapon is a point-to-point weapon, the total damage is equal to the strength of this weapon. Otherwise the damage is the sum of the damage in all cells within its radius. The damage caused to agents occupying a cell is inversely proportional to the distance from the input cell. For example, if the weapon radius is 2 and the weapon strength is 3, the distribution of the damage is shown in Figure 25.2. Therefore the total damage is 35.

1	1	1	1	1
1	2	2	2	1
1	2	3	2	1
1	2	2	2	1
1	1	1	1	1

Fig. 25.2 Damage distribution

The operational cost is a proportion of the weapon cost in this paper, which is defined in the equation 25.6.

$$C_t^o = \sum_i^N \frac{A_i \times C_i^a}{T} \tag{25.6}$$

where:
C_t^o denotes the total operational cost of blue force;
N is the total number of blue agents;
A_i denotes the ammunition of blue agent i;
C_i^a presents the weapon cost of blue agent i;
T is a constant. In this paper we set it to 100.

We use the ratio of the number of dead agents at the end of the simulation to the total number of the agents to represent the casualty rate suffered by a force (Equation 25.7), because the surveillance agents are invisible to any other adversary agents.

$$R_c = \frac{N_d}{M_o} \tag{25.7}$$

where:
R_c denotes the casualty rate of one force;
N_d represents the number of dead agents in one force at the end of the simulation;
M_o is the number of non-surveillance agents in a force.

For each scenario, we evolve for 100 generations, each of which includes 20 individuals. Each individual is evaluated for 30 times while each simulation lasts 150 time steps. The crossover rate is set to 0.8 and the mutation rate is set to 0.01. We repeat each experiment for 10 times with different seeds.

The simulation environment is 30x30 cells and the destination flag (each force has a goal to occupy this area) is located at the mid of the left of the environment. Initially the blue force is located at the upper-right corner while the red force is located at the bottom-right corner. Figure 25.3 presents an example of the initial position of both blue and red forces.

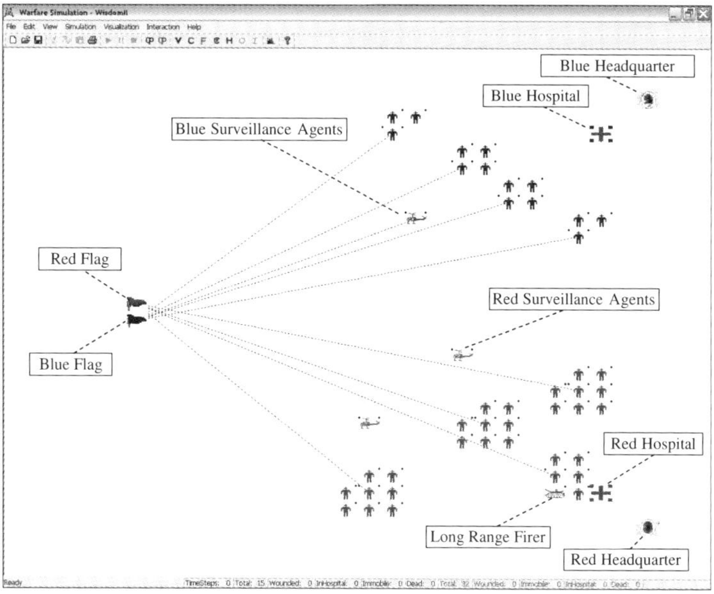

Fig. 25.3 Screen dump for the simulation

25.4 Results

Figure 25.4 depicts a 2-D projection of the Pareto-optimal set. Each bar represents the cost of capability requirements of blue. We need to point out that when interpreting the casualty rate, one needs to keep in mind that the blue force consists of 15 agents while the red force consists of 30 combatant agents. Thus, a casualty rate of 50% in each side would correspond to a loss of 7 agents in blue and 15 agents in red.

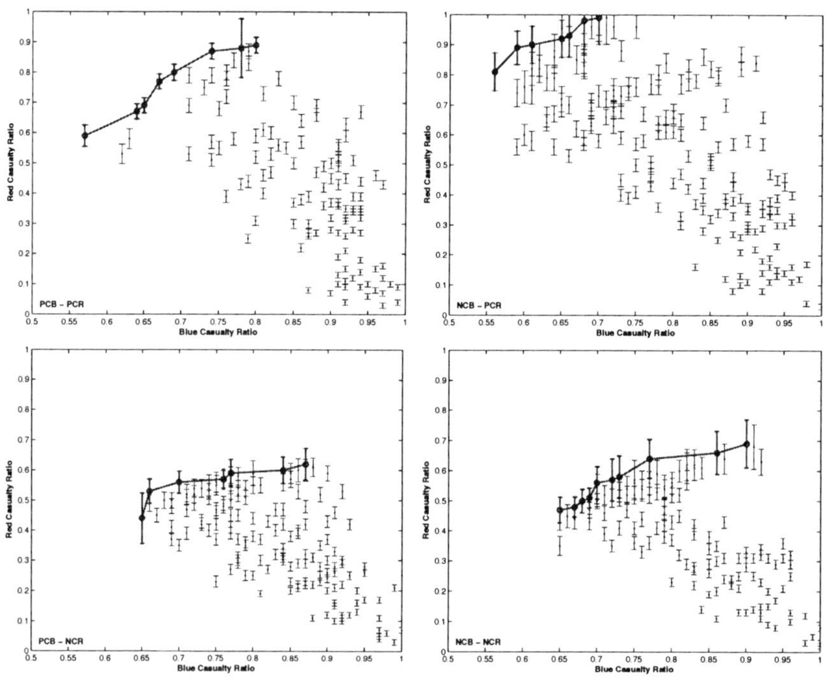

Fig. 25.4 Pareto-optimal set with blue cost. NCB: network centric blue force; NCR: network centric red force; PCB: platform centric blue force; PCR: platform centric red force

The first observation from the graph is that the only scenario where blue can completely destroy the red force in the finite time of the simulation is NCB-PCR. This clearly indicates that the NCW mode has a significant advantage over the platform based. Obviously solutions where blue is destroyed are not interesting in this context since we can avoid these situations by choosing better capabilities. However, we can also notice that there is no efficient solution where the red can destroy the blue when the red is networked and the blue is not. This is interesting since the red has fixed capabilities, thus even when the red operates in the NCW mode, changing capabilities of the blue can re-balance the equation since the blue has the flexibility to decide on what capabilities it requires for each situation. Overall, if the red operates in a platform based mode, the blue can gain an advantage through capabilities but it will gain maximum advantage through capabilities and NCW mode of operation.

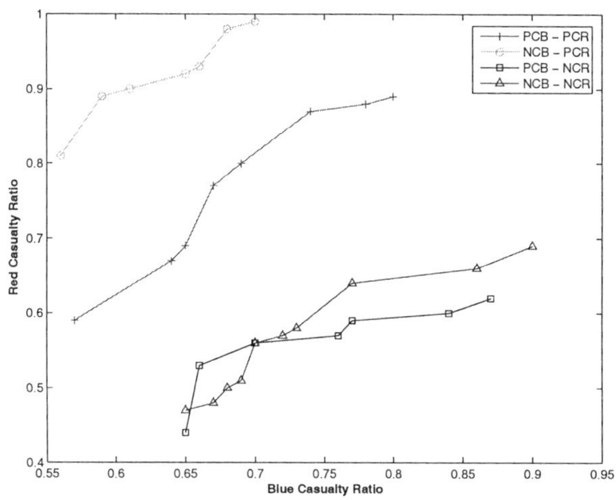

Fig. 25.5 Pareto-optimal set regarding blue casualty ratio and red casualty ratio only

If cost is not an issue, we can eliminate totally the objective concerning cost and visualize the Pareto set using two objectives only as presented in Figure 25.5. Each efficient solution in this graph represent a possible scenario of engagement. It is clear that the most efficient scenarios correspond to NCB-PCR followed by PCB-PCR. For the same level of damage in the blue, operating in the NCW mode would increase the damage in the red by 35%. Obviously, in a real life situation this does not necessary mean real damage, it just reflects the superiority gained by networking. Once the red operates in an NCW mode, the blue gains no significant advantage regardless of its mode of operation. This is not surprising as the red has double the agents of the blue, but it can be seen that higher capabilities need to compensate for that.

Figures 25.6 and 25.7 present the Pareto set using two objectives only. In the first figure, the two objectives are the cost of the blue and the casualty in the blue, while the second figure captures the casualty in the red instead. It

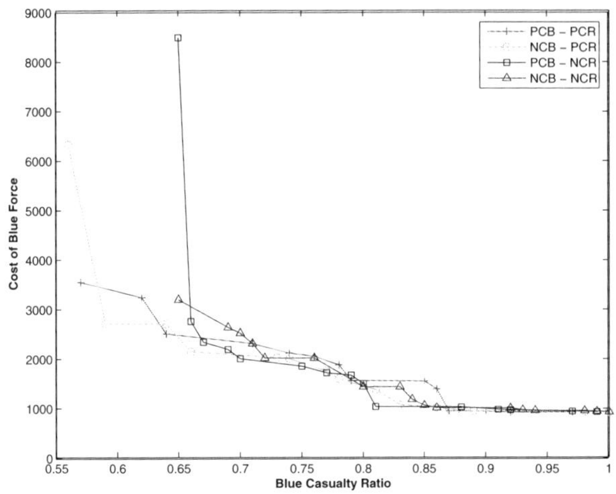

Fig. 25.6 Pareto-optimal set regarding blue cost and blue casualty ratio only

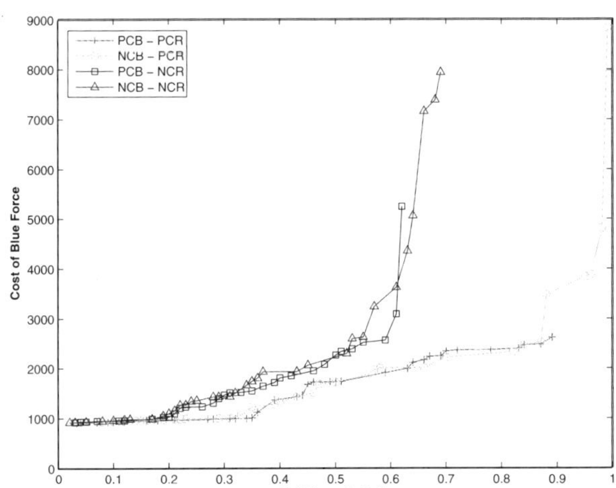

Fig. 25.7 Pareto-optimal set regarding blue cost and red casualty ratio only

is expected to see that the cost reduces as the damage in the blue increases and the damage in the red decreases. What is interesting is the slow change of cost with low damage in the red then the sudden exponential increase in cost from almost 0.5 damage in the red and upwards. It is more interesting to see that all four scenarios are close to each other, except when low damage in the blue or high damage in the red are required. Although this is somehow expected, it is interesting that the simulations confirmed this hypothesis. Overall, we can conclude that for a force to operate efficiently, we should expect a dramatic increase in cost and capability requirements, as well as a decrease in the blue damage and increase in the red damage. However, one can see that for the same level of damage (either for the blue or the red), the cost associated with NCW is less than that associated with the platform mode.

25.5 Conclusion and future work

NCW is a new concept emerging recently in the military domain. Since it is hard to provoke a real war to verify and test it, the number of the proponents is almost the same as that of the opponents. There are currently insufficient evidences for either side to convince the other. This paper attempts to use computer based simulations and red teaming to study the theory of NCW in certain scenarios. We adopt WISDOM–II, a network centric agent based simulation combat system as the simulation platform. With the aid of multi-objective optimization techniques, we find that a networked force may perform better than a platform centric force no matter whether its enemy is network centric or not. In order to perform as well as possible, the networked force must have corresponding high level capabilities. Otherwise it cannot exhibit the advantage of the networked force. We also find that it is relatively hard to beat a networked force. Being networked appears to be a force multiplier, but when both forces are networked, this factor may cancel out.

In this paper in order to make the problem not too complicated, we fix the composition and capability of the red force. It limits the potential of the red force's performance. In the future we would like to study the system dynamics and behaviors by co-evolving the composition and the capability of both blue and red forces.

Acknowledgments

This work is supported by the University of New South Wales grant PS04411 and the Australian Research Council (ARC) Centre for Complex Systems grant number CEO0348249.

Appendix

Network Diagram

Submitted by: James Watson, John Hawkins, Daniel Bradley, Dharma Dassanayake, Jim Hanan, Scott Heckbert, and Andrew Hockey.

1a. Pattern name: Network Diagram

1b. Classification: Visualization (Structure, Dynamics, Function, Micro, Macro, State Space)

2. Intent. A fundamental property of complex systems is that the interactions between micro-level components result in emergent macro level behaviour. This emergent behaviour means that it is difficult to know which components are of interest. In addition, there are large numbers of possible metrics available to analyze the properties of a complex system. Thus, a common starting point is to simply generate some form of initial visualization of components and their interactions.

A network diagram provides a spatial representation of entities and their relationships. By hiding the specifics of entities and their relationships, an initial overview of the complex system is provided. For example, a tabular description of genes and their interactions makes it difficult to follow the paths of gene activation. By depicting each gene as a node and its interactions as links between nodes, we gain an immediate impression of paths of activation. This is achieved by removing superfluous information such as gene name, etc., and by providing an intuitive spatial representation.

Furthermore, numerous views of the same system can be generated by changing what the nodes and links represent. This flexibility allows researchers to home in on the pertinent features of the system.

3. Also known as: Graph

4. Motivation. *Static System:* Characterizing the nature of co-author relationships in a given academic field is difficult when browsing a textual description of their collaborations. Visualizing authors as nodes and co-authorship as links renders large amounts of information visually accessible.

Dynamic System: When investigating state spaces, an activation diagram makes it difficult to determine properties such as cyclic behaviour, length of cycles, etc. By removing information such as individual entity states for each time step, and visualizing unique states as nodes and the transitions between them as links, such properties are immediately apparent.

5. Applicability. The Network Diagram pattern can be applied whenever you can define entities and their relationships. In many physical systems, there is a an obvious way to define the system as entities and relationships. For example, regulations between genes, interactions between people, etc. However, the Network Diagram has wider applicability than these corporal mappings. More abstract properties of a system, such as state space transitions, can also benefit from the Network Diagram visualization.

6. Structure. (Intentionally left blank).

7. Participants. Node, Links, Layout, Manipulation

8. Collaborations. Entities are represented as nodes, with their relationships represented as links between these nodes. The nodes are placed in a spatial representation defined by the layout. Manipulation provides a means of interacting with all aspects of the Network Diagram (such as layout, node states, etc.).

9. Consequences. The Network Diagram focuses on the entities and relationships of interest.

- it focuses on the relationships of interest
- layout allows different views, possibly different interpretations
- ease of view / analysis comes from reduction of information
- generality means one has to choose entities / relationships
  - this is a strength and a weakness
    - strength: flexibility
    - weakness: lack of guidance of what should be nodes and links

10. Implementation.

- there are lots of different layouts / methods of manipulation, which can largely influence usefulness
- giving user choice over layout (and manipulation) is a useful approach
- a random layout is the simplest to implement, but is generally unsuitable for (e.g.) visualizing the giant component of the network. Increasingly sophisticated network layout algorithms can incur a cost in processor time (many optimal layouts are likely to be NP-complete).

11. Sample code.

```
n = network, indexed by node
p = position of each node

clear the screen
for i = 1 to (number of nodes in n):
    p[i] = random position
    draw sphere at p[i]

for i = 1 to (number of nodes in n):
    r = list of nodes regulated by n[i]
    for j = 1 to (number of nodes in r):
        draw arrow from p[i] to p[r[j]]
```

12. Known uses. Boolean network visualization and design, social network visualization, neural network visualization and design.

13. Related patterns. Discrete State-Space Trajectory Diagram, Activation Diagram.

Synchronous System State Update

Submitted by: Jim Hanan, James Watson, Daniel Bradley, John Hawkins, Dharma Dassanayake, Scott Heckbert, Andrew Hockey.

1a. Pattern name: Synchronous System State Update

1b. Classification: Model (Behavioural)

2. Intent. In many complex systems simulations, all objects that interact in the model should be updated synchronously, reflecting the parallel nature of the abstracted system. This pattern describes a method of implementing synchronous updating by creating a new set of states from current states.

3. Also known as: Synchronous update.

4. Motivation. Given the Boolean network model shown in Figure 1, you want to analyze system behaviour when all node interactions are taken into account before states are changed.

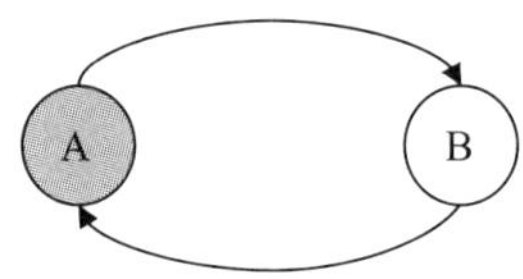

Fig. 1 Simple Boolean network with node A activated.

Using the Synchronous State Update pattern, you would create and assign values to a new set of states (one for each node) according to all the interactions between nodes in the current state (see Figure 2). The new set of states then becomes the current set.

Node	Current State
A	active
B	inactive

Node	New State
A	inactive
B	active

Fig. 2 For node A, no activation is received from its inputs, so it becomes inactive. Node B is activated by node A, so it becomes active. This process of swapping states can occur indefinitely.

By determining the new set of states for each node in the network according to all interactions among the old states, synchronous updating is achieved.

This is different to asynchronous updating, where all nodes and their interactions are not considered before updating. From the initial conditions of Figure 1, asynchronous updating could

- first update A according to B's state without considering A's input to B, so A and B become inactive forever
- first update B according to A's state without considering B's input to A, so A and B become active forever

5. Applicability. The Synchronous System State Update is useful whenever all components of a complex system are concurrently interacting on the same timescale. For example, this is often the case when you are interested in the influence of interactions between entities, and thus abstract away the time differences between these interactions and consider them all at the same point in time. Examples include some Boolean models of gene regulation, social networks, and L-systems plant development. It is best suited to processes with discrete time-steps, though continuous systems can be approximated with small time-steps.

6. Structure. (Intentionally left blank).

7. Participants. Entities, Relationships, Current states, Next states

8. Collaborations. Each entity has a corresponding current state. Next states are created according to the current states of the entities and the relationships between them. The next states become the current states at the end of updating.

9. Consequences.

- Since interactions are only considered at each time-step, this pattern abstracts away the actual time events occur.
  - For non-stochastic simulations, this causes the system to exhibit deterministic behaviour, which is often desirable for system-wide analysis.
  - If the system being modeled has asynchrony, this information is lost, unless modeled explicitly by incorporating delays and fine scale time steps.
  - May lose variability potentially created by asynchronous updating.
- Updating slows as the number of entities increases.
- Approximating continuous processes requires small time-steps, which requires longer processing times.

10. Implementation.

- Small time-step sizes can be used to approximate continuous time.
- Pointers to the new/current states lists can be manipulated to improve performance, e.g.:

```
free(current)
current = next
next = new (un-initialized) set of states
```

- Depending on the model's architecture, each entity (such as an agent) can handle its own updating and state, or a separate global process can maintain state lists and iterate through all entities. For very large numbers of entities, the global algorithm approach can be more efficient if simple representations are used for entities (e.g., a bit array for Boolean networks) instead of object instantiations.

11. Sample code.

```
C = current states of entities, indexed by node number
N = next states of entities, indexed by node number
t = current time-step

t = 1
while (t <= time-steps of interest)
    for i = 1 to number of nodes in the system
        N[i] = state based on inputs to C[i]

    C = N
    t = t + 1
```

12. Known uses. Network update [330], Game loop, L-system derivation process [244]

13. Related patterns. Asynchronous System State Update should be used when asynchronous updating is important to the model. Sequential State Update should be used when the ordering of the updates is inherent in the system being modeled (e.g., Chomsky grammars).

Discrete State-Space Trajectory Diagram

Submitted by: Daniel Bradley, John Hawkins, James Watson, Dharma Dassanayake, Scott Heckbert, Andrew Hockey.

1a. Pattern name: Discrete State-Space Trajectory Diagram

1b. Classification: Visualization (Dynamics, State Space, Function, Macro Behaviours)

2. Intent. The pattern describes a technique for visualizing a system's dynamic qualities as a discrete graph. It allows for the identification of the number, size and structure of attractors, as well as the number and length of transients leading to these attractors.

By removing the specific details about the states the system takes, a state space diagram shows an abstract map of a systems dynamics. The removal of details irrelevant to describing dynamics eases the task of analysis by visual inspection.

3. Also known as. Wuensche Diagram, Basin of attraction field

4. Motivation. In all complex systems simulations at each moment the state of the system is described by a set of variables. As the system is updated over time these variables undergo changes that are influenced by the previous state of the entire system. System dynamics can be viewed as tabular data depicting the changes in variables over time. However, it is hard to analyze system dynamics just looking at the changes in these variables, as causal relationships between variables are not readily apparent.

By removing all the details about the actual state and the actual temporal information, we can view the dynamics as a graph with nodes describing states and links describing transitions. For instance software applications can have a large number of states. Problems occur when software applications reach uncommon or unanticipated states. Being able to visualize the entire state space, and quickly comprehend the paths leading to any particular state, allows more targeted analysis. Common states can be thoroughly tested, uncommon states can be identified and artificially induced.

State space diagrams allow for numerous insights into system behaviour, in particular some states of the system can be shown to be unreachable, while others are unavoidable.

5. Applicability. Any situation in which you have a model or system which changes state over time and you want to examine the abstract dynamical qualities of these changes. For example, social network theory, gene regulatory networks, urban and agricultural water usage, and concept maps in cognition and language modeling.

6. Structure. See Figure 3.

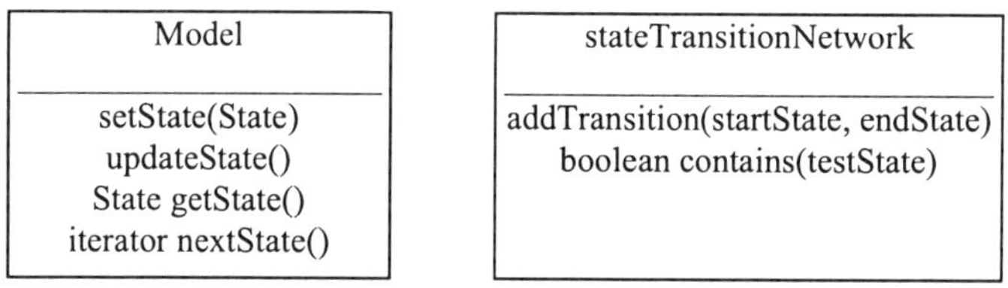

Fig. 3

7. Participants. See Figure 4.

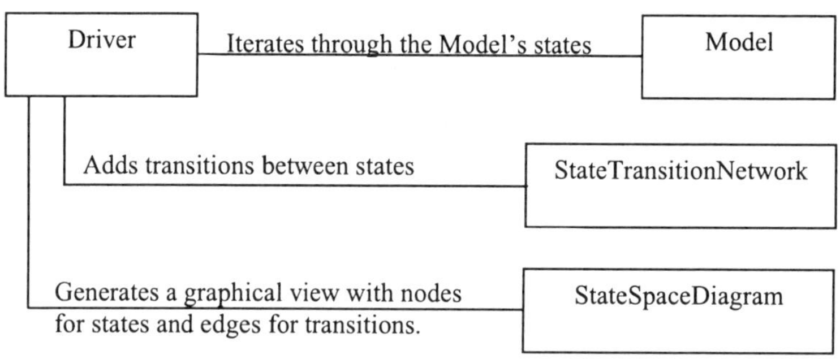

Fig. 4

8. Collaborations. Can be used with Activation Diagram to visualize the details of a given state.

9. Consequences.

- The pattern supports it's objectives by removing details unrelated to the system dynamics, then examining only states and transitions.
- There are several tradeoffs in applying this pattern. Firstly it works only for discrete state-spaces or discretely quantised continuous spaces. It is not tractable to show entire state space dynamics for systems with

a large number of possible states. However, one can perform random sampling of the system and view partial state space diagrams.

- For large diagrams the layout of nodes and connections is critical for comprehending the structure.
- There can be a very large number of possible states for a given system. In these cases, visualization of the entire state space at once can be difficult.

10. Implementation.

- There are two major methods of implementing the data structure beneath the transition network. It can be written as an adjacency list where each node points to a linked list of connections. Secondly it may be stored as a complete matrix of NxN dimensions (where N is the number of states). Non-zero entries in the matrix indicate the existence of a connection between nodes.
- With an appropriate layout, e.g., Wuensche, the whole state space can be viewed at once. For large number of states, the diagram can be made easier to read by trimming off the most outlying states from transients.
- Ghosting/colour changes allow the user to examine changes to system dynamics when some underlying aspect of the system is changed.
- It is important to recognise that the graphical model is abstract only. Any spatial information contained in the underlying model is being removed.

11. Sample code. A complete state space transition network is produced thus:

```
--------------------------------------------------------------
initialiseStateTransitionNetwork(noOfStates)
For each currentState in Model
    model.setState(currentState)
    model.updateState()
    nextState = model.getState()
    StateTransitionNetwork.addTransition(currentState,nextState)
End for
--------------------------------------------------------------
```

Similarly to explore a particular trajectory through the state space:

```
------------------------------------------------------------
initialiseStateTransitionNetwork(initialTrajectoryState)
currentState = initialTrajectoryState
nextState = model.nextState(currentState)
StateTransitionNetwork.addTransition(currentState, nextState)

While Not StateTransitionNetwork.contains(nextState)
    currentState = nextState
    model.setState(currentState)
    model.updateState()
    nextState = model.getState()
    StateTransitionNetwork.addTransition(currentState, nextState)
End for
------------------------------------------------------------
```

These code snippets cover the generation of the data structure representing the state space diagram. Once this has been generated there are a number of layout algorithms for creating the actual visualization.

12. Known uses. Boolean models of Gene networks, NK landscape problems, Visualizing basins of attraction in random Boolean networks.

13. Related patterns. Network Diagram, Neural Networks, Activation Diagram, Hypercube

Bibliography

[1] H. A. Abbass. Speeding up back propagation using multi objective evolutionary algorithms. *Neural Computation*, 15(11):2705–2726, 2003.

[2] H. A. Abbass and K. Deb. Searching under multi-evolutionary pressures. In *Proceedings of the Second International Conference on Evolutionary Multi-Criterion Optimization*, pages 391–404. Springer-Verlag, 2003.

[3] C. Adami. *Introduction to artificial life*. Springer-Verlag, Heidelberg, 1998.

[4] C. Adami. Avida (digital life laboratory), available from http://dllab. caltech. edu/avida. 2005.

[5] L. M. Adleman. Molecular computation of solutions to combinatorial problems. *Science*, 266(5187):1021–1024, 1994.

[6] L. M. Adleman. On constructing a molecular computer. In *DNA Based Computers*, volume 27 of *DIMACS Series in Discrete Mathematics and Theoretical Computer Science*, pages 1–21, 1996.

[7] D. Aerts, S. Aerts, J. Broekaert, and L. Gabora. The Violation of Bell Inequalities in the Macroworld. *Foundations of Physics*, 30:1387–1414, 2000.

[8] P. M. Agapow and N. T. B. Isaac. Macrocaic: correlates of species richness. *Diversity and distributions*, 8:41 – 43, 2002.

[9] P. M. Agapow and A. Purvis. Power of eight tree shape statistics to detect non-random diversification: a comparison by simulation of two models of cladogenesis. *Syst. Biol.*, 51(2002):866 – 872, 2002.

[10] I. J. R. Aitchison. *An informal introduction to gauge field theories*. Cambridge University Press, 1982.

[11] D. J. Albers, J. C. Sprott, and W. D. Dechert. Routes to chaos in neural networks with random weights. *International Journal of Bifurcation and Chaos*, 8(7):1463–1478, 1998.

[12] R. Albert and A. H. Barabási. Statistical mechanics of complex networks. *Review of Modern Physics*, 74:47–97, 2002.

[13] D. S. Alberts, J. J. Garstka, and F. P. Stein. *Network centric warfare: Developing and leveraging information superiority (2nd edition)*. CCRP Publication Series, 1999.

[14] M. Aldana. Boolean dynamics of networks with scale-free topology. *Physica D*, 185:45–66, 2003.

[15] P. J. Angeline and J. B. Pollack. Coevolving high-level representations. In C. Langton, editor, *Artificial Life III*, pages 55–71. Reading MA: Addison-Wesley, 1994.

[16] D. Angus and T. Hendtlass. Ant colony optimisation applied to a dynamically changing problem. In T. Hendtlass and M. Ali, editors, 14th *International Conference on Industrial and Engineering Applications of Artificial Intelligence*, volume 2358 of *Lecture Notes in Artificial Intelligence*, pages 618–627. Springer Verlag, Berlin, 2002.

[17] B. Appleton. Patterns and software: essential concepts and terminology, 2000. http://www. cmcrossroads. com/bradapp/docs/patterns-intro. html.

[18] J. C. Astor and C. Adami. A developmental model for the evolution of artificial neural networks. *Artificial Life*, 6:189–218, 2000.

[19] S. Y. Auyang. *How is Quantum Field Theory Possible?* Oxford University Press, 1995.

[20] S. Y. Auyang. *Foundations of Complex-System Theories: In Economics Evolutionary Biology and Statistical Physics*. Cambridge University Press, 1998.

[21] N. A. Baas. Emergence, hierarchies, and hyperstructures. *Artificial Life III*, pages 515–537, 1994.

[22] T. Bäck. *Evolutionary algorithms in theory and practice*. Oxford University Press, New York, 1996.

[23] T. Bäck, D. Fogel, and Z. Michalewicz. *Handbook of Evolutionary Computation*. Oxford University Press, Oxford, UK, 1997.

[24] D. Baird and R. E. Ulanowicz. The seasonal dynamics of the chesapeake bay ecosystem. *Ecological Monographs*, 59(4):329–364, 1989.

[25] P. Bak. *How Nature Works*. Oxford University Press, 1997.

[26] A. L. Barabási and R. Albert. Emergence of scaling in random networks. *Science*, 286:509–512, 1999.

[27] B. Baran. Improved antnet routing. *SIGCOMM Comput. Commun. Rev.*, 31(2):42–48, 2001.

[28] M. Barlow and A. Easton. Crocadile - an open, extensible agent-based distillation engine. *Information & Security*, 8(1):17–51, 2002.

[29] T. Barraclough and V. Savolainen. Evolutionary rates and species diversity in flowering plants. *Evolution*, 55:677 – 683, 2001.

[30] A. J. Bateman. Intra-sexual selection in Drosophila. *Heredity*, 2:349–368, 1948.

[31] J. Beasley. OR-Library. Online at `http://people.brunel.ac.uk/~mastjjb/jeb/info.html`, 2005.

[32] J. Beasley, M. Krishnamoorthy, Y. Sharaiha, and D. Abramson. The displacement problem and dynamically scheduling aircraft landings. *Journal of the Operational Research Society*, 55:54–64, 2004.

[33] J. Beasley, J. Sonander, and P. Havelock. Scheduling aircraft landings at London Heathrow using a population heuristic. *Journal of the Operational Research Society*, 52:483–493, 2001.

[34] M. A. Bedau, J. S. McCaskill, N. H. Packard, St. Rasmussen, C. Adami,

D. G. Green, T. Ikegami, K. Kaneko, and T. S. Ray. Open problems in artificial life. *Artificial Life*, 6:363–376, 2000.

[35] M. A. Bedau, E. Snyder, and C. T. Brown. A comparison of evolutionary activity in artificial evolving systems and in the biosphere. In P. Husbands and I. Harvey, editors, *ALife IV*, pages 125–134, 1997.

[36] M. A. Bedau, E. Snyder, and N. H. Packard. A classification of long-term evolutionary dynamics. *Artificial Life VI*, pages 228–237, 1998.

[37] R. Bellman. *Dynamic Programming.* Princeton University Press, Princeton, NJ.

[38] D. Benedetto, E. Caglioti, and V. Loreto. Language trees and zipping. *Phys. rev. Lett.*, 88:048702, 2002. arXiv:cond-mat/0108530.

[39] Y. Benenson, T. Paz-Ellzur, R. Adar, E. Kelnan, Z. Livneh, and E. Shapiro. Programmable and autonomous computing machine made of biomolecules. *Nature*, 414:430–434, Nov 2001.

[40] P. Bentley and S. Kumar. Three ways to grow designs: A comparison of evolved embryogenies for a design problem. In *Proc. of the Genetic and Evolutionary Conference (GECCO '99)*, pages 35–43, 1999.

[41] N. Bernstein. *The Coordination and Regulation of Movements.* Oxford: Pergamon Press, 1967.

[42] T. Bersano-Begey. Controlling exploration, diversity and escaping local optima in GP. In *Late Breaking Papers at the Genetic Programming Conference*, pages 7–10. MIT Press, 1997.

[43] N. Bertschinger and T. Natschläger. Real-time computation at the edge of chaos in recurrent neural networks. *Neural Comput.*, 16(7):1413–1436, 2004.

[44] E. Bilotta, D. Groß, T. Smith, T. Lenaerts, S. Bullock, H. H. Lund, J. Bird, R. Watson, P. Pantano, L. Pagliarini, H. A. Abbass, R. Standish, and M. Bedau, editors. *ALife VIII — workshops.* University of New South Wales, 2002.

[45] E. Bilotta, A. Lafusa, and P. Pantano. Is self-replication an embedded characteristic of artificial/living matter? In *ICAL 2003: Proceedings of the eighth international conference on Artificial life*, pages 38–48, Cambridge, MA, USA, 2003. MIT Press.

[46] E. Bilotta, A. Lafusa, and P. Pantano. Life-like self-reproducers. *Complex.*, 9(1):38–55, 2003.

[47] S. Bleuer, M. Braek, L. Thiele, and E. Zitzler. Multiobjective genetic programming: reducing bloat using SPEA2. In *Proceedings of the IEEE Congress on Evolutionary Computation*, volume 1, pages 536–543. IEEE Press, 2001.

[48] C. Blum. ACO applied to group shop scheduling: A case study on intensification and diversification. In M. Dorigo, G. Di Caro, and M. Sampels, editors, *Third International Workshop on Ant Algorithms, ANTS 2002*, volume 2463 of *Lecture Notes in Computer Science*, pages 14–27, Brussels, Belgium, 2002. Springer-Verlag.

[49] E. J. W. Boers and H. Kuiper. *Biological metaphors and the design of modular artificial neural networks.* Master's thesis, Leiden University, The

Netherlands, 1992.

[50] E. Bonabeau, M. Dorigo, and G. Theraulaz. *Swarm intelligence: from natural to artificial systems.* Oxford University Press, Inc., New York, NY, USA, 1999.

[51] E. Bonabeau, F. Henaux, S. Guerin, D. Snyers, P. Kuntz, and G. Theraulaz. Routing in telecommunications networks with ant-like agents. In *IATA '98: Proceedings of the second international workshop on Intelligent agents for telecommunication applications*, pages 60–71, 1998.

[52] J. E. Bond and B. D. Opell. Testing adaptive radiation and key innovation hypotheses. *Evolution*, 52(1998):403 – 414, 1998.

[53] A. Borgu. The challenges and limitations of 'network centric warfare'. Presented at the "Network Centric Warfare: Improving ADF Capabilities Through Network Enabled Operations" conference, September 2003.

[54] S. Boswell, N. Curtis, P. Dortmans, and N. Tri. A parametric study of factors leading to success in conflict: Potential warfighting concepts. In *Proceedings of the Land Warfare Conference*, pages 369 – 381, 2003.

[55] K. Brading and E. Castellani, editors. *Symmetries in Physics: Philosophical Reflections.* Cambridge University Press, 2003.

[56] J. Branke. *Evolutionary Optimization in Dynamic Environments.* Kluwer Academic Publishers, Dordrecht, 2001.

[57] F. Briand and J. E. Cohen. Community food webs have scale invariant structure. *Nature*, 307:264–266, 1984.

[58] L. T. Bui, J. Branke, and H. A. Abbass. Multiobjective optimization for dynamic environments. In *Proceedings of the IEEE Congress on Evolutionary Computation.* in press, 2005.

[59] R. K. Bullough and P. J. Caudrey. The Soliton and its History. In R. K. Bullough and P. J. Caudrey, editors, *Solitons*, volume 17 of *Topics in Current Physics*, chapter 1, pages 1–64. Springer-Verlag, 1980.

[60] P. Buri. Gene frequency in small populations of mutant drosophila. *Evolution*, 10:367–402, 1956.

[61] R. Burkard, S. Karisch, and F. Rendl. QAPLIB - A quadratic assignment problem library. *Journal of Global Optimization*, 10:391–403, 1997.

[62] A. Caballero. Developments in the prediction of effective population size. *Heredity*, 73:657–679, 1994.

[63] R. T. Cahill. Hadronisation of QCD. *Australian Journal of Physics*, 42:171–186, 1989.

[64] Caltech. Digital life laboratory, available from http://dllab. caltech. edu. 2005.

[65] N. A. Cambell. *Biology.* The Benjamin Cummings Publishing Company, Redwood City, CA, second edition edition, 1990.

[66] A. Cangelosi, S. Nolfi, and D. Parisi. Artificial life models of neural development. In S. Kumar and P. Bentley, editors, *On Growth, Form and Computers.* Elsevier Academic Press, 2003.

[67] A. Cangelosi, D. Parisi, and S. Nolfi. Cell division and migration in a 'genotype' for neural networks (cell division and migration in neural networks). *Network: Computation in Neural Systems*, 5:497–515, 1994.

[68] G. Di Caro and M. Dorigo. Antnet: A mobile agents approach to adaptive routing. *Technical Report IRIDIA, Universit Libre de Bruxelles*, (12), 1997.

[69] G. Di Caro and M. Dorigo. Antnet: distributed stigmergetic control for communication networks. *Journal of Artificial Intelligence Research (JAIR)*, 9:317–365, 1998.

[70] E. Castellani. On the meaning of symmetry breaking. In K. Brading and E. Castellani, editors, *Symmetries in physics: philosophical reflections*. Cambridge University Press, 2003.

[71] J. L. Casti. On system complexity: Identification, measurement and management. In John L. Casti and Anders Karlqvist, editors, *Complexity, Language and Life: Mathematical Approaches*, volume 16 of *Biomathematics*, pages 146–173. Springer-Verlag, 1986.

[72] J. L. Casti. *Would-be Worlds*. John Wiley and Sons, New York, 1997.

[73] D. Caterinicchia and M. French. Network-centric warfare: Not there yet. http://www. fcw. com/fcw/articles/2003/0609/cov-netcentric-06-09-03. asp, June 2003.

[74] A. Chakravarthy and D. Ghose. Obstacle avoidance in a dynamic environment: A collision cone approach. *IEEE Transaction on Systems, Man, and Cybernetics Part A: System and Humans*, 28(5), SEPTEMBER 1998.

[75] K. Y. Chang, G. Jan, and I. Parberry. A method for searching optimal routes with collision avoidance on raster charts. *The Journal of Navigation,The Royal Institute of Navigation*, 56, 2003.

[76] A. D. Channon and R. I. Damper. Towards the evolutionary emergence of increasingly complex advantageous behaviours. *International Journal of Systems Science, Special Issue: 'Emergent Properties of Complex Systems'*, 31:843–860, 2000.

[77] P. Chen, J. Akoka, H. Kangassalo, and B. Thalheim, editors. *Conceptual Modeling: Current Issues and Future Directions*. Lecture Notes in Computer Science. Springer-Verlag, Heidelberg, 1999.

[78] R. R. Christian and J. J. Luczkovich. Organizing and understanding a winters seagrass foodweb network through effective trophic levels. *Ecological Modelling*, 117:99–124, 1999.

[79] D. Chu and J. Rowe. Spread of Vector Borne Diseases in a Population with Spatial Structure. In *In Proceedings of PPSN VIII - Eight International Conference on Parallel Problem Solving from Nature*, number 3242 in Lecture Notes in Computer Science, pages 222–232, Birmingham, UK, September 2004. Springer-Verlag.

[80] P. Chu and J. Beasley. A genetic algorithm for the multidimensional knapsack problem. *Journal of Heuristics*, 4:63–86, 1998.

[81] J. C. Claussen. Offdiagonal complexity: A computationally quick complexity measure for graphs and networks. arXiv:q-bio. MN/0410024.

[82] T. H. Clutton-Brock. *The evolution of parental care*. Princeton University Press, Princeton, New Jersey, 1991.

[83] E. Coen, A. G. Rolland-Lagan, M. Matthews, J. A. Bangham, and P. Prusinkiewicz. The genetics of geometry. *Proceedings of the National Academy of Science, USA*, 101:4728–4735, 2004.

[84] O. Cordón, I. de Viana, and F. Herra. Analysis of the best-worse ant system and its variants on the QAP. In M. Dorigo, G. Di Caro, and M. Sampels, editors, *Third International Workshop on Ant Algorithms, ANTS 2002*, volume 2463 of *Lecture Notes in Computer Science*, pages 228–234. Springer-Verlag, Brussels, Belgium, 2002.

[85] O. Cordón, F. Herra, I. de Viana, and L. Moreno. A new ACO model integrating evolutionary concepts: The best-worse ant system. In *Second International Workshop on Ant Algorithms, ANTS 2000*, pages 22–29, Brussels, Belgium, 2000.

[86] D. Cornforth, D. G. Green, and J. Awburn. Weasel world: a simple artificial environment for investigating open-ended evolution. In *Proceedings of the 8th Asia Pacific Symposium on Intelligent and Evolutionary Systems*, pages 40–49. Monash University, Australia, 2004.

[87] J. P. Crutchfield. The calculi of emergence: computation, dynamics and induction. *Phys. D*, 75(1-3):11–54, 1994.

[88] J. P. Crutchfield and K. Young. Computation at the edge of chaos. In WH Zurek, editor, *Complexity, Entropy and the Physics of Information*, pages 117–125. Addison-Wesley, Redwood City, California, 1990.

[89] M. Daly and M. Wilson. Whom are newborn babies said to resemble? *Ethology and Sociobiology*, 3:69–78, 1982.

[90] M. Daly and M. I. Wilson. Some differential attributes of lethal assaults on small children by stepfathers versus genetic fathers. *Ethology and Sociobiology*, 15:207–217, 1994.

[91] T. J. Davies, T. G. Barraclough, M. W. Chase, P. S. Soltis, D. E. Soltis, and V. Savolainen. Darwin's abominable mystery: insights from a supertree of the angiosperms. *Proc. Natl. Acad. Sci. USA*, in press.

[92] R. Dawkins. *The Blind Watchmaker: why the evidence of evolution reveals a universe without design.* W. W. Norton, New York, 1986.

[93] R. Dawkins and T. R. Carlisle. Parental investment, mate desertion and a fallacy. *Nature*, 262:131–133, 1976.

[94] E. D. De Jong and Jordan B. Pollack. Multi-objective methods for tree size control. *Genetic Programming and Evolvable Machines*, 4(3):211–233, 2003.

[95] E. D. De Jong, R. A. Watson, and J. B. Pollack. Reducing bloat and promoting diversity using multiobjective methods. In *Proceedings of the Genetic and Evolutionary Computation Conference*, pages 11–18. Morgan Kaufmann, 2001.

[96] Y. Van de Peer, J. S. Taylora, and A. Meyer. Are all fishes ancient polyploids? *J. Struct. Funct. Genomics*, 3:65–73, 2003.

[97] A. de Queiroz. Contingent predictability in evolution: key traits and diversification. *Systematic Biology*, 56(2002):917 – 929, 2002.

[98] K. Deb. *Multi-objective optimization using evolutionary algorithms.* Wiley, Chichester, 2001.

[99] K. Deb, S. Agrawal, A. Pratab, and T. Meyarivan. A fast elitist non-dominated sorting genetic algorithm for multi-objective optimization: NSGA-II. In *Proceedings of the Parallel Problem Solving from Nature VI*

Conference, pages 849–858. Springer-Verlag, 2000.

[100] K. Deb, S. Agrawal, A. Pratap, and T. Meyarivan. A fast and elitist multiobjective genetic algorithm: Nsga-ii. *IEEE Transactions on Evolutionary Computation (IEEE-TEC)*, 6(2):182 - 197, 2002.

[101] K. Deb, M. Mohan, and S. Mishra. Towards a quick computation of well-spread pareto-optimal solutions. In *Evolutionary Multi-Criterion Optimization: Second International Conference*, pages 222–236. Springer-Verlag, 2003.

[102] W. D. Dechert and R. Gencay. Lyapunov exponents as a nonparametric diagnostic for stability analysis. *Journal of Applied Econometrics*, 7:S41–S60, 1992.

[103] P. L. Deininger, M. A. Batzer, C. A. I. Hutchinson, and M. H. Edgell. Master genes in mammalian repetitive dna amplification. *Trends Genet.*, 8:307 - 311, 1992.

[104] E. D. deJong. Representation development from pareto-coevolution. In *Proceedings of the Genetic and Evolutionary Computation Conference (GECCO)*, 2003.

[105] B. Derrida and Y. Pomeau. Random networks of automata: A simple annealed approximation. *Europhysics Letters*, 1:45–49, 1986.

[106] R. Dewar. Informational theory explanation of the fluctuation theorem, maximum entropy production and self-organized criticality in non-equilibrium stationary states. *J. Phys. A*, 36:631–641, 2003.

[107] K. P. Dial and J. M. Marzluff. Nonrandom diversification within taxonomic assemblages. *Systematic Zoology*, 38(1989):26 - 37, 1989.

[108] P. Dittrich, J. Ziegler, and W. Banzhaf. Artificial chemistries — a review. *Artificial Life*, 7:225–275, 2001.

[109] L. Dolan and K. Roberts. Plant development: pulled up by the roots. *Current Opinion in Genetics and Development*, 5:432–438, 1995.

[110] M. Dorigo and G. Di Caro. The ant colony optimization meta-heuristic. In D. Corne, M. Dorigo, and F. Glover, editors, *New Ideas in Optimization*, pages 11–32. McGraw-Hill, London, 1999.

[111] M. Dorigo, G. Di Caro, and L. Gambardella. Ant algorithms for discrete optimization. *Artificial Life*, 5(2):137–172, 1999.

[112] M. Dorigo and L. Gambardella. Ant Colony System: A cooperative learning approach to the traveling salesman problem. *IEEE Transactions on Evolutionary Computation*, 1(1):53–66, 1997.

[113] A. Dorin and J. McCormack. Self-assembling dynamical hierarchies. *Artificial Life VIII*, pages 423–428, 2002.

[114] M. Dror and W. Powell. Stochastic and dynamic models in transportation. *Operations Research*, 41:11–14, 1993.

[115] J. A. Dunne, R. J. Williams, and N. D. Martinez. Network structure and biodiversity in food webs: robustness increases with connectance. *Ecology Letters*, 5:558–567, 2002.

[116] M. S. Capcarrère E. M. A. Ronald M. Sipper. Testing for emergence in artificial life. In Dario Floreano, Jean-Daniel Nicoud, and Francesco Mondada, editors, *Advances in Artificial Life: Proc. 5th European Conf. on Artifical*

Life (ECAL'99), Switzerland, pages 13–20. Pub Springer-Verlag, 1999.

[117] B. Edmonds. *Syntactic Measures of Complexity*. PhD thesis, University of Manchester, 1999.

[118] P. Eggenberger. Cell interactions as a control tool of developmental processes for evolutionary robotics. In *Proceedings of SAB '96*, pages 440–448, 1996.

[119] P. Eggenberger. Evolving morpohologies of simulated 3D organisms based on diffferential gene expression. In P. Husbands and I. Harvey, editors, *Fourth European Conference on Artificial Life*, pages 205–213, Cambridge, MA, 1997. The MIT Press/Bradford Books.

[120] J. L. Elman. Finding structure in time. *Cognitive Science*, 14:179–211, 1990.

[121] J. M. Epstein and R. L. Axtell. *Growing Artificial Societies: Social Science from the Bottom Up*. MIT Press, Cambridge, 1996.

[122] P. Erdös and A. Rényi. On random graphs. *Publ. Math. Dubrecen*, 6:290–291, 1959.

[123] C. Eyckelhof and M. Snoek. Ant systems for a dynamic TSP: Ants caught in a traffic jam. In M. Dorigo, G. Di Caro, and M. Sampels, editors, *Third International Workshop on Ant Algorithms, ANTS 2002*, volume 2463 of *Lecture Notes in Computer Science*, pages 88–99, Brussels, Belgium, 2002. Springer-Verlag.

[124] J. Felsenstein. Phylip - phylogeny inference package. *Cladistics*, 5:164 – 166, 1989.

[125] T. Feo and M. Resende. Greedy randomised adaptive search procedures. *Journal of Global Optimization*, 51:109–133, 1995.

[126] K. Fleischer and A. H. Barr. A simulation testbed for the study of multicellular development: The multiple mechanisms of morphogenesis. In C. Langton, editor, *Artificial Life III*, pages 389–416, Redwood City, CA, 1994. Addison-Wesley.

[127] A. J. Fleming. The mechanism of leaf morphogenesis. *Planta*, 216:17–22, 2002.

[128] J. Foster and W. Holzl. *Applied Evolutionary Economics and Complex Systems*. Edward Elgar, Cheltenham, England, 2004.

[129] G. L. Fox and C. Bruce. Conditional fatherhood: Identity thoery and parental investment theory as alternative sources of explanation of fathering. *Journal of Marriage and Family*, 63(2):394–403, 2001.

[130] A. L. Fradkov. Memorandum 1447: Exploring nonlinearity by feedback, 1998.

[131] R. P. Freckleton, M. Pagel, and P. H. Harvey. Comparative methods for adaptive radiations. In K. J. Gaston T. M. Blackburn, editor, *Macroecology: Concepts and Consequences*, pages 391 – 407. Blackwell, 2003.

[132] R. M. French and A. Messinger. Genes, phenes and the baldwin effect: Learning and evolution in a simulated population. In R. Brooks and P. Maes, editors, *Artificial Life IV*, pages 277–282. the MIT Press, Cambridge, MA, 1994.

[133] D. J. Futuyma. *Evolutionary biology*. Sinauer Associates, Inc., Sunderland, 1998.

[134] L. Gabora and D. Aerts. Contextualizing Concepts using a Mathematical Generalization of the Quantum Formalizm. *Journal of Experimental and Theoretical Artificial Intelligence*, 14:327–358, 2002.

[135] D. P. Galligan. Modelling shared siturational awareness using the mana model. *Journal of Battlefield Technology*, 7(3):35–40, 2004.

[136] D. P. Galligan, M. A. Anderson, and M. K. Lauren. *MANA:Map Aware Non-uniform Automata Version 3. 0 Users Manual.* Defence Technology Agency, Devonport, Aukland, November 2003. Draft version.

[137] L. Gambardella, E. Taillard, and M. Dorigo. Ant colonies for the quadratic assignment problem. *Journal of the Operational Research Society*, 50:167–176, 1999.

[138] E. Gamma, R. Helm, R. Johnson, and J. Vlissides. *Design Patterns: elements of reusable object-oriented software.* Addison-Wesley, 1995.

[139] M. R. Gardner and W. R. Ashby. Connectance of large dynamical (cybernetic) systems: critical values for stability. *Nature*, 228:784, 1970.

[140] J. J. Garstka. Network-centric warfare offers warfighting advantage: Datalinks are the new weapon of the information age. *Signal*, 57:58–60, May 2003.

[141] N. Geard and J. Wiles. A gene network model for developing cell lineages. *Artificial Life*, 11(3):249–268, 2005.

[142] N. Geard, K. Willadsen, and J. Wiles. Perturbation analysis: A complex systems pattern. This volume.

[143] Carlos Gershenson. Updating schemes in random Boolean networks: Do they really matter? In J. Pollack, M. Bedau, P. Husbands, T. Ikegami, and R. A. Watson, editors, *Artificial Life IX Proceedings of the Ninth International Conference on the Simulation and Synthesis of Living Systems*, pages 238–243. MIT Press, 2004.

[144] S. F. Gilbert. *Developmental Biology.* Sinauer, 7 edition, 2003.

[145] M. E. Gilpin. *Group Selection in Predator-Prey Communities.* Princeton University Press, 1975.

[146] F. Glover and M. Laguna. *Tabu Search.* Kluwer Academic Publishers, Boston, MA, 1997.

[147] D. E. Goldberg and J. Richardson. Genetic algorithms with sharing for multimodal function optimization. In *Proceedings of the 2nd International Conference on Genetic Algorithms and their Applications*, pages 41–49. Lawrence Erlbaum Associates, 1987.

[148] L. Goldwasser and J. A. Roughgarden. Construction and analysis of a large caribbean food web. *Ecology*, 74(4):1216–1233, 1970.

[149] B. Gompertz. On the Nature of the Function Expressive of the Law of Human Mortality and on a New Mode of Determining Life Contingencies. *Philosophical Transactions of the Royal Society of London*, 115:513–585, 1825.

[150] C. Gondro. *Modelagem e implementacao de um modelo biologico computacional para estudos em genetica de populacoes.* Msc, Federal University of Parana, 2002.

[151] C. Gondro and J. C. M. Magalhaes. Using a computer based biological

model for teaching population genetics and evolution. In *Proceedings of the XIX International Congress of Genetics*, page 288, 2003.

[152] S. J. Gould. Bushes all the way down. *Natural History*, page 19, 1987.

[153] S. J. Gould. *Wonderful Life*. Penguin, London, 1989.

[154] S. J. Gould. *Wonderful Life*. Vintage, 2000.

[155] A. Grafen. Natural Selection, Kin Selection and Group Selection. In J. Krebs and N. Davies, editors, *Behavioural Ecology: An Evolutionary Approach*, pages 62–84, Oxford, 1984. Blackwell Scientific Publications.

[156] D. G. Green. Emergent behaviour in biological systems. *Complexity International*, 1, 1993. accessed at http://www. csu. edu. au/ci/vol1/.

[157] D. G. Green. Self-organization in complex systems. In T. R. J. Bossomaier and D. G. Green, editors, *Complex Systems*, pages 7–41. Cambridge University Press, 2000.

[158] D. G. Green, D. Newth, D. Cornforth, and M. Kirley. On evolutionary processes in natural and artificial systems. In *Proceedings of the Fifth Australia-Japan Joint Workshop on Intelligent and Evolutionary Systems*, pages 1–10. Dunedin, New Zealand: The University of Otago, 2001.

[159] G. Grimmett. *Percolation*. Springer-Verlag, Heidelberg, 1989.

[160] D. Großand T. Lenaerts. Towards a definition of dynamical hierarchies. In Bilotta et al. [44], pages 45–53.

[161] D. Gross and B. McMullin. Is it the right ansatz? *Artificial Life*, 7:355–365, 2001.

[162] F. Gruau. *Neural network synthesis using cellular encoding and the genetic algorithm*. Ph. d. thesis, l'Ecole Normale Suprieure de Lyon, 1994.

[163] F. Gruau, D. Whitley, and L. Pyeatt. A comparison between cellular encoding and direct encoding for genetic neural networks. In *Genetic Programming 1996: Proceedings of the First Annual Conference*, pages 81–89. MIT Press, 1996.

[164] S. M. Gunner. *Modelling Low Energy Processes in Quantum Chromodynamics using the Global Colour Model*. PhD thesis, School of Chemistry Physics and Earth Sciences, The Flinders University of South Australia, 2004.

[165] M. Guntsch and M. Middendorf. Applying population based ACO to dynamic optimization problems. In M. Dorigo, G. Di Caro, and M. Sampels, editors, *Third International Workshop on Ant Algorithms, ANTS 2002*, volume 2463 of *Lecture Notes in Computer Science*, pages 111–122. Springer-Verlag, Brussels, Belgium, 2002.

[166] M. Guntsch and M. Middendorf. A population based approach for ACO. In S. Cagnoni, J. Gottlieb, E. Hart, M. Middendorf, and G. Raidl, editors, *EVO Workshops 2002*, volume 2279 of *Lecture Notes in Computer Science*. Springer-Verlag, Kinsale, Ireland, 2002.

[167] M. Guntsch, M. Middendorf, and H. Schmeck. An ant colony optimization approach to dynamic TSP. In L. Spector, E. Goodman, A. Wu, W. Langdon, H. Voigt, M. Gen, S. Sen, M. Dorigo, S. Pezeshk, M. Garzon, and Ei. Burke, editors, *Proceedings of the Genetic and Evolutionary Computation Conference*, pages 860–867. Morgan Kaufmann, Washington, DC, 2001.

[168] H. Gutowitz and C. Langton. Mean field theory of the edge of chaos. In *ECAL*, pages 52–64, 1995.

[169] H. A. Gutowitz, J. D. Victor, and B. W. Knight. Local structure theory for cellular automata. *Physica D*, 28:18–48, 1987.

[170] C. Guyer and J. B. Slowinski. Comparisons of observed phylogenetic topologies. *Evolution*, 45(1991):340 - 350, 1991.

[171] Pattee H. H. Simulations. Realizations, and Theories of Life. In Margaret A. Boden, editor, *The Philosophy of Artificial Life*, pages 379–393. Oxford University Press, 1996.

[172] Pattee H. H. The physics of symbols: bridging the epistemic cut. *BioSystems*, 60:5–21, 2001.

[173] A. Habel. *Hyperedge replacement: grammars and languages.* Lecture Notes in Computer Science 643. Springer-Verlag, Berlin ; New York, 1992.

[174] M. Hagiya. Towards molecular programming — a personal report on DNA8 and molecular computing. In *Modelling in Molecular Biology*, Natural Computing Series, pages 125–140. Springer, 2004.

[175] S. J. Hall and D. Raffaelli. Food-web patterns: lessons from a species-rich web. *Journal of Animal Ecology*, 60(3):823–842, 1991.

[176] J. Hallinan and J. Wiles. Evolving genetic regulatory networks using an artificial genome. In Y. P. Chen, editor, *Proc. Second Asia-Pacific Bioinformatics Conference (APBC2004)*, volume 29 of *Conferences in Research and Practice in Information Technology*, pages 291–296. Australian Computer Society, 2004.

[177] W. D. Hamilton. The Genetical Evolution of Social Behaviour, I and II. *Journal of Theoretical Biology*, 7:1–52, 1964.

[178] W. D. Hamilton. Innate Social Aptitudes of Man: an Approach from Evolutionary Genetics. In R. Fox, editor, *Biosocial Anthropology*, pages 133–155, New York, 1975. John Wiley and Sons.

[179] D. L. Hartl and A. G. Clark. *Principles of Population Genetics.* Sinauer, Sunderland, 1997.

[180] I. Harvey. Species adaptation genetic algorithms: A basis for a continuing saga. In F. J. et al. Varela, editor, *Proceedings of First European Conference on Artificial Life.* MIT Press/Bradford Books, 1992.

[181] I. Harvey and T. Bossamaier. Time out of joint: Attractors in asynchronous random Boolean networks. In P. Husbands and I. Harvey, editors, *Fourth European Conference on Artificial Life*, Cambridge, MA, 1997. The MIT Press/Bradford Books.

[182] P. H. Harvey, A. J. L. Brown, J. M. Smith, and S. Nee. *New uses for new phylogenies.* Oxford University Press, 1996.

[183] K. Havens. Scale and structure in natural food webs. *Science*, 257(5073):1107–1109, 1992.

[184] K. A. Hawick, H. A. James, and C. J. Scogings. Web site, containing model snapshots and movie sequences, available from http://www. massey. ac. - nz/~kahawick/alife. 2005.

[185] K. A. Hawick, H. A. James, and C. J. Scogings. A zoology of emergent patterns in a predator-prey model. Technical Report CSTN-015, Massey

University, March 2005.

[186] K. A. Hawick, C. J. Scogings, and H. A. James. Defensive spiral emergence in a predator-prey model. In *Proc. Complexity 2004, Cairns, Australia*, December 2004.

[187] T. Hendtlass. TSP optimisation using multi tour ants. In R. Orchard, C. Yang, and M. Ali, editors, *Innovations in Applied Artificial Intelligence, 17th International Conference on Industrial and Engineering Applications of Artificial Intelligence and Expert Systems*, volume 3029 of *Lecture Notes in Computer Science*, pages 533–532, Ottawa, Canada, 2004. Springer.

[188] T. Hendtlass and M. Randall. A survey of ant colony and particle swarm meta-heuristics and their application to discrete optimisation problems. In *Proceedings of the Inaugual Workshop on Artificial Life*, pages 15–25, Adelaide, Australia, 2001.

[189] M. Heusse, D. Snyers, S. Gurin, and P. Kuntz. Adaptive agent-driven routing and load balancing in communication networks. *ENST de Bretagne Tech. Rep*, 1998.

[190] T. Higuchi et al. Evolvable hardware at function level. In *IEEE International Conference on Evolutionary Computation*, pages 187–192, 1997.

[191] Y. C. Ho. Perturbation analysis: Concepts and algorithms. In J. J. Swain, D. Goldsman, R. C. Crain, and J. R. Wilson, editors, *Proceedings of the 1992 Winter Simulation Conference*, Piscataway, NJ, 1992. The IEEE Press.

[192] P. Hogeweg. Shapes in the shadow: Evolutionary dynamics of morphogenesis. *Artificial Life*, 6:85–101, 2000.

[193] J. Holland. *Emergence*. Oxford University Press, Oxford, 1998.

[194] J. Holland. Echo, available from http://www. santafe. edu/projects/echo/-echo. html. 2005.

[195] J. H. Holland. *Adaptation in Natural and Artificial Systems*. The University of Michigan Press, Ann Harbor, 1975.

[196] J. H. Holland. *Adaptation in Natural and Artificial Systems*. The MIT Press, Cambridge, 1992.

[197] J. H. Holland. *Hidden Order: How Adaptation Builds Complexity*. Addison-Wesley, Reading, 1995.

[198] H. Honda, H. Yamanaka, and M. Dan-Sohkawa. A computer simulation of geometrical configurations during cell division. *Journal of Theoretical Biology*, 106:423–435, 1984.

[199] G. S. Hornby. *Generative representations for evolutionary design automation*. Ph. d. thesis, Brandeis University Dept. of Computer Science, 2003.

[200] P. Hraber, T. Jones, and S. Forrest. The ecology of echo. *Artificial Life*, 3(3):165–190, 1997.

[201] S. B. Hrdy. Infanticide among animals: A review, classification and examination of the implications for the reproductive strategies of females. *Ethology and Sociobiology*, 1:13–40, 1979.

[202] M. Huxham, S. Beaney, and D. Raffaelli. Do parasites reduce the chances of triangulation in a real food web? *OIKOS*, 76(2):284–300, 1996.

[203] I. Hwang and C. Tomlin. Protocol based cd&r for atc control. Technical

report, Hybrid System Laboratory, Department of Aeronautics & Astronautics, Stanford University, USA, July 2002.

[204] ICAO. The icao global air navigation plan for cns/atm systems. Technical Report Document 9750, UN-International Civil Aviation Organisation, Dec 2002.

[205] A. Ilachinski. Irreducible semi-autonomous adaptive combat (isaac): An artificial life approach to land combat. Research Memorandum CRM 97-61, Center for Naval Analyses, Alexandria, 1997.

[206] A. Ilachinski. Irreducible semi-autonomous adaptive combat (isaac): An artificial life approach to land combat. *Military Operations Research*, 5(3):29–46, 2000.

[207] A. Ilachinski. *Artificial War: Multiagent-Based Simulation of Combat.* World Scientific Publishing Company, Singapore, 2004.

[208] N. Jakobi. Harnessing morphogenesis. Technical Report CSRP 423, School of Cognitive and Computer Science, University of Sussex, 1995.

[209] H. A. James, C. J. Scogings, and K. A. Hawick. A framework and simulation engine for studying artificial life. *Research Letters in the Information and Mathematical Sciences*, 6:143–155, May 2004.

[210] H. A. James, C. J. Scogings, and K. A. Hawick. Parallel synchronisation issues in simulating artificial life. In *Proc. 16th IASTED Int. Conf. on Parallel and Distributed Computing and Systems (PDCS), Boston*, pages 815–820, November 2004.

[211] M. A. Janssen and W. J. M. Martens. Modeling malaria as a complex adaptive system. *Artificial Life*, 3(3):213–236, 1997.

[212] T. Kalganova. Bidirectional incremental evolution in extrinsic evolvable hardware. In *The Second NASA/DoD Workshop on Evolvable Hardware*, pages 65–74. IEEE Computer Society, 2000.

[213] T. Kalganova. An extrinsic function-level evolvable hardware approach. In *Proc. of the Third European Conference on Genetic Programming (EUROGP2000), LNCS 1802*, pages 60–75. Springer-Verlag, 2000.

[214] K. Kaneko and I. Tsuda. *Complex Systems: Chaos and Beyond.* New York: Springer-Verlag, 2000.

[215] I. Kassabalidis, M. A. El-Sharkawi, R. J. Marks II, P. Arabshahi, and A. A. Gray. Adaptive-sdr: Adaptive swarm-based distributed routing. In *World Congress on Computational Intelligence*, 2002.

[216] S. A. Kauffman. Metabolic stability and epigenesis in randomly constructed genetic nets. *Journal of Theoretical Biology*, 22:437–467, 1969.

[217] S. A. Kauffman. Gene regulation networks: a theory for their global structure and behaviours. *Current Topics in Developmental Biology*, 6:145–182, 1971.

[218] S. A. Kauffman. *The Origins of Order: Self-Organization and Selection in Evolution.* Oxford University Press, Oxford, UK, 1993.

[219] A. I. Kaufman. Be careful what you wish for: The dangers of fighting with a network centric military. *Journal of Battlefield Technology*, November 2002.

[220] H. Kawamura, M. Yamamoto, and A. Ohuchi. Improved multiple ant colonies system for traveling salesman problems. In E. Kozan and

A. Ohuchi, editors, *Operations Research/Management Science at Work*, volume 43 of *International Series in Operations Research and Management Science*, pages 41–59. Kluwer, Boston, MA, 2002.

[221] S. Kessler and N. Sinha. Shaping up: the genetic control of leaf shape. *Current Opinion in Plant Biology*, 7:65–72, 2004.

[222] G. T. Kim, H. Tsukaya, Y. Aito, and H. Uchiyama. Change in the shape of leaves and flowers upon overexpression of cytochrome p450 in arabidopsis. *Proc Natl Acad Sci USA*, 96:943309437, 1999.

[223] J. Kim. *Supervenience and Mind*. Cambrdige University, Cambridge, 1993.

[224] H. Kitano. Designing neural networks using genetic algorithms with graph generation systems. *Complex Systems*, 4(4):461–476, 1990.

[225] H. Kitano. Biological robustness. *Nature Reviews Genetics*, 5:826–837, 2004.

[226] I. Kleinschmidt. *Spatial statistical analysis, modelling and mapping of malaria in Africa*. PhD thesis, University Basel, 2001.

[227] J. R. Koza. *Genetic programming: on the programming of computers by means of natural selection*. MIT Press, Cambridge, 1992.

[228] C. Krawjewski. Phylogenetic measures of biodiversity. *Biol. Conserv.*, 69:33–39, 1994.

[229] Y. Kuniyoshi and S. Suzuki. Dynamic emergence and adaptation of behavior through embodiement as coupled chaotic field. *Proc. 17th Int. Conf. on Intelligent Robots and Systems*, pages 2042–2048, 2004.

[230] R. Lachner. Collision avoidance as a differential game, real-time approximation of optimal strategies using higher derivatives of the value function. Technical Report D-38678, Technische Universit at Claust hal Institut fur Mathematik, Clausthal- Zellerfeld, Germany, 1998.

[231] W. B. Langdon. Size fair and homologous tree crossovers for tree genetic programming. *Genetic Programming and Evolvable Machines*, 1(1/2):95–119, 2000.

[232] C. G. Langton. Computation at the edge of chaos: phase transitions and emergent computation. *Phys. D*, 42(1-3):12–37, 1990.

[233] M. K. Lauren. Modelling combat using fractals and the statistics of scaling systems. *Military Operations research*, 5(3):47–58, 2000.

[234] M. K. Lauren and R. T. Stephen. Mana: Map-aware non-uniform automata: A new zealand approach to scenario modelling. *Journal of Battlefield Technology*, 2002.

[235] D. Layzer. Growth and order in the universe. In B. H. Weber, D. J. Depew, and J. D. Smith, editors, *Entropy, Information and Evolution*, pages 23–39. MIT Press, Cambridge, Mass., 1988.

[236] J. Lee and J. Sitte. A gate-level model for morphogenetic evolvable hardware. In *Proc. of the 2004 IEEE International Conference on Field-Programmable Technology (FPT'04)*, pages 113–119, 2004.

[237] J. Lee and J. Sitte. An implementation of a morphogenetic evolvable hardware system. In *5th International Conference on Simulated Evolution And Learning (SEAL04)*, Busan, Korea, 2004. Korea Advanced Institute of Science and Technology (KAIST).

[238] R. E. Lenski, C. Ofria, T. C. Collier, and C. Adami. Genome complexity, robustness and genetic interactions in digital organisms. *Nature*, 400(6745):661–4, 1999.

[239] R. E. Lenski, C. Ofria, R. T. Pennock, and C. Adami. The evolutionary origin of complex features. *Nature*, 423:139–144, 2003.

[240] J. S. Levinton. *Genetics, Paleontology, and Macroevolution.* Cambridge, 1988.

[241] S. Levy. *Artificial Life*. Penguin Books, 1992.

[242] M. Li and P. Vitányi. *An Introduction to Kolmogorov Complexity and its Applications*. Springer, New York, 2nd edition, 1997.

[243] S. Liang, A. Zincir-Heywood, and M. Heywood. Intelligent packets for dynamic network routing using distributed genetic algorithm. In *Genetic and Evolutionary Computation Conference, GECCO*, 2002.

[244] A. Lindenmayer. Mathematical models for cellular interactions in development, parts I and II. *Journal of Theoretical Biology*, 18:280–315, 1968.

[245] R. J. Lipton. DNA solution of hard computational problems. *Science*, 268(5210):542–545, 1995.

[246] Y. Liu, X. Yao, A. Shao, and T. Higuchi. Scaling up fats evolutionary programming with cooperative coevolution. In *Proceedings of 2001 IEEE Congress on Evolutionary Computation*, Seoul, Korea, 2001. IEEE Press.

[247] A. J. Lotka. *Elements of mathematical biology.* Dover Publications, 1925.

[248] M. H. Luerssen. Graph grammar encoding and evolution of automata networks. In *Proceedings of the 28th Australasian Computer Science Conference, Newcastle, Australia*, pages 229–238. Australian Computer Society, Inc., 2005.

[249] M. H. Luerssen and D. M. W. Powers. On the artificial evolution of neural graph grammars. In *Proceedings of the 4th International Conference on Cognitive Science*, pages 369–377. University of New South Wales, 2003.

[250] M. H. Luerssen and D. M. W. Powers. Graph composition in a graph grammar-based method for automata network evolution. In *Proceedings of the IEEE Congress on Evolutionary Computation.* in press, 2005.

[251] S. Luke and L. Panait. A survey and comparison of tree generation algorithms. In *Proceedings of the Genetic and Evolutionary Computation Conference*, pages 81–88. Morgan Kaufmann, 2001.

[252] M. Lungarella and L. Berthouze. On the interplay between morphological, neural, and environmental dynamics: a robotic case-study. *Adaptive Behavior*, 10(3-4):223–241, 2002.

[253] J. E. Lycett and R. Dunbar. Abortion rates reflect optimisation of parental investment strategies. *Proceedings of the Royal Society, B. : Biological Sciences*, 266(1436):2355–2358, 1999.

[254] R. H. MacArthur. Fluctuation of animal populations and a measure of community stability. *Ecology*, 36:533–536, 1955.

[255] D. R. Maddison, D. L. Swofford, and W. P. Maddison. Nexus: an extensible file format for phylogenetic data. *Syst. Biol.*, 46:590 – 621, 1997.

[256] J. C. Magalhaes and D. Krause. Suppes predicate for genetics and natural selection. *J Theor Biol*, 209(2):141–53, 2001.

[257] S. Mallat. A theory for multiresolution signal decomposition: the wavelet representation. *IEEE Trans. Pattern Analysis and Machine Intelligence*, 11(7):674–693, 2002.

[258] C. A. Martin. On continuous symmetries and the foundations of modern physics. In Katherine Brading and Elena Castellani, editors, *Symmetries in Physics: Philosophical Reflections*, pages 29–60. Cambridge University Press, 2003.

[259] N. D. Martinez. Artifacts or attributes? effects of resolution on the little rock lake food web. *Ecological Monographs*, 61(4):367–392, 1991.

[260] N. D. Martinez, B. A. Hawkins, H. A. Dawah, and B. P. Feifarek. Effects of sampling effort on characterization of food-web structure. *Ecology*, 80(3):1044–1055, 1999.

[261] A. Martinoli, A. J. Ijspeert, and L. M. Gambardella. A probabilistic model for understanding and comparing collective aggregation mechansims. In *Proc. of the 5th European Conf. on Artificial Life, ECAL'99*, 1999.

[262] S. Mascaro, K. B. Korb, and A. E. Nicholson. Suicide as an evolutionary stable strategy. Technical Report 89, School of Computer Science and Software Engineering, Monash University, Victoria, Australia, 2001.

[263] S. Mascaro, K. B. Korb, and A. E. Nicholson. Suicide as an Evolutionary Stable Strategy. In *Advances in Artificial Life, 6th European Conference*, pages 358–361, Prague, Czech Republic, 2001.

[264] S. Mascaro, K. B. Korb, and A. E. Nicholson. Alife investigation of parental investment in reproductive strategies. In *ALIFE VIII - Proceedings of The 8th International Conference*, pages 358–361, Sydney, 2002.

[265] J. Matsukidaira, J. Satsuma, D. Takahashi, T. Tokihiro, and M. Torii. Toda-type cellular automaton and its n-soliton solution. *Physics Letters A*, 225:287–295, 1997.

[266] R. M. May. Will a large complex system be stable? *Nature*, 238:413–414, 1972.

[267] R. M. May. *Stability and complexity in model ecosystems.* Princeton Landmarks in Biology. Princeton University Press, 2001.

[268] J. Maynard Smith. Group Selection. *Quarterly Review of Biology*, 51:277–283, 1976.

[269] E. Mayr. *The Growth of Biological Thought.* Harvard University Press, Cambridge, 1982.

[270] K. S. McCann. The diversity-stability debate. *Nature*, 450:228–255, 2000.

[271] B. D. McKay. Practical graph isomorphism. *Congressus Numerantium*, 30:45–87, 1981.

[272] R. I. McKay and H. A. Abbass. Anticorrelation measures in genetic programming. In *Australasia-Japan Workshop on Intelligent and Evolutionary Systems*, pages 45–51, 2001.

[273] N. McPhee and N. Hopper. Analysis of genetic diversity through population history. In *Proceedings of the Genetic and Evolutionary Computation Conference*, pages 1112–1120. Morgan Kaufmann, 1999.

[274] C. Melhuish, J. Welsby, and C. Edwards. Using templates for defensive wall building with autonomous mobile ant-like robots. In *Proc. of Towards*

Intelligent Mobile Robots (TIMR'99), 1999.

[275] J. Memmott, N. D. Martinez, and J. E. Cohen. Predators, parasitoids and pathogens: species richness, trophic generality and body sizes in a natural food web. *Journal of Animal Ecology*, 69(1):1–15, 2000.

[276] N. D. Mermin. Hidden variables and the two theorems of John Bell. *Reviews of Modern Physics*, 65(3):803–815, 1993.

[277] B. Meyer. On the convergence behaviour of ant colony search. In *Proceedings of the 7th Asia-Pacific Conference on Complex Systems*, pages 153–167, 2004.

[278] G. F. Miller, P. M. Todd, and S. U. Hegde. Designing neural networks using genetic algorithms. In *Proceedings of the Third International Conference on Genetic Algorithms and Their Applications*, pages 379–384. Morgan Kaufmann, 1989.

[279] J. F. Miller and P. Thomson. Cartesian genetic programming. In *Proceedings of the Third European Conference on Genetic Programming (EuroGP2000)*, pages 121–132. Springer-Verlag, 2000.

[280] R. Milo, S. Itzkovitz, N. Kashtan, R. Levitt, S. Shen-Orr, I. Ayzenshtat, and M. Shefferand U. Alon. Superfamilies of evolved and designed networks. *Science*, 303(5663):1538–1542, 2004.

[281] J. Mitteldorf. Aging Selected for its Own Sake. *Evolutionary Ecology Research*, 7:1–17, 2004.

[282] J. Mitteldorf. Chaotic Population Dynamics and the Evolution of Aging. In *ALIFE IX - Proc. of The 9th International Conference on the Simulation and Synthesis of Living Systems*, pages 346–352, 2004.

[283] A. Ø. Mooers and S. B. Heard. Inferring evolutionary process from the phylogenetic tree shape. *Quarterly Review of Biology*, 72:31 – 54, 1997.

[284] D. E. Moriarty and R. Miikkulainen. Efficient reinforcement learning through symbiotic evolution. *Machine Learning*, 22(1-3):11–32, 1996.

[285] H. Murao, H. Tamaki, and S. Kitamura. A coevolutionary approach to adapt the genotype-phenotype map in genetic algorithms. In *Proceedings of the 2002 Congress on Evolutionary Computation*. IEEE Press, 2002.

[286] Y. Nakamichi and T. Arita. Diversity control in ant colony optimization. In H.A. Abbass, editor, *Proceedings of the Inaugural Workshop on Artificial Life*, pages 69–79, Adelaide, Australia, 2001.

[287] S. Nee, R. M. May, and P. H. Harvey. The tempo and mode of evolution revealed from molecular phylogenies. *Phil. Trans. R. Soc. Lond. B*, 344:305 - 311, 1994.

[288] D. Newth. *Building Blocks and Modules — some mechanisms for adaptation in complex systems and evolutionary computation*. Ph. d. thesis, School of Environmental and Information Sciences, Charles Sturt University, 2002.

[289] D. Newth. Food-web complexity: latitudinal trends, topological properties and stability. In In Stonier et al., editor, *Complex Systems 2004: Proceedings of the 7th Asia-Pacific Conference on Complex Systems*, pages 700–712. Cairns Convention Centre, Cairns, Australia, 2004.

[290] D. Newth, J. E. Lawrence, and D. G. Green. Emergent structure in dynamic systems. In A. Namatame et al., editor, *Complex Systems 2002: The Pro-*

ceedings of the Sixth International Conference on Complex Systems, pages 229–237. National Defense Force Academy Japan, Japan, 2002.

[291] W. P. Niedringhaus. Stream option manager (som), automated integration of aircraft separation, merging, stream management, and other air traffic control functions. *IEEE Transaction on Systems, Man, and Cybernetics*, 25(9), SEPTEMBER 1995.

[292] A. Nishikawa, M. Yamamura, and M. Hagiya. DNA computation simulator based on abstract bases. *Soft Computing*, 5(1):25–38, 2001.

[293] S. Nolfi, O. Miglino, and D. Parisi. Phenotypic plasticity in evolving neural networks. Technical Report PCIA-95-05, Department of Cognitive Processes and Artificial Intelligence, Institute of Psychology C. N. R. - Rome , Italy, 1994.

[294] Radio Technical Committee on Aeronautics. Free Flight implementation, Final Report of RTCA Task Force 3. Technical report, RTCA Inc,, Washington, DC, 1995.

[295] Who Study Group on Malaria. Vector control for malaria and other mosquito borne diseases. Technical Report 857, World Health Organisation, 1995.

[296] D. Orrel and L. A. Smith. Visualising bifurcations in high dimensional systems: The spectral bifurcation diagram. *Int. J. of Bifurcation and Chaos*, 13(10):3015–3027, 2003.

[297] M. J. Osborne. *An Introduction to Game Theory.* Oxford University Press, 2004.

[298] E. Ott, C. Grebogi, and J. Yorke. Controlling chaos. *Phys. Rev. Lett.*, 64(11):1196–1199, 1990.

[299] J. Paredis. Coevolutionary computation. *Artificial Life*, 2(4):355–375, 1995.

[300] C. A. C. Parker, H. Zhang, and C. R. Kube. Blind bulldozing: multiple robot nest construction. In *Proc. of 2003 IEEE/RSJ International Conference on Intelligent Robots and Systems (IROS'03)*, 2003.

[301] H. H. Pattee. Artificial life needs a real epistemology. In F. Moran, A. Moreno, J. J. Merelo, and P. Chacon, editors, *Advances in Artificial Life*, Third European Conference on Artificial Life, Granada, Spain, June 4–6, 1995, pages 23–38. Springer, 1995.

[302] M. E. Peskin and D. V. Schroeder. *An Introduction to Quantum Field Theory.* Addison-Wesley Publishing Company, 1995.

[303] R. Pfeifer and C. Scheier. *Understanding Intelligence.* Cambridge, MA: MIT Press, 1999.

[304] S. L. Pimm. Food web design and the effect of species deletion. *OIKOS*, 35:139–149, 1980.

[305] S. L. Pimm. The complexity and stability of ecosystems. *Nature*, 307:321–326, 1984.

[306] G. A. Polis. Complex trophic interactions in deserts: An empirical critique of food-web theory. *Nature*, 138(1):123–155, 1991.

[307] M. A. Potter. *The Design and Analysis of a Computational Model of Cooperative Coevolution.* PhD thesis, George Mason University, Faiffax, Virginia, USA, 1997.

[308] M. A. Potter and K. A. De Jong. A cooperative coevolutionary approach to function optimization. In *Proceedigns of the third parallel problem solving from nature, PPSN*, pages 249–257, Jerusalem, Israel, 1994. Springer Verlag.

[309] M. Prandini, J. Lygeros, A. Nilim, and S. Sastry. A probabilistic framework for aircraft conflict detection. In *In Proc. AIAA Guidance, Navigat. , Contr. Conf.*, volume 99, pages 1047–1057, Portland, 1999.

[310] G. R. Price. Selection and Covariance. *Nature*, 227:520–521, 1970.

[311] K. Price. An introduction to differential evolution. In D. Corne, M. Dorigo, and F. Glover, editors, *New ideas in optimization*, pages 79–108. McGraw-Hill, London, 1999.

[312] I. Prigogine. *From Being into Becoming.* W. H. Freeman, San Francisco, 1980.

[313] P. Prusinkiewicz. Modeling plant growth and development. *Current Opinion in Plant Biology*, 7:79–83, 2004.

[314] A. Purvis. Using interspecific phylogenies to test macroevolutionary hypotheses. In P. H. Harvey, A. J. Leigh Brown, J. Maynard Smith, and S. Nee, editors, *New uses for new phylogenies*, pages 153 – 168. O. U. P., 1996.

[315] A. Purvis, P. M. Agapow, J. L. Gittleman, and G. M. Mace. Non-random extinction increases the loss of evolutionary history. *Science*, 288:328–330, 2000.

[316] A. Purvis, P. H. Harvey, A. Rambaut, and P. M. Agapow. Phylogeny shape. *Science*, submitted.

[317] A. Purvis, A. Katzourakis, and P. M. Agapow. Evaluating phylogenetic tree shape: Two modifications to fusco and cronk's method. *J. theor. Biol.*, 214(2002):99 – 103, 2002.

[318] A. Purvis and A. Rambaut. Comparative analysis by independent contrasts. *Science*, 11:247 – 251, 1995.

[319] O. G. Pybus, A. Rambaut, and P. H. Harvey. An integrated framework for the inference of population history from reconstructed genealogies. *Genetics*, 155:1429–1437, 2000.

[320] L. R. Rabiner, A. E. Rosenberg, and S. E. Levinson. Considerations in dynamic time warping algorithms for discrete word recognition. In *IEEE Trans. Acoust. Speech & Signal Proc.*, volume 26, pages 575–582, 1978.

[321] M. Randall. A general framework for constructive meta-heuristics. In E. Kozan and A. Ohuchi, editors, *Operations Research/Management Science at Work*, volume 43 of *International Series in Operations Research and Management Science*, pages 111–128. Kluwer Academic Publishers, Boston, MA, 2002.

[322] M. Randall. Scheduling aircraft landings using ant colony optimisation. In *Sixth IASTED International Conference Artificial Intelligence and Soft Computing (ASC 2002)*, pages 129–133, Banff, Alberta, Canada, 2002.

[323] M. Randall. A systematic strategy to incorporate intensification and diversification into ant colony optimisation. In H.A. Abbass and J. Wiles, editors, *Proceedings of the Australian Conference on Artificial Life*, pages

199–208, Canberra, Australia, 2003.

[324] M. Randall. Near parameter free ant colony optimisation. In M. Dorigo, M. Bairattari, C. Blum, L. Gambardella, F. Monadada, and T. Stützle, editors, *Lecture Notes in Computer Science*, volume 3172, pages 374–381, Berlin, 2004. Springer-Verlag.

[325] M. Randall and E. Tonkes. Intensification and diversification strategies in ant colony optimisation. *Complexity International*, 9, 2002.

[326] T. Ray. Tierra, available from http://www. isd. adr. co. jp/~ray/tierra. 2005.

[327] T. S. Ray. An approach to the synthesis of life. In C. Langton, C. Taylor, J. D. Farmer, and S. Rasmussen, editors, *Artificial Life II, Santa Fe Institute Studies in the Sciences of Complexity*, volume XI, pages 371–408. Addison-Wesley, Reading, 1991.

[328] T. S. Ray. Evolution and optimization of digital organisms. In Billingsley K. R. et al., editor, *Scientific Excellence in Supercomputing: The IBM 1990 Contest Prize Papers, Athens, GA, 30602:*, pages 489–531. The Baldwin Press, The University of Georgia, 1991.

[329] J. H. Reif. Parallel molecular computation. In *Proceedings of the 7th Annual ACM Symposium on Parallel Algorithms and Architectures, SPAA '95*, pages 213–223, 1995.

[330] T. Reil. Dynamics of gene expression in an artificial genome - implications for biological and artificial ontogeny. In D. Floreano, F. Mondada, and J. D. Nicoud, editors, *Proceedings of the 5th European Conference on Artificial Life*, pages 457–466. Springer Verlag, 1999.

[331] G. Reinelt. TSPLIB - A traveling salesman problem library. *ORSA Journal on Computing*, 3:376–384, 1991.

[332] R. E. Ricklefs. Global diversification rates of passerine birds. *Proc. R. Soc. London B*, 270(2003):2285 – 2292, 2003.

[333] R. E. Ricklefs and S. S. Renner. Species richness within families of flowering plants. *Evolution*, 48:1619 – 1636, 1994.

[334] M. Ridley. *Evolution*. Blackwell Publishers, Oxford, 3rd edition, 2003.

[335] A. Roberts. Stability of a feasible random ecosystem. *Nature*, 251:607–808, 1974.

[336] D. Roggen, D. Floreano, and C. Mattiussi. A morphogenetic evolutionary system: Phylogenesis of the poetic circuit. In *Proc. of the 5th International Conference on Evolvable Systems: From Biology to Hardware ICES 2003, LNCS 2606*, pages 153–164. Springer, 2003.

[337] J. P. Rosca. Entropy-driven adaptive representation. In *Proceedings of the Workshop on Genetic Programming: From Theory to Real-World Applications*, pages 23–32, Tahoe City, 9 1995.

[338] R. Rosen. *Life Itself: A comprehensive inquiry into the nature, origin, and fabrication of life.* Complexity in Ecological Systems Series. Columbia University Press, New York, 1991.

[339] S. Roweis, E. Winfree, R. Burgoyne, N. V. Chelyapov, M. F. Goodman, P. W. K. Rothemund, and L. M. Adleman. A sticker based model for DNA computation. In *DNA Based Computers II*, volume 44 of *DIMACS Series*

in *Discrete Mathematics and Theoretical Computer Science*, pages 1–27, 1999.

[340] T. Rudge and J. Haseloff. A computational model of cellular morphogenesis in plants. To appear in *VIIIth European Conference on Artificial Life*, Canterbury, UK, 2005.

[341] N. Sadeh and A. Kott. Models and techniques for dynamic demand-responsive transportation planning. Technical Report CMU-RI-TR-96-09, Robotics Institute, Carnegie Mellon University, 1996.

[342] K. Sakamoto, H. Gouzu, K. Komiya, D. Kiga, S. Yokoyama, T. Yokomori, and M. Hagiya. Molecular computation by DNA hairpin formation. *Science*, 288:1223–1226, 2000.

[343] I. Salazar-Ciudad and J. Jernvall. How different types of pattern formation mechanisms affect the evolution of form and development. *Evolution and Development*, 6:6–16, 2004.

[344] I. Salazar-Ciudad, S. A. Newman, and R. V. Solé. Phenotypic and dynamical transitions in model gene networks I. Emergence of patterns and genotype-phenotype relationships. *Evolution and Development*, 3:84–94, 2001.

[345] M. J. Sanderson and M. J. Donoghue. Shifts in diversification rate with the origin of angiosperms. *Science*, 264:1590 – 1593, 1994.

[346] K. Sastry, M. Pelikan, and D. E. Goldberg. Decomposable problems, niching, and scalability of multiobjective estimation of distribution algorithms. IlliGAL Report No. 2005004, Illinois Genetic Algorithms Laboratory, University of Illinois at Urbana-Champaign, IL, 2005.

[347] R. Sato and R. Ramachandran, editors. *Conservation Laws and Symmetry: Applications to Economics and Finance*. Kluwer Academic Publishers, 1990.

[348] J. H. Scherrer. Risks and vulnerabilities of network-centric forces: Insights from the science of complexity. Newport r. i., Naval War College, February 2003.

[349] R. Schoonderwoerd, O. E. Holland, J. L. Bruten, and L. J. M. Rothkrantz. Ant-based load balancing in telecommunications networks. *Adaptive Behavior*, (2):169–207, 1996.

[350] W. Shakespeare. *The Tragedy of Hamlet, Prince of Denmark*. Project Gutenburg, 1999.

[351] S. Shen-Orr, R. Milo, S. Mangan, and U. Alon. Network motifs in the transcriptional regulation network of escherichia coli. *Nature Genetics*, 31:64–68, 2002.

[352] S. Y. Shin, H. Y. Jang, M. H. Tak, and B. T. Zhang. Simulation of DNA hybridization chain reaction based on thermodynamics and artificial chemistry. In *Preliminary Proceedings of the 10th International Meeting on DNA Computing*, June 2004.

[353] H. J. Sims and K. J. McConway. Non-stochastic variation of species-level diversification rates within angiosperms. *Evolution*, 57:460 – 479, 2003.

[354] J. M. W. Slack. *From egg to embryo*. Cambridge University Press, Cambridge, 2 edition, 1991.

[355] R. Smith. Open dynamics engine. http://ode. org/.

[356] E. Sober and D. S. Wilson. *Unto Others.* Havard University Press, Cambridge, 1998.

[357] R. V. Solé, B. Luque, and S. Kauffman. Phase transitions in random networks with multiple states. Working Paper 00-02-011, Santa Fe Institute, 2000.

[358] R. V. Solé and J. M. Montoya. Complexity and fragility in ecological networks. *Proceedings of the Royal Society of London, Series B.*, 268:2039–2045, 2001.

[359] R. V. Solé, I. Salazar-Ciudad, and J. Garcia-Fernández. Common pattern formation, modularity and phase transitions in a gene network model of morphogenesis. *Physica A*, 305:640–654, 2002.

[360] M. W. Spong and M. Vidyasagar. *Robot Dynamics and Control.* New York: Wiley, 1989.

[361] J. C. Sprott. Numerical calculation of largest Lyapunov exponent. `http://sprott. physics. wisc. edu/chaos/lyapexp. htm`, 2004.

[362] E. Stam. Does imbalance in phylogenies reflect only bias? *Evolution*, 56(2002):1292 – 1295, 2002.

[363] R. K. Standish. On complexity and emergence. *Complexity International*, 9, 2001. arXiv:nlin. AO/0101006.

[364] R. K. Standish. Open-ended artificial evolution. *International Journal of Computational Intelligence and Applications*, 3:167, 2003. arXiv:nlin. AO/0210027.

[365] R. K. Standish. The influence of parsimony and randomness on complexity growth in Tierra. In Bedau et al., editors, *Life IX Workshop and Tutorial Proceedings*, 2004.

[366] K. O. Stanley and R. Miikkulainen. Evolving neural networks through augmenting topologies. *Evolutionary Computation*, 10(2):99–127, 2002.

[367] K. O. Stanley and R. Miikkulainen. A taxonomy for artificial embryogeny. *Artificial Life*, 9(2):93–130, 2003.

[368] S. M. Stanley. *Macroevolution.* John Hopkins, 1998.

[369] R. L. Stewart and R. A. Russell. Emergent structures built by a minimalist autonomous robot using a swarm-inspired template mechanism. In *Proc. of The First Australian Conference on Artificial Life (ACAL'03)*, 2003.

[370] R. L. Stewart and R. A. Russell. Building a loose wall structure with a robotic swarm using a spatio-temporal varying template. In *Proc. of 2004 IEEE/RSJ International Conference on Intelligent Robots and Systems (IROS04)*, 2004.

[371] R. L. Stewart and R. A. Russell. A distributed feedback mechanism to regulate wall construction by a robotic swarm. unpublished, 2005.

[372] S. Strogatz. *Nonlinear Dynamics and Chaos: With Applications to Physics, Biology, Chemistry, and Engineering.* Cambridge, MA: Perseus Books Publishing, 1994.

[373] T. Stützle and M. Dorigo. ACO algorithms for the quadratic assignment problem. In D. Corne, M. Dorigo, and F. Glover, editors, *New Ideas in Optimization*, pages 33–50. McGraw-Hill, London, 1999.

[374] T. Stützle and H. Hoos. Improvements on the Ant-System: Introducing the $\mathcal{MAX} - \mathcal{MIN}$ Ant System. In *Proceedings of the Third International Conference on Artificial Neural Networks and Genetic Algorithms*, 1997.

[375] T. Stützle and H. Hoos. $\mathcal{MAX} - \mathcal{MIN}$ Ant System. *Future Generation Computer Systems*, 16:889–914, 2000.

[376] D. Subramanian, P. Druschel, and J. Chen. Ants and reinforcement learning: A case study in routing in dynamic networks. In *The Fifteenth International Joint Conference on Artificial Intelligence*, pages 832–839, 1997.

[377] P. N. Suganthan, N. Hansen, J. J. Liang, K. Deb, Y. P. Chen, A. Auger, and S. Tiwari. Problem definitions and evaluation criteria for the cec 2005 special session on real-parameter optimization. Technical Report 2005005, Nanyang Technological University, Singapore, and KanGAL, IIT Kanpur, India, 2005.

[378] G. Taga. Emergence of bipedal locomotion through entrainment among the neuro-musculo-skeletal system and the environment. *Physica D*, 75(1-3):190–208, 1994.

[379] S. Takatuzi. *Estudo Comparativo Entre Metodologias Para Estimar Populacao Efetiva (Ne) Usando Simulacoes.* Msc, Federal University of Parana, 2004.

[380] J. Teo and H. A. Abbass. Automatic generation of controllers for embodied legged organism pareto evolutionary multi-objective approach. *Journal of Evolutionary Computation,MIT Press*, 12(3), 2004.

[381] E. Thelen and L. Smith. *A Dynamic Systems Approach to the Development of Cognition and Action.* Cambridge, MA: MIT Press, 1995.

[382] G. Theraulaz, J. Gautrais, S. Camazine, and J. L. Deneubourg. The formation of spatial patterns in social insects: from simple behaviours to complex structures. *Phil. Trans. R. Soc. Lond.*, A(361):1263–1282, 2003.

[383] J. M. T. Thompson and H. B. Stewart. *Nonlinear Dynamics and Chaos.* New York: Wiley, 2002.

[384] R. Thornhill and C. T. Palmer. *A Natural History of Rape.* MIT Press, Cambridge, Massachusetts, 2000.

[385] R. Thornhill and N. W. Thornhill. Human rape: An evolutionary analysis. *Ethology and Sociobiology*, 4(3):137–173, 1983.

[386] B. H. Tiffney. Seed size, dispersal syndromes, and the rise of the angiosperms. *Ann. Mo. Bot. Gard.*, 71:551 - 576, 1984.

[387] B. H. Tiffney and S. J. Mazer. Angiosperm growth habit, dispersal, and diversification reconsidered. *Evol. Ecol.*, 9:93 - 117, 1995.

[388] A. Toffolo and E. Benini. Genetic diversity as an objective in multi-objective evolutionary algorithms. *Evolutionary Computation*, 11(2):151–168, 2003.

[389] K. Tominaga. A formal model based on affinity among elements for describing behavior of complex systems. Technical Report UIUCDCS-R-2004-2413, Department of Computer Science, University of Illinois at Urbana-Champaign, March 2004.

[390] K. Tominaga. Modelling DNA computation by an artificial chemistry based on pattern matching and recombination. In *Proceedings of the Workshop*

on Artificial Chemistry and Its Applications, part of the 9th International Conference on the Simulation and Synthesis of Living Systems (ALIFE9), 2004.

[391] K. Tominaga. Describing protein synthesis and a cell cycle of an imaginary cell using a simple artificial chemistry. In *Proceedings of the Workshop on Artificial Chemistry and Its Applications, part of the 8th European Conference, ECAL 2005*, 2005. In press.

[392] J. Torresen. A divide-and-conquer approach to evolvable hardware. In *2nd International Conference on Evolvable Systems: from biology to hardware (ICES 98), LNCS 1478*, pages 57–65. Springer-Verlag, 1998.

[393] M. Toussaint. *The evolution of genetic representations and modular neural adaptation*. Ph. d. thesis, Institut für Neuroinformatik, Ruhr-Universität Bochum, 2003.

[394] C. R. Townsend, R. M. Thompson, A. R. McIntosh, C. Kilroy, E. Edwards, and M. R. Scarsbrook. Disturbance, resource supply, and food-web architecture in streams. *Ecology Letters*, 1(3):200–209, 1998.

[395] K. Tregonning and A. Roberts. Complex systems which evolve towards homeostasis. *Nature*, 281:563–564, 1979.

[396] R. L. Trivers. Parental investment and sexual selection. In Bernard Campbell, editor, *Sexual Selection and the Descent of Man*, pages 136–179. Heinemann, London, 1972.

[397] R. L. Trivers. Parent-offspring conflict. *American Zoologist*, 11:249–264, 1974.

[398] R. L. Trivers and D. E. Willard. Natural selection of parental ability to vary the sex ratio of offspring. *Science*, 179(4068):90–92, 1973.

[399] H. Tsukaya. Organ shapen and size: a lesson from studies of leaf morphogenesis. *Current Opinion in Plant Biology*, 6:57–62, 2003.

[400] P. Turney, D. Whitley, and R. Anderson. Evolution, learning and instinct: 100 years of the baldwin effect. *Evolutionary Computation*, 4(3), 1997.

[401] L. M. Van Valen. Group selection and the evolution of dispersal. *Evolution*, 25, 1971.

[402] V. K. Vassilev, T. C. Fogarty, and J. F. Mille. Information characteristics and the structure of landscapes. *Evolutionary Computation*, 1(8):31–60, 2000.

[403] V. K. Vassilev and J. F. Miller. Scalability problems of digital circuit evolution: Evolvability and efficient designs. In *Proc. of the 2nd NASA/DoD Workshop on Evolvable Hardware*. IEEE, 2000.

[404] G. Wagner, G. Booth, and H. Bagheri-Chaichian. A population genetic theory of canalization. *Evolution*, 51:329–347, 1997.

[405] J. A. Walker and J. F. Miller. Evolution and acquisition of modules in cartesian genetic programming. In *Proceedings of the 7th European Conference on Genetic Programming*, pages 187–197. Springer-Verlag, 2004.

[406] P. H. Warren. Spatial and temporal variation in the structure of a freshwater food web. *OIKOS*, 55:299–311, 1989.

[407] D. J. Watts. *Small Worlds*. Princeton Studies in Complexity. Princeton University Press, 1999.

[408] D. J. Watts and S. H. Strogatz. Collective dynamics of 'small-world networks. *Nature*, 393:440–442, 1998.

[409] J. Wawerla, G. S. Sukhatme, and M. J. Matarić. Collective construction with multiple robots. In *Proc. of 2002 IEEE/RSJ International Conference on Intelligent Robots and Systems (IROS'02)*, 2002.

[410] H. Wedde, M. Farooq, and Y. Zhang. Beehive: An efficient fault-tolerant routing algorithm inspired by honey bee behavior. In *4th International Workshop, ANTS 2004*, 2004.

[411] A. Weismann. *Essays upon Heredity and Kindred Biological Problems*. Clarendon Press, London, 1889.

[412] J. Werfel, Y. Bar-Yam, and R. Nagpal. Construction by robot swarms using extended stigmergy. Technical Report AI Memo AIM-2005-011, MIT Computer Science and Artificial Intelligence Laboratory, 2005.

[413] M. J. West-Eberhard. Phenotypic Plasticity and the Origins of Diversity. *Annual Review of Ecology and Systematics*, 20:249–278, 1989.

[414] G. White. The mathematical agent - a complex adaptive system representation in bactowars. In *First workshop on complex adaptive systems for defence*, Adelaide, Australia, 2004.

[415] T. White, B. Pagurek, and F. Oppacher. ASGA: Improving the ant system by integration with genetic algorithms. In *Genetic Programming 1998: Proceedings of the Third Annual Conference*, pages 610–617, 22-25 1998.

[416] A. Wieland. Evolving neural network controllers for unstable systems. In *Proceedings of the International Joint Conference on Neural Networks*, pages 667–673. IEEE Press, 1991.

[417] J. Wiles and J. Watson. Patterns in complex systems modeling. *Lecture Notes in Computer Science*, 3578 (June):532–539, 2005.

[418] G. Williams. *Sex and Evolution*. Princeton University Press, Princeton, New Jersey, 1975.

[419] R. J. Williams and N. D. Martinez. Simple rules yield complex food webs. *Nature*, 404:180–183, 2000.

[420] C. Wilson. Network centric warfare: Background and oversight issues for congress. Report for congress, CRS, June 2004.

[421] D. S. Wilson. *Natural Selection of Populations & Communities*. Benjamin/Cummings, Menlo Park, 1980.

[422] J. R. Wilson. Network-centric warfare 21st century. *Military & Aerospace Electronics*, January 2000.

[423] M. Wilson, C. Melhuish, A. B. Sendova-Franks, and S. Scholes. Algorithms for building annular structures with minimalist robots inspired by brood sorting in ant colonies. *Autonomous Robots*, 17(2-3):115–136, 2004.

[424] S. Wilson. The animat path to ai. In J. A. Meyer and S. Wilson, editors, *From Animals to Animats 1: Proceedings of The First International Conference on Simulation of Adaptive Behavior*, pages 15–21. The MIT Press/Bradford Books, Cambridge, MA, 1991.

[425] E. Winfree, T. Eng, and G. Rozenberg. String tile models for DNA computing by self-assembly. In Anne Condon and Grzegorz Rozenberg, editors, *DNA Computing*, volume 2054 of *LNCS*, pages 63–88, Berlin, 2000.

Springer.

[426] A. Wolf, J. B. Swift, H. L. Swinney, and J. A. Vastano. Determining Lyapunov exponents from a time series. *Physica D*, 16:285–317, 1985.

[427] S. Wolfram. Cellular automata as models of complexity. *Nature*, 311:419–424, 1984.

[428] L. Wolpert. *The Principles of Development*. Oxford University Press, Oxford, UK, 1998.

[429] J. R. Woodward. Modularity in genetic programming. In *Genetic Programming, Proceedings of EuroGP'2003*, pages 254–263. Springer-Verlag, 2003.

[430] A. Wuensche. The ghost in the machine: basin of attraction fields of random boolean networks. In C. G. Langton, editor, *Artificial Life III*, Sante Fe Institute Studies in the Sciences of Complexity, Reading, MA, 1994. Addison-Wesley.

[431] A. Wuensche. Classifying cellular automata automatically: finding gliders, filtering, and relating space-time patterns, attractor basins, and the z parameter. *Complex.*, 4(3):47–66, 1999.

[432] V. C. Wynne-Edwards. *Animal Dispersion in Relation to Social Behavior*. Oliver and Boyd, Edinburgh, 1962.

[433] J. Wyrzykowska, S. Pien, W. H. Shen, and A. J. Fleming. Manipulation of leaf shape by modulation of cell division. *Development*, 129:957–964, 2002.

[434] T. Yamamoto and Y. Kuniyoshi. Global dynamics: a new concept for design of dynamical behavior. *Proc. 2nd Int. Workshop on Epigenetic Robotics*, pages 177–180, 2002.

[435] A. Yang. Understanding network centric warfare. *ASOR BULLETIN*, 23(4):2 – 6, December 2004.

[436] A. Yang, H. A. Abbass, and R. Sarker. Landscape dynamics in multi-agent simulation combat systems. In *Proceedings of 17th Joint Australian Conference on Artificial Intelligence, LNCS*, Cairns, Australia, December 2004. Springer-Verlag.

[437] A. Yang, H. A. Abbass, and R. Sarker. Characterizing warfare in red teaming. *IEEE Transactions on Systems, Man, Cybernetics, Part B*, 2005. accepted to appear.

[438] A. Yang, H. A. Abbass, and R. Sarker. Wisdom-II: A network centric model for warfare. In *Ninth International Conference on Knowledge-Based Intelligent Information & Engineering Systems (KES 2005), LNCS*, Melbourne, Australia, September 2005.

[439] X. Yao, Y. Liu, and G. Lin. Evolutionary programming made faster. *IEEE Transactions on Evolutionary Computation*, 3(2):82–102, 1999.

[440] T. Yokomori. YAC: Yet another computation model of self-assembly. In *DNA Based Computers V*, volume 54 of *DIMACS Series in Discrete Mathematics and Theoretical Computer Science*, pages 153–167, 2000.

[441] T. Yokomori and S. Kobayashi. DNA-EC: A model of DNA-computing based on equality checking. In *DNA Based Computers III*, volume 48 of *DIMACS Series in Discrete Mathematics and Theoretical Computer Science*, pages 347–360, 1999.

[442] S. Rasmussen, N. A. Baas, B. Mayer, M. Nilsson, and M. W. Olesen. Ansatz for Dynamical Hierarchies. *Artificial Life*, 7:329–353, 2001.
[443] S. Rasmussen, N. A. Baas, B. Mayer, and M. Nilsson. Defence of the Ansatz for Dynamical Hierarchies. *Artificial Life*, 7:367–373, 2001.

Index